Frontiers in Civil Engineering

Volume 1

Stability of Geotechnical Structures: Theoretical and Numerical Analysis

Authored By

Y.M. Cheng

Department of Civil and Environmental Engineering, Hong Kong Polytechnic University, Hong Kong

H. Wong

LTDS (UMR 5513), ENTPE, Université de Lyon, France

C.J. Leo

School of Computing, Engineering and Mathematics, University of Western Sydney, Locked Bag 1797, Penrith NSW 2751, Australia

&

C.K. Lau

Fong On Geotechnics Limited, Hong Kong

Frontiers in Civil Engineering

Stability Analysis of Geotechnical Structures: Theoretical and Numerical Approaches

Volume # 1

Authors: Y.M. Cheng, H. Wong, C.J. Leo & C.K. Lau

ISSN (Online): 2468-4708

ISSN: Print: 2468-4694

ISBN (eBook): 978-1-68108-303-2

ISBN (Print): 978-1-68108-304-9

PREFACE

Due to the difficulty to define the initial stress and boundary conditions, the loading paths as well as the constitutive model of geomaterials, stability analysis has always been a very important discipline in geotechnical engineering. Towards this, engineers will assess the ultimate conditions where the strength of the system is fully mobilized. The facture of failure or the collapse load will then be assessed based on the ultimate analysis which is considered to be unaffected by the initial conditions of the system.

This book will introduce the fundamental concepts and applications of plasticity theory, limit equilibrium, and limit analysis in geotechnical engineering. These concepts will be illustrated using analytical examples whenever possible in order to enhance understanding at a fundamental level and can also be used to make preliminary estimates of geotechnical stability. In parallel, suitable numerical methods and advanced computational tools will be introduced for the engineers to solve theoretical and geotechnical problems of practical interests which require greater detailed consideration.

This book deals with the challenging subject matter in a systematic fashion, from a theoretical standpoint to practice in the real world. For this reason, the book is divided into 3 parts. In part 1, the fundamental concepts in plasticity, limit equilibrium, limit analysis and instability for geomaterials are presented as a first step in introducing readers as the theoretical basis. Analytical and semi-analytical solutions are then discussed in Part 2, with liberal use of illustrative examples, as a further step to shed insights and reinforce the underlying principles embodied in the theory. Finally in Part 3, examples utilising advanced computational tools like the finite element and discrete element methods are covered for the purpose of elucidating the complexity of dealing with stability problems of the real world using numerical approaches.

A particular feature of this book is that it stresses the rigorous formulation as much as the computational techniques to tackle stability problems. It is noted that the solution of these problems is far from trivial. The search of failure load and the corresponding failure mechanism involve the constrained optimisation of discontinuous objective functions containing multiple optimum points. In short, this book is an attempt to present within a single volume the fundamentals as well as the practical developments of stability analysis in geotechnical engineering in an easily accessible manner. Most of the materials are either based on the research works from the authors or the teaching materials to the postgraduate students.

This book is aimed at researchers and engineers working in the field of geotechnical engineering having to make design decisions concerning the stability and the risk of failure of geotechnical structures. These include natural and man-made slopes, dams, shallow or deep foundations, soil retaining structures, embankments, road or railway tunnels, large scale underground structures including for underground storage of nuclear wastes or CO_2 sequestration.

ACKNOWLEDGEMENTS

The authors would like to acknowledgement the support from the project "Theory, Analysis and Monitoring of Landslides and Debris Flows" (account ZVCR) for parts of the works in this book.

CONFLICT OF INTEREST

There is not any conflict of interest for the content of this book with any project, universities or parties.

Dr. Y.M. Cheng
Department of Civil and Environmental Engineering
Hong Kong Polytechnic University
Hong Kong

Dr. H. Wong
LTDS (UMR 5513), ENTPE
Université de Lyon
France

Dr. C.J. Leo
School of Computing, Engineering and Mathematics
University of Western Sydney, Locked Bag 1797
Penrith NSW 2751
Australia

Dr. C.K. Lau
Fong On Geotechnics Limited
Hong Kong

Frontiers in Civil Engineering

Volume 1

Stability of Geotechnical Structures: Theoretical and Numerical Analysis

CHAPTER 1

Introduction

Abstract: Stability analysis has always been a critical issue in geotechnical engineering. Many of the design works by engineers are actually based on the ultimate condition which can be assessed without the need of the initial condition and sophisticated constitutive models. A brief introduction about the stability analysis will be given in this chapter.

Keywords: Close-form solution, Finite element, Limit analysis, Limit equilibrium, Stability analysis.

1.1. INTRODUCTION

Due to the difficulty to define the initial stress and boundary conditions, the loading paths as well as the constitutive model of geomaterials, stability analysis has always been a very important discipline in geotechnical engineering. Towards this, engineers will assess the ultimate conditions where the strength of the system is fully mobilized without the detailed information about initial condition. In fact, it is extremely difficult and expensive to determine the initial condition for a general problem, even if the ground condition is simple, and no continuous results will be obtained even if explorations are carried out. The initial condition can be affected by the soil formation process, development and loading of the site, ground water and other surrounding effects and possibly many other factors. The failure or the collapse load will be assessed based on the ultimate analysis which is considered to be unaffected by the initial conditions of the system. Although this approach appears to be highly simplified without considering the initial condition as well as the constitutive model of soil, it appears to be indispensable for most of the practical problems. Furthermore, the engineers

Y. M. Cheng, H. Wong, C. J. Leo & C. K. Lau

are now well familiar with this approach, and many design and analysis have been carried out successfully with this approach.

This book will introduce the fundamental concepts and applications of stability analysis with particular reference to plasticity theory, limit equilibrium, limit analysis, finite element and discrete element methods in geotechnical engineering. Suitable numerical methods and advanced computational tools will be introduced for the engineers to solve theoretical and geotechnical problems of practical interests which require greater detailed consideration.

There are various advancement on stability analysis over last 40 years, and many results are developed and used for a variety of problems. It is impossible to cover all these results in the present work. This book is mainly based on the research and teaching materials by the authors, but sufficient background about the works by other researchers are also given for the readers.

1.2. BACKGROUND

Due to the growth of population and economic activities, terraces are created for buildings and infrastructures like quays, canals, railways and roads. Man-made cut and fill slopes have to be formed to facilitate such developments. In the past, the stability of many slopes, foundations and retaining structures are assessed by simple rules of thumb, due to the lack of adequate fundamental knowledge and computing power. There are various attempts to improve the rules of thumb approach in 20 century. One of the earliest attempts was by a French engineer Alexander Collin (Collin, 1846) which is not really better than the rules of thumb. In 1916, a series of quay failures had occurred in Sweden, and the Swedes had developed one of the earliest methods to assess slope stability using the method of slices and limit equilibrium method. The method is now called the Swedish

Method (or the Ordinary Method) of Slices (Fellenius, 1927) which is still used in limited sense up to present. A number of subsequent refinements to the method were made later: Taylor's stability chart (Taylor, 1937) based on moment equilibrium; Bishop's Simplified Method of Slices (Bishop, 1955) which is also based on the moment equilibrium; Janbu's method which is extended to the non-circular slip (Janbu, 1973); Morgenstern & Price (1965) which try to ensure forcing moments and forces to be simultaneously achieved; Spencer's parallel inter-slice forces (1967) as a special case of the Morgenstern & Price's method; and Sarma's method which is based on a horizontal earthquake approach (1973). These various methods are now basically unified under the Modern Generalized Method of Slices (GMS) (*e.g.* Low *et al.*, 1998).

In the classical Limit Equilibrium approach, the user has to define a slip surface before the stability analysis. There are different techniques to ensure a critical slip surface can indeed be identified. The finite element method (Griffiths & Lane, 1999) or the equivalent finite difference method (Cundall & Strack, 1979) are some of the modern computational methods which are used to evaluate the stability problems directly using the strength reduction algorithm (Dawson *et al.*, 1999). Zhang (1999) and others have proposed the rigid finite element method which is still limited to research purpose for the stability analysis up to now. The advantage of these methods is that there is no need to assume any inter-slice forces or slip surface which have to be prescribed in the classical limit equilibrium or limit analysis, but there are also various limitations to these methods.

Currently, most of the engineers are using computer methods which are commonly limit equilibrium or finite element method to solve different kinds of stability problems. However, every numerical method has its own assumptions and limitations. It is therefore necessary for the engineers and researchers to be fully aware of them so that the methods can be used within its limitations in real design situation. There is however one fundamental question which is realistic but is difficult to be overcome. All the observed failures are invariably 3-D in nature, but virtually all the engineering calculations for routine design are based on plane strain analysis and design. There are noticeable difference between the shear strengths for 3-D and 2-D (plane strain) conditions. For example, typical sand can mobilize up to 6 degrees higher in the frictional angle under plane strain condition, as compared with that for 3-D or axi-symmetric strain (Bishop, 1972). Currently, the two key issues in practical analysis are: using 3-D strength data but 2-D model, and thus rendering the existing practice highly dubious. However, the increase in shear strength in plane strain usually far outweighs the higher FOS in a 3-D analysis, hence the current practice will be conservative in a practical sense. This is probably the reason why in nature, all slopes fail in 3-D, and there is practical no 2D slope failure even though the FOS for 2D analysis is smaller than the corresponding result for 3D analysis. Besides that, most of the 3D limit equilibrium methods also suffer from various limitations. Based on the works by Cheng and Yip (2007), more rigorous 3D limit equilibrium methods have been developed which have overcome the critical problem of sliding direction. Even though 3-D slope stability analysis has been well established, there is still practical difficulty for the engineers to adopt 3D analysis, as it is also expensive to carry sufficient borehole assessments to determine a sufficiently good 3D geological profile for the analysis correctly.

1.3. CLOSED-FORM SOLUTIONS

For some simple and special cases with simple loading and geometry, closed-form but non-trivial solutions can be found. These are very important results (except for practical application) because they are academically pleasing, and these solutions provide the benchmark results to calibrate other computational methods. Engineers, particularly those younger ones, tend to rely heavily on the use of computer programs without a good feel on the engineering problems they are facing in their work. There are discussions within the Hong Kong Institution of Engineers, and the current phenomenon where engineers tend to over-rely on the black box type solution and computer programs without simple but reliable engineering sense should be avoided.

For example, for a circular slip failure with $c \neq 0$ and $\phi = 0$, if we take moment at the center of rotation, the factor of safety will be obtained easily which is the classical Swedish method. The factor of safety from the simple Swedish method should be exactly equal to that from the Bishop method (1955) for this case. On the other hand, the more refined Morgenstern-Price method will fail to converge easily for this case (depends on the angle of inclination) while the Sarma method (1973) will give result very close to that from the Swedish method. Engineers should be aware of these simple results which are usually in assessing the applicability of different computer programs and computational methods. Apart from this case, the classical bearing capacity and earth pressure problem closed-form solutions may also be used to calibrate and verify a code calculation. It is interesting to note that most of the computer programs fail to reproduce the exact solutions for these simple cases, unless advanced theory is employed which will be discussed in chapter 3. A bearing capacity problem can be seen as a slope with a zero slope angle but with substantial surcharge loading. The beauty of this classical problem is that it is relatively easy to extend the 2D solution to 3-D or at least axi-symmetric case where closed-form solution also exists. For example, the well-known ultimate bearing capacity 5.14Cu for 2-D case can be extended to 5.69 Cu for axi-symmetric case (Shield, 1955), where Cu is the undrained shear strength of the soil. A computer program should yield FOS = 1.0 if the surcharge loadings are set to 5.14 Cu and 5.69 Cu respectively for the corresponding case (but practically lesser for most of the computer codes). Likewise, similar bearing capacity solutions also exist for frictional material in both plane strain and axi-symmetric strain (Cox, 1962 or Bolton & Lau, 1993). Virtually all commercial programs fail to reproduce these classical solutions, and the limit of application of each computer program and the numerical procedures should be assessed by the engineers. The extremum principle by Cheng *et al.,* (2010) has demonstrated that the limitation of all existing programs and methods are: (1) use of iterative analysis, (2) assumption on the interslice force function. These limitations can be removed by the use of double QR method and extermum principle by Cheng (2003) and Cheng *et al.,* (2010).

Similarly, the earth pressure problems can also be checked for the computer code. The problem can be considered as a slope which has a slope angle of 90 degrees. Alternatively, it can be viewed as a bearing capacity problem where the ground in front of the foundation is not horizontal. By applying an active or passive pressure at the vertical face, the computer program should yield FOS=1.0 for both active and passive cases which will be illustrated in sec 3.9. Likewise, the problem can be extended to 3-D or more precisely axi-symmetric case for a shaft stability problem (Kwong, J S M 1991), from which the lateral earth pressure for a circular excavation can be determined.

Our argument is that all computational methods and programs should be benchmarked and validated against these "closed-form" solutions for comparison. The comparison would reveal both the strengths and limitations of the methods or programs so that. The basic understanding of these problems are hence vital to the assessment of the analysis, in particular, the use of computer programs. The author has found that many computer programs actually fail to reproduce the exact solutions, for which the engineers should be aware of.

1.4. ENGINEERING JUDGMENT

Engineering judgment is one of the most valuable assets of an engineer because engineering it is an integration of art and science. However, the best engineers always use their engineering judgment sparingly. To us, engineering judgment is really a euphemism for a jump in the dark. We should be able to use simple and understandable tools to carry out back-of-the-envelope type calculations to assist us to assess and evaluate the computation tools and design situations we are facing, so that we can develop a good feel about the problem and make our final judgment in the analysis.

1.5. LAYOUT OF THIS BOOK

This book is written by four authors: Y.M. Cheng, Henry Wong, C.J. Leo and C.K. Lau. The content covers the basic theory of lower and upper bound analysis, which are followed by the more specific methods such as limit equilibrium, limit analysis and plasticity theory. Finite element and distinct element methods are introduced as the numerical method. Besides the basic theory, there are plenty of examples and applications to various problems in this book. Some simple computer programs are also introduced to help the readers to understand and apply the limit analysis method in lateral earth pressure theory. Engineers and researchers will find this book useful for many geotechnical problems.

REFERENCES

Bishop A.W. (1955), The use of the slip circle in the stability analysis of earth slopes, Geotechnique, 5(1), 7–17.

Bishop A.W. (1972), Shear strength parameters for undisturbed and remoulded soils specimens, in: R.H.G. Parry (ed.), Stress–Strain Behaviour of Soils, pp. 3–58, London: Foulis.

Bolton M.D. and Lau C.K. (1993), Vertical bearing capacity factors for circular and strip footings on a Mohr–Coulomb soil, Canadian Geotechnical Journal, 30(6), 1024–1033.

Cheng Y.M. (2003), Locations of critical failure surface and some further studies on slope stability analysis, Computers and Geotechnics, 30(3), 255–267.

Cheng Y.M. and Yip C.J. (2007), Three-dimensional asymmetrical slope stability analysis – Extension of

Bishop's, Janbu's, and Morgenstern–Price's techniques, Journal of Geotechnical and Geoenvironmental Engineering, 133(12), 1544–1555.

Cheng Y.M., Zhao Z.H. and Sun Y.J. (2010), Evaluation of interslice force function and discussion on convergence in slope stability analysis by the lower bound method, Journal of Geotechnical and Geoenvironmental Engineering, ASCE, 136(8):1103-1113.

Collin A. (1846), Recherches Expérimentales sur les Glissements Spontanés des Terrains Argileux, accompagnées de Considerations sur Quelques Principes de la Méchanique Terrestre. Carilian-Goeury and Dalmont, Paris.

Cox A.D. (1962), Axially-symmetric plastic deformation in soils – II, Indentation of ponderable soils, International Journal of Mechanical Sciences, 4, 371–380.

Cundall P.A. and Strack O.D.L. (1979), A discrete numerical model for granular assemblies, Geotechnique, 29(1), 47–65.

Dawson E.M., Roth W.H. and Drescher A. (1999), Slope stability analysis by strength reduction. Geotechnique, 49(6), 835–840.

Fellenius W. (1927), Erdstatische Berechnungen mit Reibung und Kohasion (in German), Berlin: Ernst.

Griffiths D.V. and Lane P.A. (1999), Slope stability analysis by finite elements, Geotechnique, 49(3), 387–403.

Janbu N. (1973), Slope stability computations, in: R.C. Hirschfield and S.J. Poulos (eds.), Embankment-Dam Engineering, pp. 47–86, John Wiley.

Kwong J.S.M. (1991), Shaft Stability, PhD thesis, King's College London.

Low B.K., Gilbert R.B. and Wright S.G. (1998), Slope reliability analysis using generalized method of slices, Journal of Geotechnical and Geoenvironmental Engineering, 124(4), 350–362.

Morgenstern N.R. and Price V.E. (1965), The analysis of stability of general slip surface, Geotechnique, 15(1), 79–93.

Petterson K.E. (1955), The early history of circular sliding surfaces, Geotechnique, 5, 275–296.

Sarma S.K. (1973), Stability analysis of embankments and slopes, Geotechnique 23(3), 423–433.

Shield R.T. (1955), On the plastic flow of metals under conditions of axial symmetry, Proceedings of the Royal Society of London Series A, 233, 267–287.

Spencer E. (1967), A method of analysis of the stability of embankments assuming parallel inter-slice forces,

Geotechnique, 17, 11–26.

Taylor D.W. (1937), Stability of earth slopes, Journal of the Boston Society of Civil Engineers, 24, 197–246.

Zhang X. (1999), Slope stability analysis based on the rigid finite element method, Geotechnique, 49(5), 585–593.

CHAPTER 2

Upper and Lower Bound Approaches

Abstract: In this chapter, the basic theory about the lower and upper approaches will be introduced. These approaches will then be applied to various types of problems, and the applicability of the lower and upper bound approaches will be illustrated through many types of problems, for which analytical/rigorous solutions are available.

Keywords: Failure mechanism, Lower bound, Upper bound, Yield.

2.1. INTRODUCTION

To determine the collapse load of a structure composed of an elastic-plastic material and subject to a set of external forces (volume forces, surface tractions), there are essentially two different approaches: (1) an incremental load path analysis taking into account the complete stress-strain behaviour or (2) a direct limit analysis only taking into account the failure condition.

The first approach consists of performing an incremental elastic-plastic analysis up to failure to determine the collapse load. The material behaviour in terms of the stress-strain relationship up to failure is accounted for. Numerical computations are usually performed using computer programs based on finite elements or finite differences. Here failure is considered to take place when a representative displacement tends to increase without bound. Examples of this approach using finite element method can be found in Nagtegaal *et al.,* (1974), Chen (1975), Sload and Randolph (1982), Alehossein *et al.,* (1992), Yu *et al.,* (1993) and Potts and Zdravkovic (1999). The main inconveniences of this approach are that: it requires accounting for the complete stress-strain relations hence the amount of computations

Y. M. Cheng, H. Wong, C. J. Leo & C. K. Lau

involved is usually large; it needs the material properties for each material component which are difficult and expensive to obtain; failure load is determined when the computer program diverges and stops to operate normally.

Alternatively, a second approach uses the bound theorems which only takes into account the strength criterion of the material, but not the stress-strain behaviour. The material strength is supposed invariant during straining in order that the method works; locally failure and yield are synonymous and is described by a yield function. This approach leads directly to the determination of the collapse load, requiring much less amount of computation hence cost-effective, which constitutes its main advantage. Note that in civil engineering problems, structural collapse is normally the main concern. Therefore, the non-consideration of displacement and strain fields is in many cases not an important issue, especially at the preliminary design stage where a choice needs to be made among different alternatives.

Chen (1975) and Chen and Liu (1990) used this approach to geotechnical problems. A more theoretical treatment of this subject can be found in the monograph of Salençon (1983). For simple geometries and homogeneous material domains, analytical solutions are available. A very classical and illustrative problem concerns the tunnel face stability for which many researchers have made contributions, for example: Davis *et al.,* (1980), Chambon and Corté (1994), Leca and Dormieux (1990), Subrin and Wong (2002), Wong and Subrin (2006), only to cite a few. For complex geometries and heterogeneous material domains, numerical procedures are necessary. A brief account on numerical bound analysis based on finite elements and linear programming can be found in Yu (2006), which summarized the contributions of previous authors like of Lysmer (1970), Bottero *et al.,* (1980), Sloan (1988, 1989), Sloan and Kleeman (1995). These works are based on a linearization of the strength criterion. More recently, non-linear programming was introduced: D.Z. Li and Y.M. Cheng (2013).

In this chapter, emphasis will be put on a clear presentation of fundamental tools and concepts. We therefore limit ourselves to simple geometries where analytical solutions are accessible. For numerical approach, the reader can refer to the work of Yu (2006).

2.2 MATERIAL STRENGTH AND ITS MATHEMATICAL DESCRIPTION

As aforementioned, the bound-approach to the determination of collapse loads is based on the consideration of material strengths, discarding the stress-strain relation. It is therefore appropriate to begin with the mathematical description of the admissibility condition of stress fields, at the same time fixing notations for future references.

At any material point inside the physical domain, in order that the stress be supportable by the material, it must remain inside a domain of "admissible stresses" G as shown in Fig. (**2.1**) which is a subdomain of the six-dimensional Euclidean space $\mathbb{R}^6$ for 3D problems. Restricting the analysis to isotropic materials, the spatial orientation of the stress tensor does not count, only 3 independent scalar invariants representing the stress intensities intervene, for example the 3 principal stresses. The domain G can then be represented as a subdomain of the three-dimensional Euclidean space $\mathbb{R}^3$. Thermodynamic stability conditions require this domain to be convex, which is commonly described using a convex function $f(\boldsymbol{\sigma})$ such that:

$$G = \{\boldsymbol{\sigma} \in \mathbb{R}^3 : f(\boldsymbol{\sigma}) \leq 0\} \quad \textbf{(2.1)}$$

If the physical domain is not homogeneous, then G would be a function of space coordinates: $G = G(\boldsymbol{x})$, and so is the case of the yield function: $f = f(\boldsymbol{\sigma}, \boldsymbol{x})$. To simplify the presentation, we will omit this dependence on x and continue to write $f = f(\boldsymbol{\sigma})$, unless this dependence is explicitly required.

Note that a scalar function $f(\boldsymbol{\sigma})$ is convex if and only if for any 2 given points $\boldsymbol{\sigma}_A$ and $\boldsymbol{\sigma}_B$, and for any positive scalar λ such that $0 < \lambda < 1$, we have $f(\lambda\boldsymbol{\sigma}_A + (1-\lambda)\boldsymbol{\sigma}_B) < \lambda f(\boldsymbol{\sigma}_A) + (1-\lambda)f(\boldsymbol{\sigma}_B)$. In consequence, for any $\boldsymbol{\sigma}_A$ and $\boldsymbol{\sigma}_B$ inside G: $f(\lambda\boldsymbol{\sigma}_A + (1-\lambda)\boldsymbol{\sigma}_B) < \lambda f(\boldsymbol{\sigma}_A) + (1-\lambda)f(\boldsymbol{\sigma}_B) < 0$. Hence any point between $\boldsymbol{\sigma}_A$ and $\boldsymbol{\sigma}_B$ is also inside G thereby establishing the convexity of the latter domain. Fig. (**2.1**) provides a graphical illustration of the above ideas.

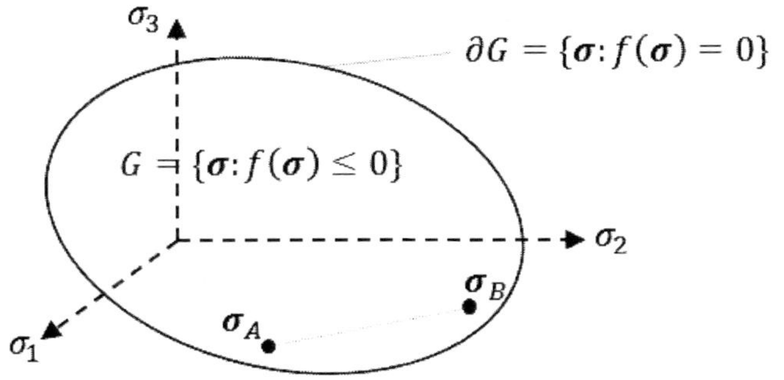

Fig. (2.1). The domain of admissible stresses *G*, the yield function $f(\boldsymbol{\sigma})$, and the convexity.

The bounding surface of *G*, noted ∂G, represents the limiting cases where the stress can just be supported and is defined by:

$$\partial G = \{\boldsymbol{\sigma} \in \mathbb{R}^3 : f(\boldsymbol{\sigma}) = 0\} \qquad \textbf{(2.2)}$$

The above definition implies that the boundary ∂G is part of the domain G. For the bound-approach to work, the domain G must be invariant with respect to deformations, implying absence of strain-hardening behaviour. Change of geometry due to straining is also discarded so that the equations of static equilibrium are considered at the initial, undeformed configuration. Ideally, the approach to be developed in this chapter aims at determining the collapse load of structures composed of materials having a rigid-plastic behaviour, so that locally the failure criterion can be identified with the yield criterion and the convex function $f(\boldsymbol{\sigma})$ used to define the domain *G* will be here-after referred as the "yield function".

2.3. PROBLEM SETTING AND OBJECTIVE

2.3.1. Geometry and Loadings

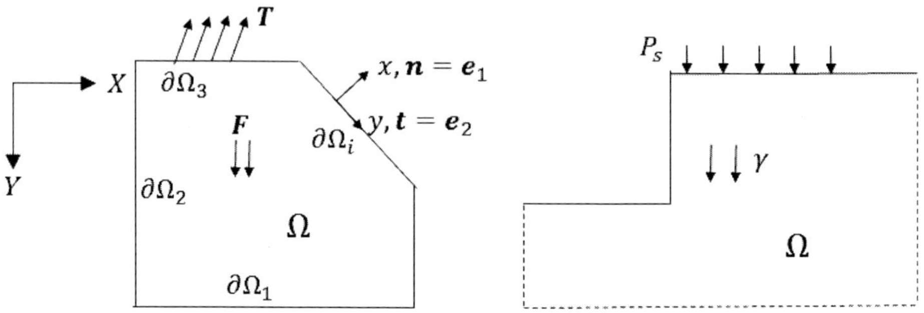

(a) Generic configuration (b) Particular case of a soil slope

Fig. (2.2). Notation for the physical domain, boundary surface and applied forces.

Consider a solid body, or equivalently a "physical domain" (*i.e.* the structure under analysis), contained in the geometric region Ω in (Fig. **2.2**). The couple (X, Y) represents the global coordinates while (x, y) the local axes for a subsurface $\partial\Omega_i$. $\boldsymbol{n} = \boldsymbol{e}_1$ is the outward unit normal while $\boldsymbol{t} = \boldsymbol{e}_2$ is the unit tangential vector. To simplify the presentation, only 2D problems are considered here, but the developments and notations can be generalized to 3D problems without difficulty. On this body, two types of loadings are applied:

(1) A volume force field $\boldsymbol{F}$ – in most cases the gravity (*i.e.* $\boldsymbol{F} = \boldsymbol{g}$) – is imposed on every material element inside Ω;

(2) Surface tractions $\boldsymbol{T}$ applied on the boundary $\partial\Omega$.

Note that only part of the surface tractions are prescribed quantities, termed the "applied surface loads", whereas others arise from the support conditions (for example a rigid support imposing a zero normal displacement on the subsurface $\partial\Omega_1$) and are *a priori* unknown quantities. The prescribed surface forces and velocities (*i.e.* support conditions) constitute the boundary conditions which are generally defined on a partition of the boundary into M subsurfaces: $\partial\Omega = \bigcup_{i=1}^{M} \partial\Omega_i$.

In classical continuum mechanics textbooks, on a given subsurface $\partial\Omega_i$, either a surface force is imposed, leading to: $\boldsymbol{\sigma} \cdot \boldsymbol{n} = \boldsymbol{T}$, or a given velocity is imposed: $\boldsymbol{v} = \boldsymbol{U}$. Here, $\boldsymbol{\sigma}$ and $\boldsymbol{v}$ are unknown fields to be determined, whereas $\boldsymbol{T}$ and $\boldsymbol{U}$ denote given quantities.

However, in limit analysis, very often we have to consider more complicated "mixed" boundary conditions. For a specific subsurface, $\partial\Omega_i$, and relative to its normal and tangential directions $\{\boldsymbol{e}_1, \boldsymbol{e}_2\} = \{\boldsymbol{n}, \boldsymbol{t}\}$, it often happens that a component of the surface traction is applied in one direction (static boundary condition) and a velocity component (kinematic boundary condition) is imposed on the other, complementary direction. More precisely, denote by $\mathcal{N}_t$ the set of couples (i, k) corresponding to static boundary conditions imposed on the subsurface $\partial\Omega_i$ relative to the direction k:

$$(i, k) \in \mathcal{N}_t \Rightarrow \partial\Omega_i\colon \boldsymbol{n} \cdot \boldsymbol{\sigma} \cdot \boldsymbol{e}_k = T_k \quad \textbf{(2.3)}$$

Note that $\boldsymbol{n} \cdot \boldsymbol{\sigma} \cdot \boldsymbol{e}_k = \boldsymbol{e}_1 \cdot \boldsymbol{\sigma} \cdot \boldsymbol{e}_k = \sigma_{1k}$ relative to the local coordinate system $\{\boldsymbol{e}_1, \boldsymbol{e}_2\}$, and T_k is a given external surface force. Evidently, for 3D problems, we would need to define 2 orthogonal tangential directions.

In parallel, denote by $\mathcal{N}_u$ the set of couples (i, m) corresponding to kinematic boundary conditons imposed on the subsurface $\partial\Omega_i$ relative to the direction different from those defined above:

$$(i, m) \in \mathcal{N}_u \Rightarrow \partial\Omega_i\colon \boldsymbol{v} \cdot \boldsymbol{e}_m = U_m \quad \textbf{(2.4)}$$

where U_m is a given velocity component. A couple (i,j) cannot belong to both $\mathcal{N}_t$ and $\mathcal{N}_u$. Together, the two sets contain all the $2M$ boundary conditions for a 2D problem ($3M$ for a 3D problem). Note that the 2 directions $\{\boldsymbol{e}_1, \boldsymbol{e}_2\}$ in question depend on the orientation of $\partial\Omega_i$ hence can vary from one boundary subsurface to another.

As an example, Fig. (**2.3**) shows a case where a rigid foundation is in contact with the soil mass. The physical domain is the semi-infinite half space.

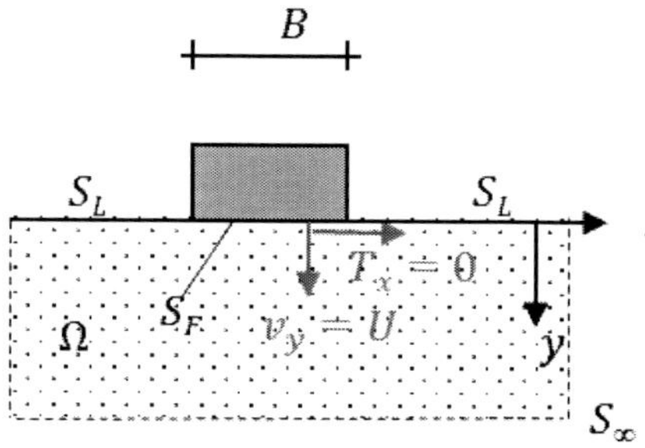

Fig. (2.3). "Mixed" boundary condition on a given boundary subsurface.

The foundation of width B penetrates into the ground, with an imposed vertical velocity $v_y = U$ over the subsurface S_F, constituting a kinematic boundary condition in the y-direction. Along the x-direction, we can for example consider either an applied horizontal force or the case of a smooth foundation with zero horizontal loading $-\boldsymbol{e}_y \cdot \boldsymbol{\sigma} \cdot \boldsymbol{e}_x = T_x = 0$ ($i.e. \boldsymbol{n} = -\boldsymbol{e}_y$). Note that in case of a foundation sitting on an inclined surface, the boundary condition would need to be defined relative to a pair of local inclined axes.

The spatial locations of the applied forces T_k and $\boldsymbol{F}$ are considered known, and their magnitudes are supposed entirely defined by a finite number N of scalar variables $\boldsymbol{Q} = (Q_1, Q_2, \dots Q_N)$, referred hereafter as load factors, so that we can write:

$$T_k = T_k(\boldsymbol{Q}) \text{ for } (i,k) \in \mathcal{N}_t\,; \quad \boldsymbol{F} = \boldsymbol{F}(\boldsymbol{Q}) \qquad \textbf{(2.5)}$$

For simplicity, we denote the ensemble of volume and surface loadings by $\boldsymbol{\mathcal{F}}(\boldsymbol{Q}) = (T_k(\boldsymbol{Q}), \boldsymbol{F}(\boldsymbol{Q}))$. For example, in a slope stability problem as shown in Fig. **(2.2(b))**, the loading can include a constant surface pressure P_s on a portion of the ground surface: $\boldsymbol{T} = -P_s\boldsymbol{n}$ and the volume force: $\boldsymbol{F} = \boldsymbol{\gamma} = \rho\boldsymbol{g}$ where ρ is the mass density and $\boldsymbol{g}$ is the gravity. We will use the notation $\gamma = ||\boldsymbol{\gamma}||$ and $g = ||\boldsymbol{g}||$ to denote the norm of vectors. In this case we have $N = 2$ and $\boldsymbol{Q} = (Q_1, Q_2) = (P_s, \gamma)$ as shown in Fig. (**2.4**). Since the load factors $\boldsymbol{Q}$ determine completely the external applied loads, it can be referred to as "the external loads" without risk of confusion. In this case, any particular loading can be considered as a point $\boldsymbol{Q}$ in the N-dimensional Euclidean space.

Local failure at a point $\boldsymbol{x}$ takes place when the stress there reaches the boundary of G: $f(\boldsymbol{\sigma}(\boldsymbol{x})) = 0$. General failure can only occur when the failure zone is sufficiently generalised such that it connects a non-trivial set of boundary points of the physical domain so that a "failure mechanism" can be formed.

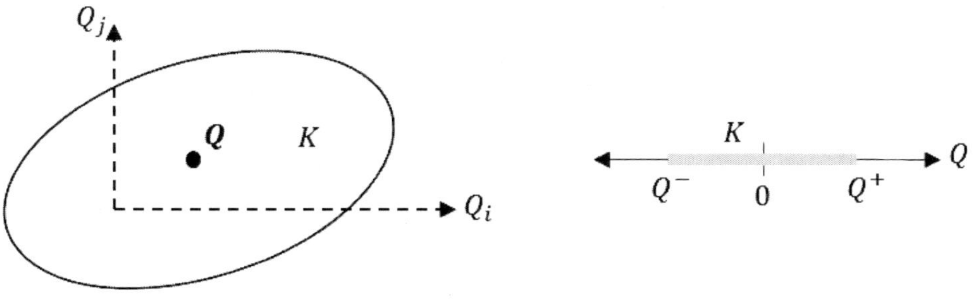

(a) Case $N = 2$ (b) Case $N = 1$

Fig. (2.4). The domain of admissible load factors.

2.3.2. Objective of Analysis: Domain of Supportable Loads

The objective is to determine a domain K of all supportable loading parameters $\boldsymbol{Q}$ such that any generic point $\boldsymbol{Q}$ inside K is "supportable" by the structure. Fig. (**2.4a**) illustrates graphically the two-dimensional case with $N = 2$ whereas Fig. (**2.4b**) the one-dimensional case with $N = 1$. Clearly this amounts to determine the boundary of K. In the particular case $N = 1$, the domain K is simply a segment of the line of real numbers, bounded by the pair of limit loads (Q^-, Q^+), and the problem is reduced to the determination of these two scalars.

By "supportable", it means that at least one stress field $\boldsymbol{\sigma}$ can be found which can equilibrate the external force fields $\boldsymbol{\mathcal{F}}(\boldsymbol{Q})$ – in other words satisfies the equilibrium equation – while the strength criterion is nowhere violated. The notion of "supportable" will be defined in more details within the context of the static theorem in the next section.

For future references and discussions, it is useful to summarize the main hypotheses and the corresponding implications:

H1. Infinitesimal displacements and strains: this implies a ***linear*** displacement strain relation $\boldsymbol{\varepsilon} = \widehat{\nabla}\boldsymbol{u}$, where $\widehat{\nabla}$ stands for the symmetrized gradient operation ($\widehat{\nabla} = (\nabla + \nabla^t)/2$) where the exponent "t" means the "transpose of". In component notation $\varepsilon_{ij} = \left(\partial_i u_j + \partial_j u_i\right)/2$, in which ∂_i means differentiation with respect to the x_i-cooordinate. In this case the actual and the initial configurations, Ω_t and Ω_0, are infinitely close, so that the differential operations relative to Lagrangian and Eulerian coordinates become identical to the first order.

H2. We will limit ourselves to elastic-perfectly plastic materials, with a fixed domain of resistance G which does not evolve during elastic-plastic straining and possess an infinite ductility. In other words, we exclude materials with a hardening behaviour. Furthermore, recall that the condition of thermodynamic stability requires G to be convex, or at least semi-convex.

H3. In the sequel, we will also assume that the functional dependence of external loads on the load factors, expressed by the functions $\boldsymbol{\mathcal{F}}(\boldsymbol{Q}) = (\boldsymbol{F}(\boldsymbol{Q}), T_k(\boldsymbol{Q}))$ are ***linear***. This linearity is a very important assumption in the construction of the theory.

H4. In this chapter, we will restrict ourselves to isotropic materials where the strength is defined by the criteria of either Mohr-Coulomb (c, ϕ) or Tresca $(c, \phi = 0)$. Note that this correspond to the quasi-totality of cases encountered in classic civil engineering problems.

2.4 LOWER BOUND OR STATIC APPROACH

The static theorem states that a given loading $\boldsymbol{Q}$ can be supported if an internal stress field $\boldsymbol{\sigma}$ can be found which satisfies the following two conditions: (1) the external and internal force fields are in static equilibrium with each other (we will simply write $\boldsymbol{\sigma}$ SA $\boldsymbol{Q}$) and (2) the internal force field $\boldsymbol{\sigma}$ satisfies the strength criterion at all material points. However, there is an important difference concerning condition (1) compared to classic continuum mechanics, as we will see shortly.

2.4.1 Statically Admissible Stress Fields

Classically, in a solid body as shown in Fig. (**2.5**), external forces defined by $\boldsymbol{Q}$ are in static equilibrium with the internal force field $\boldsymbol{\sigma}$ if the following two conditions are satisfied:

$$SA1: \forall \boldsymbol{x} \in \Omega: div(\boldsymbol{\sigma}) + \boldsymbol{F}(\boldsymbol{Q}) = \boldsymbol{0} \quad \textbf{(2.6)}$$

$$SA2: \forall (i,k) \in \mathcal{N}_t, \forall \boldsymbol{x} \in \partial\Omega_i: \boldsymbol{n} \cdot \boldsymbol{\sigma} \cdot \boldsymbol{e}_k = T_k(\boldsymbol{Q}) \quad \textbf{(2.7)}$$

The first condition expresses static equilibrium for points inside the domain Ω and the second for points on the boundary $\partial\Omega$ between the internal force field $\boldsymbol{\sigma}$ and the external applied loads $\boldsymbol{Q}$. In the case of limit analysis, an additional condition intervenes. At the onset of instability for which we try to determine the condition of appearance, the stress field can become piece-wise continuous: jump-discontinuities can arise at one or more surfaces crossing the physical domain. This is illustrated by Fig. (**2.5**) for the simplified case of one single surface of discontinuity Σ.

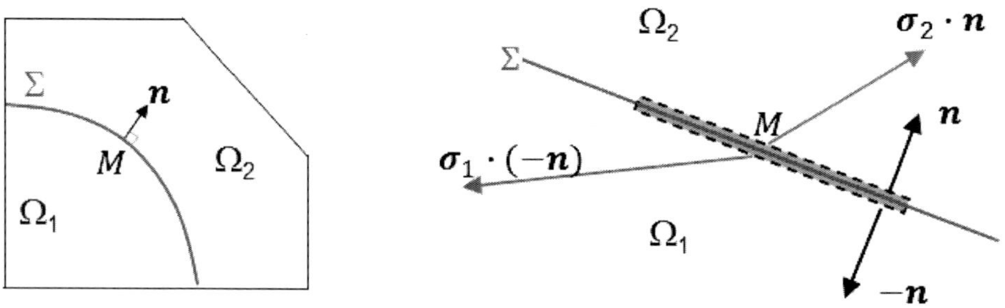

Fig. (2.5). Equilibrium across a surface of discontinuity.

The right hand figure shows an enlarged view around a generic point M on the surface Σ. Suppose we are at the critical moment when instability is about to trigger-off, hence the stress field becomes piece-wise continuous, while equilibrium still holds momentarily. Consider a very small but finite material disk, infinitely thin, of area dS whose mid-plane coincides with the surface Σ. Its upper surface inside Ω_2 is subject to a stress vector of $\boldsymbol{\sigma}_2 \cdot \boldsymbol{n}$ while the lower surface to a stress vector of $\boldsymbol{\sigma}_1 \cdot (-\boldsymbol{n})$. Imagine the thickness be arbitrarily small, so that the resulting force acting on the lateral surface can be neglected. Static equilibrium of this element then leads to:

$$SA3: [\![\boldsymbol{\sigma}]\!] \cdot \boldsymbol{n} \equiv (\boldsymbol{\sigma}_2 - \boldsymbol{\sigma}_1) \cdot \boldsymbol{n} = \boldsymbol{0} \quad \textbf{(2.8)}$$

Where the symbol $[\![.]\!]$ denotes the jump of a quantity in the direction of the unit normal $\boldsymbol{n}$. Note that this third condition SA3 is of equal importance as SA1 and SA2, and it must be satisfied in order to ensure that static equilibrium does hold "everywhere" when considering problems of limiting equilibrium.

When these conditions are satisfied, we say that the stress field is *statically admissible* (SA) for the loading $\boldsymbol{Q}$, or in an abstract form: $\boldsymbol{\sigma}\ SA\ \boldsymbol{Q}$. When the context is clear, we can simply write $\boldsymbol{\sigma}\ SA$.

2.4.2. Plastically Admissible Stress Fields

For a stress field to be admissible, it must be compatible with the strength criterion at all points $\boldsymbol{x}$ inside the physical domain. In mathematical terms, we write:

$$\forall \boldsymbol{x} \in \Omega: f\big(\boldsymbol{\sigma}(\boldsymbol{x})\big) < 0 \qquad \textbf{(2.9)}$$

bearing in mind that the domain G, hence the function f can both depend on $\boldsymbol{x}$ in case of heterogeneous material properties. When the above condition is satisfied, we say that the stress field is *plastically admissible* (PA) on account of the similarity between the failure criterion and the plastic yield condition. To make the notations compact, we will write: $\boldsymbol{\sigma}\ PA$.

2.4.3. The Static Theorem

In consequence, the static theorem can be cast into a compact form:

$$\exists \boldsymbol{\sigma}\ SA\ \boldsymbol{Q} \ \text{ and } \ \boldsymbol{\sigma}\ PA \Leftrightarrow \boldsymbol{Q} \in K \qquad \textbf{(2.10)}$$

In words, it says that for a given load $\boldsymbol{Q}$, if a stress field $\boldsymbol{\sigma}$ can be found such that it can equilibrate the load $\boldsymbol{Q}$ and at the same time satisfy the strength criterion everywhere, then the load can be supported, hence is inside the domain of supportable loads K.

This theorem is considered as an axiom in the present context and no proof will be given on its validity. Within the framework of the hypotheses H1-H4, it can be shown that an ideal elastic-plastic analysis assuming infinite ductility will give the same limit load as the ideal solution exhibited by the bound-approach.

A practical application of the static theorem can conceptually be divided into two steps. The first step consists of postulating a family of stress fields $\boldsymbol{\sigma}(\boldsymbol{a},\boldsymbol{x})$ depending on a set of unknown parameters $\boldsymbol{a}$ while all satisfy a priori the static admissibility conditions SA1, SA2 and SA3, in other words $\boldsymbol{\sigma}(\boldsymbol{a},\boldsymbol{x})$ SA $\boldsymbol{Q}$. The equilibrium conditions allow to express $\boldsymbol{a}$ in terms of $\boldsymbol{Q}$: $\boldsymbol{a}=\boldsymbol{a}(\boldsymbol{Q})$. We can therefore write $\boldsymbol{\sigma}(\boldsymbol{Q},\boldsymbol{x})$. The second step consists of imposing the plastic admissibility condition which writes: $\forall \boldsymbol{x} \in \Omega: f(\boldsymbol{\sigma}(\boldsymbol{Q},\boldsymbol{x})) \leq 0$.

More precisely, we should write $f(\boldsymbol{\sigma}(\boldsymbol{Q},\boldsymbol{x}),\boldsymbol{x}) \leq 0$ bearing in mind that the material strength can be heterogeneous. To simplify notations, we will suppose this point implicitly understood. The last condition leads to an optimization problem on the load factor $\boldsymbol{Q}$:

$$\mathbb{F}(\boldsymbol{Q}) = \max_{\boldsymbol{x}\in\Omega} f(\boldsymbol{\sigma}(\boldsymbol{Q},\boldsymbol{x})) \leq 0 \qquad \textbf{(2.11)}$$

The static bound is determined by the limiting case where the yield condition is just reached at a subset of the physical domain:

$$\mathbb{F}(\boldsymbol{Q}^{stat}) = 0 \qquad \textbf{(2.12)}$$

This last equation gives an approximate determination, denoted by ∂K^{stat}, of the true boundary ∂K, because the maximization is carried out on a finite-dimensional subset (often a small subset) of all the statically admissible stress fields, represented by the functions $\boldsymbol{\sigma}(\boldsymbol{a},\boldsymbol{x})$, depending on a finite number of parameters $\boldsymbol{a}$. A given loading $\boldsymbol{Q}^*$ which verifies

$\max_{x \in \Omega} f(\boldsymbol{\sigma}_1(\boldsymbol{Q}^*, \boldsymbol{x})) = 0$ for a family $\boldsymbol{\sigma}_1(\boldsymbol{a}_1, \boldsymbol{x})$ may appear to be on the boundary ∂K according to that family. However, it is possible that another family of stress fields $\boldsymbol{\sigma}_2(\boldsymbol{a}_2, \boldsymbol{x})$ with more refined variations (probably also a larger number of parameters $\boldsymbol{a}_2$), can lead to $\max_{x \in \Omega} f(\boldsymbol{\sigma}_2(\boldsymbol{Q}^*, \boldsymbol{x})) < 0$, implying that $\boldsymbol{Q}^*$ is in fact inside K. Hence any ∂K^{stat} determined from the static approach must be entirely inside ∂K, or equivalently:

$$K^{stat} \subset K \tag{2.13}$$

This is why we also call the static bounds the interior bounds.

2.4.4. Example of Applications: Stability of a Long Wall Mining Tunnel

As a first example, let us consider a transverse tunnel containing the excavation face in a typical configuration of long wall mining, as shown in Fig. (**2.6**) below. Fig. (**2.7**) shows an idealized geometry accounted for in subsequent analysis.

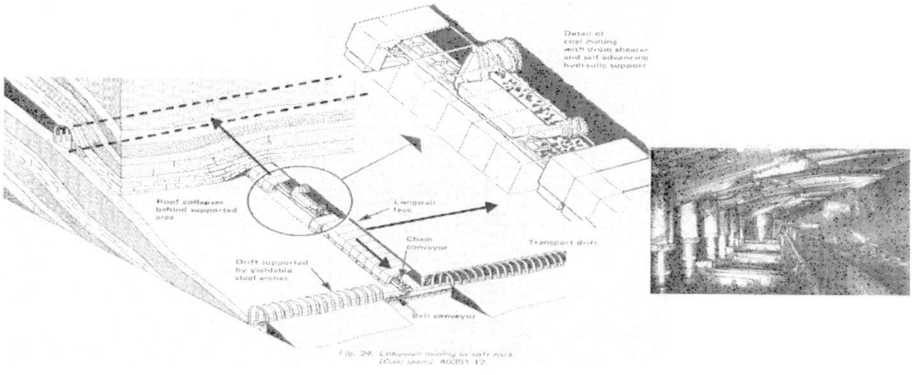

Fig. (2.6). Schematic illustration of a long wall mining.

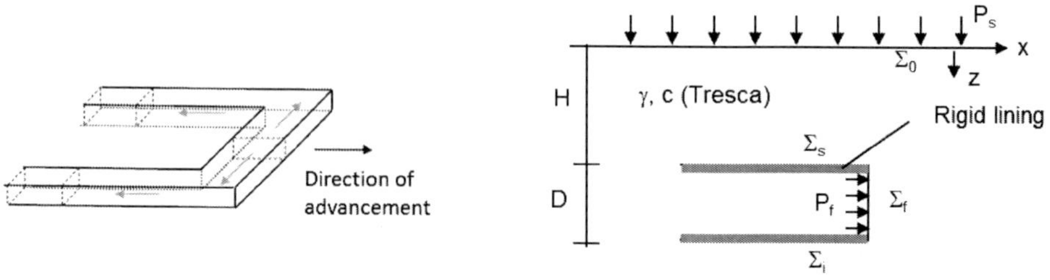

(a) Schematic 3D representation
(b) cross-section of transverse tunnel containing excavation face

Fig. (2.7). Geometric idealization of the long wall mining system.

Along the gallery in the transversal direction containing the excavation front (let us call it the excavation tunnel), coal extracted is evacuated by a conveyer belt to the two tunnels perpendicular to the excavation face (let us call them the transportation tunnels). The coal is then transported along these two transportation tunnels to the outside. The ground is supposed to exhibit an elastic-perfectly plastic behaviour, with a cohesion c and a zero friction angle $\phi = 0$. Failure is defined by Tresca's criterion. The ground above the tunnel is of thickness H while D is the height of the tunnel (Fig. **2.7**). We propose here to analyze the condition of failure. Theoretically, the three parameters (P_f, P_s, γ) can all be considered as loading parameters, so that K would be 3D. Conceptually however, and in order to fix ideas, we will take the face pressure as the only variable and all other parameters are supposed constants.

To facilitate the display of static admissibility conditions, we start by summarizing the static boundary conditions:

(a) Ground surface Σ_0: $\boldsymbol{\sigma} \cdot \boldsymbol{n} = \boldsymbol{\sigma} \cdot \begin{Bmatrix} 0 \\ -1 \end{Bmatrix} = \begin{Bmatrix} -\sigma_{xz} \\ -\sigma_{zz} \end{Bmatrix} = \begin{Bmatrix} 0 \\ P_s \end{Bmatrix}$, with P_s independent of x

(b) Tunnel floor Σ_i and ceiling Σ_s: $\sigma_{xz} = 0$

(c) Excavation face Σ_f: : $\boldsymbol{\sigma}\cdot\boldsymbol{n}=\boldsymbol{\sigma}\cdot\begin{Bmatrix}-1\\0\end{Bmatrix}=\begin{Bmatrix}-\sigma_{xx}\\-\sigma_{xz}\end{Bmatrix}=\begin{Bmatrix}P_f\\0\end{Bmatrix}$, with

P_f independent of z

Condition (b) implies a perfectly smooth soil-lining contact with zero shear stress. Condition (c) can correspond to a uniform air pressure. Another way to look at this is that in reality there is zero pressure (atmospheric pressure taken as reference datum), but we would like to situate this real condition to the pressure theoretically required to prevent collapse in order to appreciate the safety margin.

The other boundary conditions, of kinematic nature, are listed below:

- vertical velocity null at Σ_i and Σ_s: $v_z = 0$
- far from the excavation tunnel, velocity null: $v_x = v_z = 0$

These kinematic boundary conditions do not intervene when applying the static theorem.

On account of the static boundary conditions (a) – (c) above and that gravity is the only body force (*i.e.* $\boldsymbol{F}=\{0,\gamma\}$), we postulate a piece-wise continuous stress field composing of three zones as shown in Fig. (**2.8**).

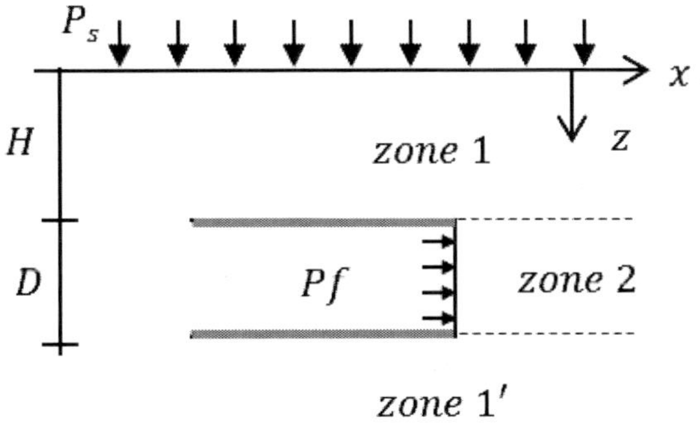

Fig. (2.8). A statically admissible stress field composing of 3 zones.

The expressions of the stress tensor in each of these 3 zones as shown in Fig. (**2.8**) are:

$$\text{zone 1 and 1}': \boldsymbol{\sigma} = \begin{bmatrix} \sigma_{xx}(z) & 0 \\ 0 & -(P_s + \gamma z) \end{bmatrix}; \quad \text{zone 2}: \boldsymbol{\sigma} = \begin{bmatrix} -P_f & 0 \\ 0 & -(P_s + \gamma z) \end{bmatrix} \quad \textbf{(2.14)}$$

We have $div(\boldsymbol{\sigma}(\boldsymbol{x})) + \boldsymbol{F} = \{0, -\gamma\} + \{0, \gamma\} = \boldsymbol{0}$ hence the condition SA1 is verified. It can also readily be seen that the stress field defined above verifies the static boundary conditions (a), (b) and (c) above; hence the condition SA2 is also verified. At the interface between zone 1 and zone 2, where $z = H$, and $\boldsymbol{n} = \boldsymbol{e}_z = \{0,1\}$, we have $\boldsymbol{\sigma}^1 \cdot \boldsymbol{n} = \boldsymbol{\sigma}^2 \cdot \boldsymbol{n} = \{0, -(P_s + \gamma H)\}$. Similarly, at the interface between zone 2 and zone 1', we can show that $\boldsymbol{\sigma}^{1'} \cdot \boldsymbol{n} = \boldsymbol{\sigma}^2 \cdot \boldsymbol{n} = \{0, -(P_s + \gamma(H + D))\}$, hence condition SA3 is verified too, allowing the conclusion that the stress field is indeed statically admissible. The next step is to enforce the plastic admissibility on this stress field, which would lead us to the desired result.

Recall that for a purely cohesive material obeying Tresca's criterion, the difference between the principal stresses must be less than twice the cohesion: $|\sigma_I - \sigma_{III}| < 2c$. This criterion can be automatically be verified in both zone 1 and 1' by choosing $\sigma_{xx}(z) = -(P_s + \gamma z)$, implying a hydrostatic stress state in these zones.

It remains to enforce the plastic admissibility in zone 2. Given the diagonal form of the stress tensor, the principal stresses are simply the diagonal members. However, their order is unknown. Therefore plastic admissibility implies that:

$$\forall\, z \in (H, H + D): \ |P_f - (P_s + \gamma z)| < 2c \quad \textbf{(2.15)}$$

This is equivalent to write:

$$\forall\, z \in (H, H + D): \ -2c < P_f - (P_s + \gamma z) < 2c \quad \textbf{(2.16)}$$

To fix ideas, suppose only P_f can vary and all other parameters are fixed. The left hand side (LHS) inequality can be simplified to:

$$\forall\, z \in (H, H + D) \colon\ P_f > P_s + \gamma z - 2c \tag{2.17}$$

This leads to a maximization problem:

$$P_f > \max_{z \in (H, H+D)} P_s + \gamma z - 2c \tag{2.18}$$

Leading to a lower bound for P_f:

$$P_f > P^-_{f,stat} = P_s + \gamma(H + D) - 2c \tag{2.19}$$

The exponent "-" indicates that it is a lower bound while the index "stat" reminds us that this lower bound is obtained by the static theorem, by anticipation of the bounds to be obtained using the kinematic theorem. Following the same steps, we get from the right hand side (RHS) the upper bound:

$$P_f < P^+_{f,stat} = P_s + \gamma H + 2c \tag{2.20}$$

Summarizing, the structure is stable if P_f is confined to an interval (*i.e.* a 1D domain) denoted by K^{stat} bounded by 2 extreme values (the static bounds):

$$\begin{aligned} &P_f \in K^{stat} \implies \text{structure potentially stable} \\ &K^{stat} = (P_s + \gamma(H + D) - 2c\,,\ P_s + \gamma H + 2c) \end{aligned} \tag{2.21}$$

It is important to note that according to the static theorem, whenever it is possible to exhibit a stress field $\boldsymbol{\sigma}$ SA and PL, the corresponding loading $\boldsymbol{Q}$ can be supported, or in other words lies strictly inside the domain of admissible loads K. This means that bounds obtained using the static theorem are interior to the exact solution domain K. Note that the limit loads have been determined by supposing an infinite ductility (see assumption **H2**), so that a complete load redistribution can take place when yielding has occurred locally. Otherwise, failure can take place at a smaller loadings. That is why we often qualify the stability condition as "potentially" stable. This point being implicitly understood, we will omit this qualification in the rest of this chapter.

To make this point as clear as possible, suppose all the load factors but one are fixed (like P_f in the above example), and we are interested by determining its lower and upper bounds. Since for any field $\boldsymbol{\sigma}$ SA and PL, the corresponding load factor must lie inside K, the bounds determined using the static theorem must be less than the true upper bound and greater than the true lower bound. If by considering a second stress field we obtain an upper bound which is larger than the one determined using the previous stress field, then this new values is a better estimate of the upper bound. Following the same logic, a smaller lower bound is a better estimate within the context of the static theorem.

However, it is not an easy task to exhibit a family of stress fields *a priori* statically admissible, defined by a small number of undetermined coefficients and exhibiting a "flexible geometry", for example with a progressive rotation of principal directions. For this reason, most analytical solutions are constructed using the kinematic theorem (the kinematic approach) which will be presented in later sections.

2.4.5. Physical Interpretation of the Two Bounds

Fig. (**2.9**) gives a physical interpretation of the two bounds.

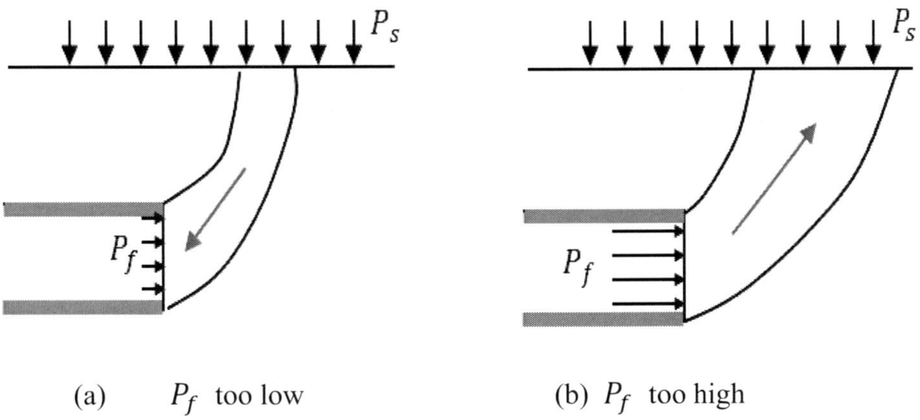

Fig. (2.9). Failure modes for the two extreme cases.

When the face pressure is too low (*i.e.* falls below the true lower bound), collapse may take place such that the mass of soil enters the cavity. On the contrary, when it is too high (*i.e.* goes above the true upper bound), it also leads to failure *via* a "blow-out" mechanism.

We also have to ensure that the domain of stability k^{stat} is not empty by verifying that $P^{+}_{f,stat} > P^{-}_{f,stat}$. A short computation leads to the relation:

$$D < \frac{4c}{\gamma} \quad \textbf{(2.22)}$$

In other words, the height of the cavity must be limited to ensure stability. Otherwise, a rotational failure as shown in Fig. (**2.10**) occurs, due to gravity effects, whatever the values of P_s and P_f. Since these pressures do not perform any work in the rotational failure, they do not intervene. This failure mode can be easily treated by the kinematic approach to be developed later (very difficult by the static approach!).

Fig. (2.10). Failure mode in case of an excessive height.

2.4.6 Improvement of the Static Bounds

To improve the previous estimates of bounds, we have to exhibit more complicated stress fields. This can be done by increasing the number of zones in the definition of a piece-wise continuous stress field as shown in (Fig. **2.11**). One example of this computation was the work of Davis *et al.,* (1980) in computing the lower bound, as shown in the figure below.

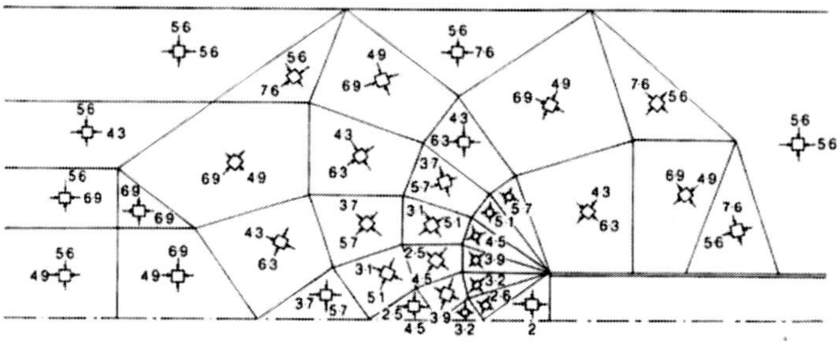

Fig. (2.11). Division of the physical domain in Davis *et al.,* (1980).

2.5. Upper Bound or Kinematic Approach

The static approach allows to obtain a pessimistic estimation of the domain of supportable loadings, in the sense that $K^{stat} \subset K$. This is no small achievement given that only the conditions of static equilibrium and the strength criterion need to be accounted for, saving a lot of computation effort compared to an incremental elastic-plastic approach. However, depending on the quality of the stress field displayed (which is a difficult task), the interior bounds obtained can be significantly far (or not!) from the ideal solutions. The problem is that, the ideal solution being unknown to the engineering, the quality of the static bounds cannot be assessed. This motivates the development of the kinematic approach, complementary to its static counterpart, whose objective is precisely to furnish an exterior estimate K^{kin} which satisfies the inclusion property:

$$K^{stat} \subset K \subset K^{kin} \qquad \textbf{(2.23)}$$

In the particular 1D case with one single independent load factor Q, this amount to the following conditions on the static and kinematic bounds:

$$Q^-_{kin} < Q^- < Q^-_{stat}\ ; \qquad Q^+_{stat} < Q^+ < Q^+_{kin} \qquad \textbf{(2.24)}$$

The ideal solutions (the "true" bounds) Q^- and Q^+ are still unknown, but the "gap" between the static and kinematic bounds: $Q^-_{stat} - Q^-_{kin}$ and $Q^+_{kin} - Q^+_{stat}$ obtained indicates the quality of the approximate solutions obtained.

2.5.1. Virtual Work Theorem

Before beginning the presentation, let us recall briefly the theorem of virtual work indispensable for the development of the kinematic approach. The theorem of virtual work allows us to recast the condition of static equilibrium (*i.e.* conditions SA1, SA2 and SA3 in section 2.3 in the form of algebraic and partial differential equations) into an integral form, more suitable for a formulation using virtual velocity fields.

Consider a stress field $\boldsymbol{\sigma}$ in equilibrium with a given set of external volume and surface forces denoted by $\boldsymbol{\mathcal{F}} = (\boldsymbol{F}, \boldsymbol{T})$, in other words verifies conditions SA1, SA2 and SA3 (*i.e.* $\boldsymbol{\sigma}$ SA $\boldsymbol{\mathcal{F}}$), and a virtual velocity field $\boldsymbol{v}$ which does not necessarily have any bearing with the true failure mechanism but must satisfy the kinematic boundary conditions (we will say that $\boldsymbol{v}$ is kinematically admissible or simply $\boldsymbol{v}$ KA). Multiplication of the equilibrium equation $div(\boldsymbol{\sigma}) + \boldsymbol{F} = \boldsymbol{0}$ by this virtual velocity, followed by an integration over the entire physical domain Ω leads to the following equality:

$$\int_{\Omega} \boldsymbol{v} \cdot (div(\boldsymbol{\sigma}) + \boldsymbol{F})\, d\Omega = \sum_{i} \int_{\Omega_i} \boldsymbol{v} \cdot (div(\boldsymbol{\sigma}) + \boldsymbol{F})\, d\Omega = 0 \qquad \textbf{(2.25)}$$

The volume integration has to take into account that both the stress field and the velocity field can be discontinuous over some (the same for both fields) internal surfaces. To simplify the presentation, only the case of a single surface of discontinuity Σ partitioning Ω into two halves is shown here in (Fig. **2.12**), but the result applies to more general cases.

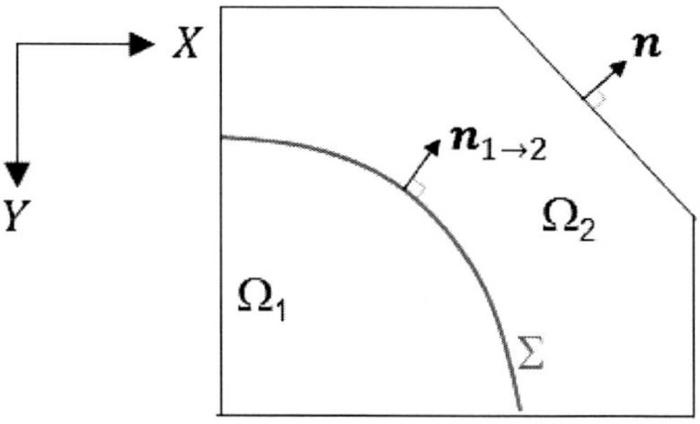

Fig. (2.12). Physical domain partitioned by a single surface of discontinuity.

On account of the identity: $\boldsymbol{v} \cdot div(\boldsymbol{\sigma}) = div(\boldsymbol{v} \cdot \boldsymbol{\sigma}) - \boldsymbol{\sigma}: \boldsymbol{d}(\boldsymbol{v})$, where $\boldsymbol{d}(\boldsymbol{v}) = \widehat{\nabla} \boldsymbol{v}$ is the strain rate field derived from the velocity field $\boldsymbol{v}$, and using Gauss' divergence theorem, we can develop the volume integral of $\boldsymbol{v} \cdot div(\boldsymbol{\sigma})$:

$$\int_{\Omega} \boldsymbol{v} \cdot div(\boldsymbol{\sigma}) = \sum_{i} \int_{\partial\Omega_i} \boldsymbol{v} \cdot \boldsymbol{\sigma} \cdot \boldsymbol{n}\, dS - \int_{\Omega} \boldsymbol{\sigma}: \boldsymbol{d}(\boldsymbol{v})\, d\Omega \qquad \textbf{(2.26)}$$

We now concentrate our attention on the first term on the right hand side. This is the sum of the surface integrals of the two subdomains Ω_1 and Ω_2. Noting that $\partial\Omega_1$ consist of half of the boundary of Ω plus the surface of discontinuity Σ, whereas $\partial\Omega_2$ is the sum of the other half of the boundary of Ω and Σ, we deduce that:

$$\sum_i \int_{\partial\Omega_i} \boldsymbol{v}\cdot\boldsymbol{\sigma}\cdot\boldsymbol{n}\,dS$$
$$= \int_{\partial\Omega} \boldsymbol{v}\cdot\boldsymbol{\sigma}\cdot\boldsymbol{n}\,dS \quad \textbf{(2.27)}$$
$$- \int_{\Sigma} (\boldsymbol{v}^2\cdot\boldsymbol{\sigma}^2\cdot\boldsymbol{n}_{1\to2} - \boldsymbol{v}^1\cdot\boldsymbol{\sigma}^1\cdot\boldsymbol{n}_{1\to2})\,d\Sigma$$

where $\boldsymbol{n}_{1\to2}$ is the unit normal vector of Σ pointing to the direction of Ω_2. Recall that $\boldsymbol{\sigma}$ has to satisfy the static condition SA3, which implies $\boldsymbol{\sigma}^2\cdot\boldsymbol{n}_{1\to2} = \boldsymbol{\sigma}^1\cdot\boldsymbol{n}_{1\to2}$ that we note simply $\boldsymbol{\sigma}\cdot\boldsymbol{n}_{1\to2}$. Moreover, noting that on the external boundary of Ω, we have $\boldsymbol{\sigma}\cdot\boldsymbol{n} = \boldsymbol{T}$, we derive finally the following relation, called the **theorem of virtual work** (if $\boldsymbol{v}$ is a displacement field) or virtual power (if $\boldsymbol{v}$ is a velocity field):

$$\boldsymbol{\sigma}\ \text{SA}\ \boldsymbol{\mathcal{F}} = \Leftrightarrow\ \forall \boldsymbol{v}\ \text{KA:} \int_{\Omega} \boldsymbol{v}\cdot\boldsymbol{F}\,d\Omega + \int_{\partial\Omega} \boldsymbol{v}\cdot\boldsymbol{T}\,dS$$
$$= \int_{\Omega} \boldsymbol{\sigma}:\boldsymbol{d}(\boldsymbol{v})\,d\Omega + \int_{\Sigma} [\![\boldsymbol{v}]\!]\cdot\boldsymbol{\sigma}\cdot\boldsymbol{n}\,d\Sigma \quad \textbf{(2.28)}$$

Note that whether $\boldsymbol{v}$ should be a (virtual !) velocity or displacement field has no importance since they are simply proportional to each other *via* an arbitrary time increment which would cancel out in the above equation due to the linearity of the construction. In the above we have used the notation $[\![\boldsymbol{v}]\!] = (\boldsymbol{v}^2 - \boldsymbol{v}^1)$ to denote the velocity jump in the direction of the unit normal vector $\boldsymbol{n}_{1\to2}$, which has been simply noted $\boldsymbol{n}$.

The LHS of the above equality represents the mechanical power furnished by the external volume and surface forces in an arbitrary but kinematically admissible velocity field $\boldsymbol{v}$, and can be put symbolically in the form $P^{ext}(\boldsymbol{\mathcal{F}},\boldsymbol{v})$. In parallel, the RHS represents the power furnished by the internal force field, on the same velocity field and can be written $P^{int}(\boldsymbol{\sigma},\boldsymbol{v})$. The theorem says that if the internal and external force fields are in equilibrium with each other, they must furnish the same mechanical power for any arbitrary but kinematically admissible velocity field. Conversely, if the internal and external force fields furnish an identical power for any arbitrary velocity field, then they must be in equilibrium with each other. In other words, the equality of mechanical power furnished by internal and external force fields on any kinematically admissible velocity field is equivalent to the 3 conditions of static equilibrium SA1, SA2 and SA3:

$$\boldsymbol{\sigma}\ \text{SA}\ \boldsymbol{\mathcal{F}} \quad \Leftrightarrow \quad \forall \boldsymbol{v}\ \text{KA}:\ P^e(\boldsymbol{\mathcal{F}},\boldsymbol{v}) = P^i(\boldsymbol{\sigma},\boldsymbol{v})$$

$$P^e(\boldsymbol{\mathcal{F}},\boldsymbol{v}) = \int_\Omega \boldsymbol{v}\cdot\boldsymbol{F}\,d\Omega + \int_{\partial\Omega} \boldsymbol{v}\cdot\boldsymbol{T}\,dS \tag{2.29}$$

$$P^i(\boldsymbol{\sigma},\boldsymbol{v}) = \int_\Omega \boldsymbol{\sigma}:\boldsymbol{d}(\boldsymbol{v})\,d\Omega + \int_\Sigma [\![\boldsymbol{v}]\!]\cdot\boldsymbol{\sigma}\cdot\boldsymbol{n}\,d\Sigma$$

Notice that both the functionals (functions of functions) $P^e(\boldsymbol{\mathcal{F}},\boldsymbol{v})$ and $P^i(\boldsymbol{\sigma},\boldsymbol{v})$ are bilinear relative to their arguments. Moreover, it is clear that the dependence of $\boldsymbol{\mathcal{F}}$ on $\boldsymbol{Q}$ is also linear, therefore, we can write $P^e(\boldsymbol{Q},\boldsymbol{v})$ instead of $P^e(\boldsymbol{\mathcal{F}},\boldsymbol{v})$ so that:

$$\boldsymbol{\sigma}\ \text{SA}\ \boldsymbol{Q} \quad \Leftrightarrow \quad \forall \boldsymbol{v}\ \text{KA}:\ P^e(\boldsymbol{Q},\boldsymbol{v}) = P^i(\boldsymbol{\sigma},\boldsymbol{v}) \tag{2.30}$$

Another important remark concerns the practical evaluation of the mechanical power developed by the surface forces $\int_{\partial\Omega} \boldsymbol{v}\cdot\boldsymbol{T}\,dS$. For a known stress field, $\boldsymbol{T}$ can of course be computed from the stress field by $\boldsymbol{T} = \boldsymbol{\sigma}\cdot\boldsymbol{n}$. However, in a practical computation, $\boldsymbol{\sigma}$ is not known in advance and we would like to bring in the prescribed static boundary conditions. Hence, on those portions of the boundary surface $\partial\Omega_i$ and for the directions k where a surface traction T_k is imposed (*i.e.* $(i,k) \in \mathcal{N}_t$), we compute the contribution to the integral using $v_k T_k$ (no sum on index k). For the complementary directions (*i.e.* $(i,j) \in \mathcal{N}_v$), we will compute T_j from $\boldsymbol{\sigma}$ using $T_j = \boldsymbol{n} \cdot \boldsymbol{\sigma} \cdot \boldsymbol{e}_j$.

2.5.2. The maximum Resisting Power and the π-Functions

A crucial step in the construction of the kinematic theorem is the introduction of an upper bound to the internal mechanical power $P^i(\boldsymbol{\sigma}, \boldsymbol{v})$, referred here-after as the resisting power, denoted by $P^r(\boldsymbol{v})$. By definition we can write:

$$P^r(\boldsymbol{v}) = \max_{\boldsymbol{\sigma}\ PA} P^i(\boldsymbol{\sigma}, \boldsymbol{v}) = \int_\Omega \pi_v(\boldsymbol{d})\, d\Omega + \int_\Sigma \pi_s([\![\boldsymbol{v}]\!], \boldsymbol{n})\, d\Sigma \qquad \textbf{(2.31)}$$

Where the two functions inside the integrals are defined by:

$$\pi_v(\boldsymbol{d}) = \max_{\boldsymbol{\sigma} \in G} \boldsymbol{\sigma} : \boldsymbol{d} \quad ; \quad \pi_s([\![\boldsymbol{v}]\!], \boldsymbol{n}) = \max_{\boldsymbol{\sigma} \in G} [\![\boldsymbol{v}]\!] \cdot \boldsymbol{\sigma} \cdot \boldsymbol{n} \qquad \textbf{(2.32)}$$

By construction, this functional $P^r(\boldsymbol{v})$ depends linearly on its argument $\boldsymbol{v}$ and represents the maximum internal power that can be derived from a given velocity field by choosing a stress field that satisfies only the strength criterion everywhere, but no other conditions, notably the equilibrium or continuity conditions. This leads us to the following important inequality which paves the way for the construction of the kinematic theorem:

$$\forall\, \boldsymbol{v}\ \text{KA}, \forall \boldsymbol{\sigma}\ \text{PA}: \ P^r(\boldsymbol{v}) \geq P^i(\boldsymbol{\sigma}, \boldsymbol{v}) \qquad \textbf{(2.33)}$$

Evidently it is equivalent to write: $\pi_v(\boldsymbol{d}) = \max_{f(\boldsymbol{\sigma}) \leq 0} \boldsymbol{\sigma} : \boldsymbol{d}$ and $\pi_s([\![\boldsymbol{v}]\!], \boldsymbol{n}) = \max_{f(\boldsymbol{\sigma}) \leq 0} [\![\boldsymbol{v}]\!] \cdot \boldsymbol{\sigma} \cdot \boldsymbol{n}$. In fact since $\boldsymbol{\sigma} : \boldsymbol{d}$ and $[\![\boldsymbol{v}]\!] \cdot \boldsymbol{\sigma} \cdot \boldsymbol{n}$ are linear relative to $\boldsymbol{\sigma}$, and G contains the origin, the optimum stress must lie on the boundary of G, hence we can also write: $\pi_v(\boldsymbol{d}) = \max_{f(\sigma)=0} \boldsymbol{\sigma}:\boldsymbol{d}$;

$\pi_S([\![\boldsymbol{v}]\!], \boldsymbol{n}) = \max_{f(\sigma)=0} [\![\boldsymbol{v}]\!] \cdot \boldsymbol{\sigma} \cdot \boldsymbol{n}$.

It is evident that the determination of π_v leads immediately to that of π_S. The symmetry of $\boldsymbol{\sigma}$ implies that we can write $[\![\boldsymbol{v}]\!] \cdot \boldsymbol{\sigma} \cdot \boldsymbol{n} = \boldsymbol{\sigma}:\boldsymbol{t}$ where $\boldsymbol{t} = [\![\boldsymbol{v}]\!] \widehat{\otimes} \boldsymbol{n} \equiv ([\![\boldsymbol{v}]\!] \otimes \boldsymbol{n} + \boldsymbol{n} \otimes [\![\boldsymbol{v}]\!])/2$ is the symmetrized tensor product of $[\![\boldsymbol{v}]\!]$ and $\boldsymbol{n}$, of components $t_{ij} = \left([\![v_i]\!] n_j + n_i [\![v_j]\!]\right)/2$. Hence:

$$\pi_S([\![\boldsymbol{v}]\!], \boldsymbol{n}) = \max_{\sigma \in G} \boldsymbol{\sigma}: [\![\boldsymbol{v}]\!] \widehat{\otimes} \boldsymbol{n} = \pi_V\left([\![\boldsymbol{v}]\!] \widehat{\otimes} \boldsymbol{n}\right) \quad \textbf{(2.34)}$$

The function $\pi_v(\boldsymbol{d})$ is associated to the domain G just like the yield function $f(\boldsymbol{\sigma})$. In fact, the domain G can equivalently be expressed as:

$$G = \{\boldsymbol{\sigma} \in \mathbb{R}^3 \colon \pi(\boldsymbol{d}) \geq \boldsymbol{\sigma}:\boldsymbol{d} \text{ for } \forall \boldsymbol{d} \in \mathbb{R}^3\} \quad \textbf{(2.35)}$$

Note that since both $\pi_V(\boldsymbol{d})$ and $\boldsymbol{\sigma}:\boldsymbol{d}$ are linear with respect to $\boldsymbol{d}$ (hence linear with respect to $\boldsymbol{v}$), we can limit the optimization to the subset of unit vectors $||\boldsymbol{d}|| = 1$, so that:

$$G = \{\boldsymbol{\sigma} \in \mathbb{R}^3 \colon \pi_V(\boldsymbol{d}) \geq \boldsymbol{\sigma}:\boldsymbol{d} \text{ for } \forall \boldsymbol{d} \in \mathbb{R}^3 \text{ such that } ||\boldsymbol{d}|| = 1\} \quad \textbf{(2.36)}$$

The function π_V has actually very simple geometric interpretations, as will be illustrated in the two following examples.

Example 1 of π_V-function: a circular domain

To see the geometric meaning of the function π_V, lets us consider the simpler case where $\boldsymbol{\sigma}$ and $\boldsymbol{d}$ are 2-component vectors in a 2D space, and that G is a circle of radius R centered upon the origin, as shown in Fig. (**2.13**). Generalization to 3D space is evident.

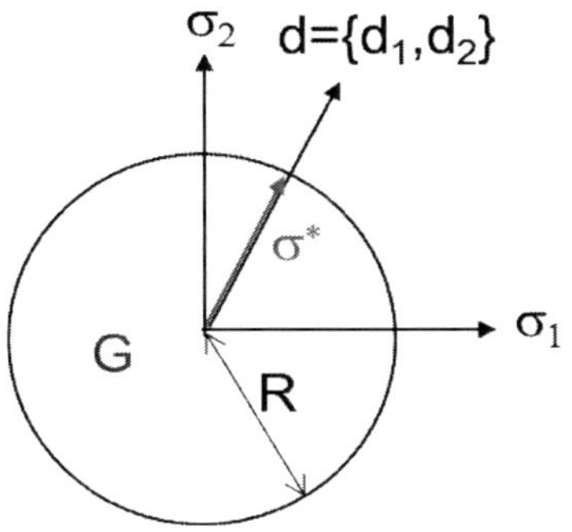

Fig. (2.13). Example of the π_V-function for a circular domain G.

For this particular case, it is easy to show that the optimum stress is the one collinear to $\boldsymbol{d}$ and lies on the boundary:

$$\pi_V(\boldsymbol{d}) = \boldsymbol{\sigma}^*:\boldsymbol{d}\,; \quad \boldsymbol{\sigma}^* = \frac{\boldsymbol{d}}{||\boldsymbol{d}||}R \qquad \textbf{(2.37)}$$

Example 2 of π_V-function: an elliptic domain

To show more clearly the geometric relations, let us consider a second case, where G is an ellipse centered upon the origin, of semi-widths a_1 and a_2 along the horizontal and vertical axes (or axis 1 and 2), as shown in Fig. (**2.14**).

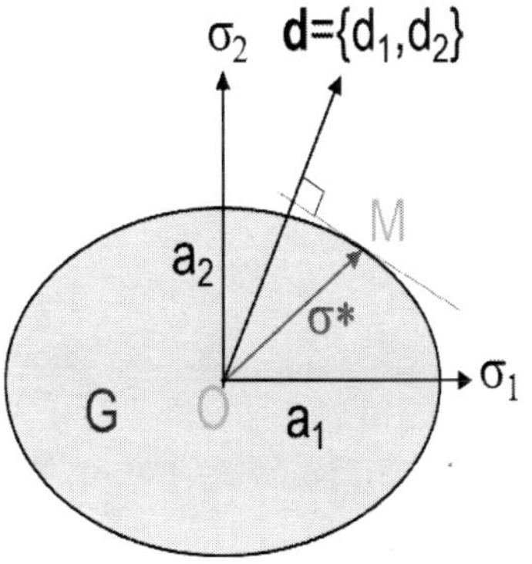

Fig. (2.14). Example of the π_V-function for an elliptical domain of admissible stresses.

Mathematically, this domain can be represented by:

$$G = \{\boldsymbol{\sigma} \in \mathbb{R}^2 : f(\boldsymbol{\sigma}) < 0\}\,; \quad f(\boldsymbol{\sigma}) = \left(\frac{\sigma_1}{a_1}\right)^2 + \left(\frac{\sigma_2}{a_2}\right)^2 - 1 \tag{2.38}$$

To find $\pi_{\mathrm{v}}(\boldsymbol{d}) = \max_{f(\boldsymbol{\sigma})=0} \boldsymbol{\sigma} : \boldsymbol{d}$, one can for example use the method of Lagrange multipliers to maximize the augmented function $h(\sigma_1, \sigma_2, \lambda) = \sigma_1 d_1 + \sigma_2 d_2 + \lambda\left[\left(\frac{\sigma_1}{a_1}\right)^2 + \left(\frac{\sigma_2}{a_2}\right)^2 - 1\right]$ by requiring the partial derivatives of h relative to each of its arguments be zero:

$$\frac{\partial h}{\partial \sigma_1} = \frac{\partial h}{\partial \sigma_2} = \frac{\partial h}{\partial \lambda} = 0 \tag{2.39}$$

The result is:

$$\boldsymbol{\sigma}^* = \frac{1}{\sqrt{(a_1\sigma_1)^2 + (a_2\sigma_2)^2}} \begin{pmatrix} a_1^2 d_1 \\ a_2^2 d_2 \end{pmatrix}\,; \quad \pi_{\mathrm{v}}(\boldsymbol{d}) = \sqrt{(a_1\sigma_1)^2 + (a_2\sigma_2)^2} \tag{2.40}$$

Geometrically, on a line perpendicular to $\boldsymbol{d}$, any point $\boldsymbol{\sigma}$ on it will yield a constant product $\boldsymbol{\sigma} : \boldsymbol{d}$, hence the optimum solution $\boldsymbol{\sigma}^*$ is obtained by determining the tangent point to ∂G of a line orthogonal to $\boldsymbol{d}$, which would yield a positive product.

2.5.3. π-Functions in Practical Applications

For practical applications, we would have to consider the two case important cases where the material strength is defined by the criteria of Tresca and Mohr-Coulomb. Note that only these two criteria are encountered in practical problems in classical civil engineering problems. For both of these criteria, the domain of admissible stresses are infinite and semi-convex. The determination of the corresponding π_{v}-functions are lengthy and complicated, and therefore will not be presented. Only the result will be shown here-below.

2.5.3.1 Criterion of Tresca

Recall that Tresca's yield criterion is defined by a single parameter - the cohesion c:

$$G = \{\boldsymbol{\sigma} \in \mathbb{R}^3 : f(\boldsymbol{\sigma}) = |\sigma_I - \sigma_{III}| - 2c < 0\} \qquad \textbf{(2.41)}$$

where $\sigma_I \geq \sigma_{II} \geq \sigma_{III}$ are the ordered principal stresses. In the space of (unordered) principal stresses $(\sigma_1, \sigma_2, \sigma_3)$, G has the form of a hexagonal cylinder with its axis coinciding with the hydrostatic axis defined by the equation: $\sigma_1 = \sigma_2 = \sigma_3$, semi-convex and of infinite dimension. The boundary of G is formed by 6 planar faces defined by the six equations:

$$\sigma_i - \sigma_j = \pm 2c \; ; \quad i \neq j \quad ; \quad i,j \in \{1,2,3\} \qquad \textbf{(2.42)}$$

The π_{v}-function for the diffused strain writes:

$$\pi_{\mathrm{v}}(\boldsymbol{d}) = \begin{cases} +\infty & \text{if } Tr(\boldsymbol{d}) \neq 0 \\ c(|d_1| + |d_2| + |d_3|) & \text{if } Tr(\boldsymbol{d}) = 0 \end{cases} \qquad \textbf{(2.43)}$$

where $Tr(\boldsymbol{d}) = d_{11} + d_{22} + d_{33} = d_1 + d_2 + d_3$ is the volumetric strain rate and (d_1, d_2, d_3) are the eigenvalues of the tensor $\boldsymbol{d}$. The components d_{i3} and d_3 do not intervene in 2D problems.

In parallel, the π-function for a surface of discontinuity writes:

$$\pi_{\mathrm{s}}(\boldsymbol{n}, [\![\boldsymbol{v}]\!]) = \begin{cases} +\infty & \text{if } \boldsymbol{n} \cdot [\![\boldsymbol{v}]\!] \neq 0 \\ c\|[\![\boldsymbol{v}]\!]\| & \text{if } \boldsymbol{n} \cdot [\![\boldsymbol{v}]\!] = 0 \end{cases} \qquad \textbf{(2.44)}$$

The above functions impose very important restrictions on the set of "possible candidates" of failure mechanism. If a potential failure mechanism involves one or more zones of diffused deformations, then the deformations must take place at constant volume; otherwise the resisting

power is infinite and failure cannot occur with the given velocity field.
In parallel, when a potential failure mechanism involves surfaces of discontinuities (or failure surfaces), then the velocity jump across any of them must take place parallel to it: $[\![\boldsymbol{v}]\!] \parallel \Sigma$, in order to keep the resisting power finite. The above conditions must hold $a\ priori$ so that a velocity field can be considered as a potential failure mechanism.

2.5.3.2 Criterion of Mohr-Coulomb

The Mohr-Coulomb criterion is defined by 2 parameters: cohesion c and friction angle ϕ:

$$G = \left\{ \boldsymbol{\sigma} \in \mathbb{R}^3 \colon f(\boldsymbol{\sigma}) = K_p \sigma_I - \sigma_{III} - 2c \frac{cos\phi}{1 - sin\phi} < 0 \right\} \quad \textbf{(2.45)}$$

The π_{v}-function for diffused strains writes:

$$\pi_{\mathrm{v}}(\boldsymbol{d}) = \begin{cases} +\infty & \text{if } Tr(\boldsymbol{d}) < (|d_1| + |d_2| + |d_3|)sin\phi \\ \dfrac{c}{tan\phi} Tr(\boldsymbol{d}) & \text{if } Tr(\boldsymbol{d}) \geq (|d_1| + |d_2| + |d_3|)sin\phi \end{cases} \quad \textbf{(2.46)}$$

In parallel, the π_{s}-function for a surface of discontinuity writes:

$$\pi_{\mathrm{s}}(\boldsymbol{n}, [\![\boldsymbol{v}]\!]) = \begin{cases} +\infty & \text{if } [\![\boldsymbol{v}]\!] \cdot \boldsymbol{n} < |[\![\boldsymbol{v}]\!]|sin\phi \\ \dfrac{c}{tan\phi} [\![\boldsymbol{v}]\!] \cdot \boldsymbol{n} & \text{if } [\![\boldsymbol{v}]\!] \cdot \boldsymbol{n} \geq |[\![\boldsymbol{v}]\!]|sin\phi \end{cases} \quad \textbf{(2.47)}$$

Therefore, when a potential failure mechanism involves diffused deformations, the volumetric strain rate must be larger than some critical value defined by the sum of the absolute values of its principal strains. In parallel, when a potential failure mechanism involves surfaces of discontinuity, then the acute angle θ between the velocity-jump and the tangential direction of the surface must be at least equal to the friction angle. In other words:

$$\theta \geq \phi \tag{2.48}$$

This condition is illustrated in Fig. (**2.15**) by the construction of a hodograph.

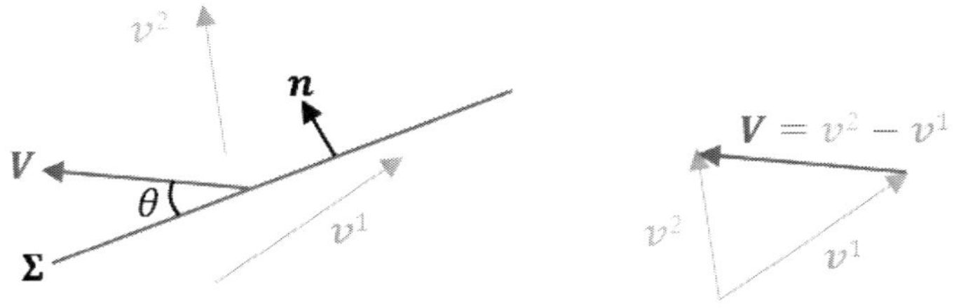

(a) Velocity fields on both sides of failure surface (b) Velocity hodograph

Fig. (2.15). Condition on velocity-jump across a failure surface in case of a Mohr-Coulomb material.

In the above figure is also shown the hodograph of velocities which is a very useful tool to impose the compatibility condition on the velocity field. Note that the normal $\boldsymbol{n}$ must point in the direction of subdomain Ω_2 and that the jump must be computed in the order $[\![\boldsymbol{v}]\!] = \boldsymbol{v_2} - \boldsymbol{v_1}$ so that the velocity-jump corresponds to a separation of the two subdomains and not an inter-penetration which is physically not viable.

We are now ready to introduce the kinematic theorem.

2.5.4. The Kinematic Theorem

For a given set of load factors $\boldsymbol{Q}$ defining completely all external loads $\boldsymbol{\mathcal{F}}(\boldsymbol{Q})$, the static theorem says that if one can find a stress field SA and PA, then this loading $\boldsymbol{Q}$ can be supported. Mathematically we write:

$$\exists \boldsymbol{\sigma} : \boldsymbol{\sigma}\, SA\, \boldsymbol{Q} \text{ and PA} \Longrightarrow \boldsymbol{Q} \in K \tag{2.49}$$

Logically, if such a stress field cannot be found, then $\boldsymbol{Q}$ cannot be supported and is outside K:

$$\nexists\boldsymbol{\sigma} : \boldsymbol{\sigma}\ SA\ \boldsymbol{Q} \text{ and PA} \Rightarrow \boldsymbol{Q} \notin K \tag{2.50}$$

The condition of non-existence of such a stress field would be difficult to establish without the virtual work theorem. With this theorem, we can now rewrite the condition of static admissibility, leading to:

$$\nexists\boldsymbol{\sigma} : \ P^e(\boldsymbol{Q},\boldsymbol{v}) = P^i(\boldsymbol{\sigma},\boldsymbol{v}) \text{ for } \forall\boldsymbol{v} \text{ KA} \quad \text{and } \boldsymbol{\sigma} \text{ PA} \quad \Rightarrow \quad \boldsymbol{Q} \notin K \tag{2.51}$$

In this form, the incompatibility between the static and plastic admissibility conditions is easy to display using the functional $P^r(\boldsymbol{v})$ that we have just constructed. Suppose that we find a particular velocity field $\boldsymbol{v}_1$ which satisfies the kinematic boundary conditions, and such that the power furnished by the external loads surpasses that of the internal resisting power: $P^e(\boldsymbol{Q},\boldsymbol{v}_1) > P^r(\boldsymbol{v}_1)$. Given that by construction, $P^r(\boldsymbol{v}_1) \geq P^i(\boldsymbol{\sigma},\boldsymbol{v}_1)\ \forall\boldsymbol{\sigma}$ PA, hence $P^e(\boldsymbol{Q},\boldsymbol{v}_1) > P^r(\boldsymbol{v}_1) \geq P^i(\boldsymbol{\sigma},\boldsymbol{v}_1)\ \forall\boldsymbol{\sigma}$ PA. We therefore deduce that it would be impossible to find a stress field which satisfies the strength criterion everywhere and at the same time can ensure the equality $P^e(\boldsymbol{Q},\boldsymbol{v}) = P^i(\boldsymbol{\sigma},\boldsymbol{v})$ for **ALL** kinematically admissible velocity fields (since we know that this equality cannot hold for the particular field $\boldsymbol{v}_1$).

In other words, the condition: $P^e(\boldsymbol{Q},\boldsymbol{v}_1) > P^r(\boldsymbol{v}_1)$ for any particular kinematically compatible velocity field $\boldsymbol{v}_1$ suffices to establish that the loading $\boldsymbol{Q}$ is not supportable, hence outside the domain K. This is the kinematic theorem:

$$\exists\boldsymbol{v}_1 \text{ KA} : \ P^e(\boldsymbol{Q},\boldsymbol{v}_1) > P^r(\boldsymbol{v}_1) \quad \Rightarrow \quad \boldsymbol{Q} \notin K \tag{2.52}$$

Note that the velocity field $\boldsymbol{v}_1$ represents a possible failure mechanism. Instead of the above, we can equivalently use the contrapositive:

$$\boldsymbol{Q} \in K \quad \Rightarrow \quad P^e(\boldsymbol{Q},\boldsymbol{v}) \leq P^r(\boldsymbol{v}) \quad \forall\boldsymbol{v} \text{ KA} \tag{2.53}$$

In words, for the external loadings $\boldsymbol{Q}$ to be supportable, the power they furnish must be less than the internal resisting power for all kinematically compatible velocity fields.

A practical application of the kinematic theorem can conceptually be divided into two steps. The first step consists of postulating a family of velocity fields: $\boldsymbol{v} = \boldsymbol{v}(\boldsymbol{b})$ defined by a set of unknown parameters $\boldsymbol{b}$. Some restrictions must be imposed on $\boldsymbol{b}$ so that $\boldsymbol{v}(\boldsymbol{b})$ is kinematically admissible and physically viable. We will use the following notation to denote this kinematically admissible set:

$$\mathbb{D}^v = \{\, \boldsymbol{b} = \{b_1, b_2, \dots, b_M\} \in \mathbb{R}^M \,:\, \boldsymbol{v}(\boldsymbol{b}) \text{ KA}\} \qquad \textbf{(2.54)}$$

The second step consists of imposing the in equality $P^e(\boldsymbol{Q}, \boldsymbol{v}(\boldsymbol{b})) \leq P^r(\boldsymbol{v}(\boldsymbol{b}))\ \forall \boldsymbol{b} \in \mathbb{D}^v$, leading to a constraint on the load factors $\boldsymbol{Q}$. For any particular element $\boldsymbol{b}$ of $\mathbb{D}^v$, the equality $P^e(\boldsymbol{Q}, \boldsymbol{v}(\boldsymbol{b})) = P^r(\boldsymbol{v}(\boldsymbol{b}))$ would yield a particular kinematic bound $\boldsymbol{Q}(\boldsymbol{b})$ outside K. An optimization intervenes here in order to determine the most optimum set $\boldsymbol{b}_{opt}$, leading to an optimum kinematic bound $\boldsymbol{Q}^{kin}$ which is the least "exterior" (*i.e.* as close as possible to the boundary of K) of all $\boldsymbol{Q}(\boldsymbol{b})$.

Examples in the next section will illustrate and clarify these ideas.

2.5.5. Example of Applications: Stability of a Long Wall Mining Tunnel

To illustrate the applicability of the kinematic theorem and also the idea of interior and exterior bounds, Let us reconsider the same example of a long wall mining tunnel already treated by the static theorem. The first step of the analysis is to choose a velocity field representing a possible failure mechanism. As a first example, we consider a sliding failure mechanism as shown in Fig. (**2.16**), which consists of a rigid block descending with a velocity $\boldsymbol{v}$ inclining at an angle $\boldsymbol{\alpha}$ relative the horizontal axis. Using the notation $\boldsymbol{v} = \|\boldsymbol{v}\| > \mathbf{0}$ to denote the amplitude of $\boldsymbol{v}$, we have $\boldsymbol{v} = \boldsymbol{v}\{-\boldsymbol{cos}(\boldsymbol{a}), \boldsymbol{sin}(\boldsymbol{a})\}$ relative to the axes $(\boldsymbol{x}, \boldsymbol{z})$. Note that this is a family of velocity fields defined by two scalars $\boldsymbol{b} = (\boldsymbol{v}, \boldsymbol{a})$.

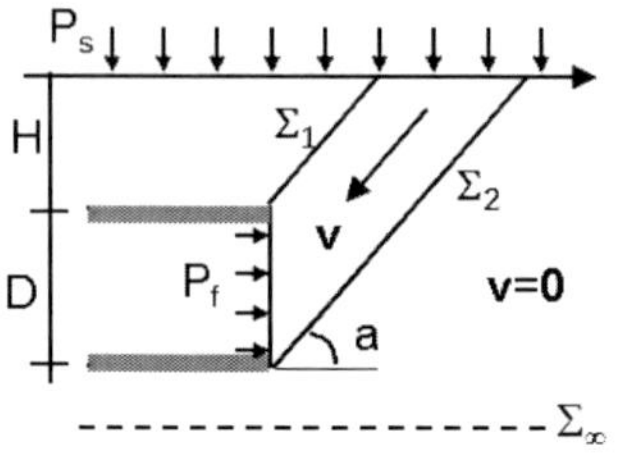

Fig, (2.16). Geometric configuration and failure mechanism considered.

This failure mechanism depends on an undetermined parameter "a", which is the inclination of the sling block relative to the horizontal axis. This allows some form of optimization to determine the most critical (or optimum) failure mechanism. For this velocity field, we need to compute the external mechanical power and the internal resisting power. It is easy to see that the two surfaces of discontinuity measure respectively $\Sigma_1 = \frac{H}{sin(a)}$ and $\Sigma_2 = \frac{H+D}{sin(a)}$, and the volume of the sliding block, of trapezoidal form, is $V = \frac{D(H+D/2)}{tan(a)}$. This block, of density γ undergoes a rigid movement of velocity inclined by an angle a to the horizontal axis. Therefore the power developed by gravity force is:

$$\int_V \boldsymbol{v} \cdot \boldsymbol{F}\, d\Omega = \gamma \left(H + \frac{D}{2}\right) v\, D \cos(a) \qquad \textbf{(2.55)}$$

There are two contributions from the pressures P_f and P_s that act respectively on surfaces of dimensions D and $D/\tan(a)$. They add up to:

$$\int_{\partial\Omega} \boldsymbol{v} \cdot \boldsymbol{T}\, d\Omega = (P_s - P_f)\, v\, D\, cos(a) \qquad \textbf{(2.56)}$$

Hence the external power writes:

$$P^e(\boldsymbol{Q},\boldsymbol{v}) = \left[\gamma \left(H + \frac{D}{2}\right) + P_s - P_f\right] v\, D\, cos(a) \tag{2.57}$$

We now have to evaluate the internal resisting power $P^r(\boldsymbol{v})$. There is no volume deformation, hence $\pi_v = 0$. For a Tresca Material and a sliding interface failure, $\pi_s([\![\boldsymbol{v}]\!],\boldsymbol{n}) = c||[\![\boldsymbol{v}]\!]|| = c\, v$. Hence:

$$P^r(\boldsymbol{v}) = c\, v(\Sigma_1 + \Sigma_2) = c\, v\, \frac{2H + D}{sin(a)} \tag{2.58}$$

For the load to be supportable, the inequality $P^e(\boldsymbol{Q},\boldsymbol{v}) \leq P^r(\boldsymbol{v})$ must hold for all admissible velocity fields, which means here for all admissible values of the angle a. By inspection, this parameter can vary between 0 and $\pi/2$. After some computations, we get the condition:

$$P_f \geq P_s + \gamma \left(H + \frac{D}{2}\right) - c\frac{\frac{2H}{D} + 1}{sin(a)cos(a)} \quad \forall a \in (0, \frac{\pi}{2}) \tag{2.59}$$

This leads to the optimization problem:

$$P_f \geq \max_{a\in(0,\frac{\pi}{2})}\left[P_s + \gamma \left(H + \frac{D}{2}\right) - c\frac{\frac{2H}{D} + 1}{sin(a)cos(a)}\right] \tag{2.60}$$

Which yields a lower bound for P_f, for the optimum solution $a = \pi/4$:

$$P_f \geq P^-_{f,kin} = P_s + \gamma \left(H + \frac{D}{2}\right) - 2c\left(\frac{2H}{D} + 1\right) \tag{2.61}$$

To find the upper bound, we have to consider the case of failure caused by an excess of face pressure pushing the ground upwards. The same failure mechanism as shown in Fig. (**2.13**) can be reused, but with a velocity in the reversed direction, in other words: $\boldsymbol{v} = v\{cos(a), -sin(a)\}$ with $v > 0$.

Using the same procedure, we arrived the following upper bound:

$$P_f < P_{f,kin}^{+} = P_s + \gamma\left(H + \frac{D}{2}\right) + 2c\left(\frac{2H}{D} + 1\right) \quad \textbf{(2.62)}$$

Summarizing, we arrive at the following exterior-bounds of K:

$$K^{kin} = \left(P_s + \gamma\left(H + \frac{D}{2}\right) - 2c\left(\frac{2H}{D} + 1\right), P_s + \gamma\left(H + \frac{D}{2}\right) + 2c\left(\frac{2H}{D} + 1\right)\right) \quad \textbf{(2.63)}$$

Comparing the above result with K^{stat} found in the last section, it can be seen that the lower and upper kinematic bounds are respectively lower and higher than the corresponding static bounds, establishing the inclusion relation: $K^{stat} \subset K^{kin}$, consistently with the theoretical construction.

2.6. FURTHER WORKED EXAMPLES

2.6.1. Stability of a Shallow Foundation on a Weightless Cohesive Soil

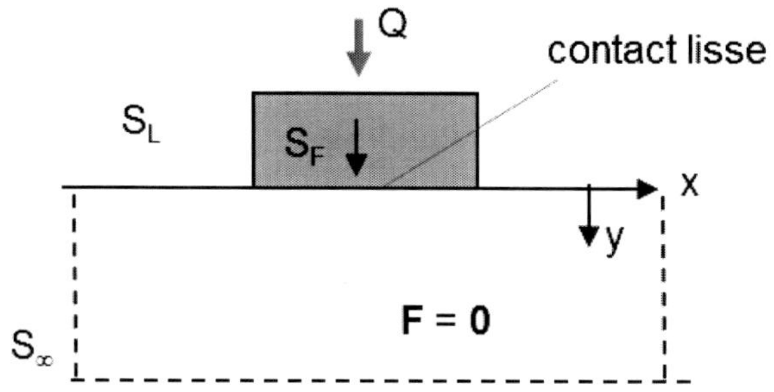

Fig. (2.17). "Mixed" boundary condition on a given boundary subsurface.

Consider a shallow infinitely long strip footing, of width B, sitting on top of a weightless ($\gamma = 0$) and purely cohesive soil as shown in Fig. (**2.17**), of cohesion c, obeying the Tresca's criterion. The vertical load per unit length Q leads to an average pressure of $\frac{Q}{B}$ applied at the ground surface S_F. The contact soil-foundation is assumed smooth so that no shear stress is transmitted. In other words, the stress vector is $(0, \frac{Q}{B})$ at the surface S_F and is null at the free surface S_L on both sides of the foundation. At far field over the surface S_∞, the velocity is null. We seek to determine the limit load per unit length Q_+ of the strip foundation. Only positive values of Q (*i.e.* compression) are considered. We will first consider the static approach, then compare the result obtained by the kinematic approach.

2.6.1.1 Static Approach

Consider the stress field defined relative to 3 subdomains, as shown in Fig. (**2.18**).

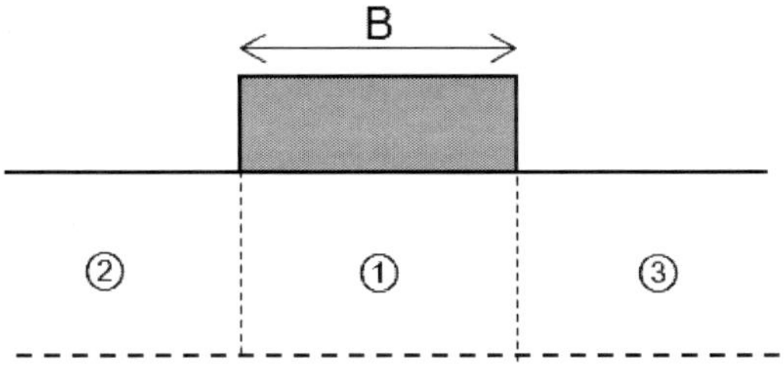

Fig. (2.18). A statically admissible stress field defined in 3 subdomains.

The expressions of the stress tensor in the three subdomains are:

$$\text{zone 1: } \boldsymbol{\sigma}^1 = \begin{bmatrix} \sigma_{xx} & 0 \\ 0 & -\frac{Q}{B} \end{bmatrix}; \quad \text{zones 2 and 3: } \boldsymbol{\sigma}^2 = \boldsymbol{\sigma}^3 = \begin{bmatrix} \sigma_{xx} & 0 \\ 0 & 0 \end{bmatrix} \qquad \textbf{(2.64)}$$

where σ_{xx} is a constant (recall that gravity effects are neglected so that the soil is weightless). It is trivial to verify that this stress field satisfies the conditions SA1, SA2 and SA3. It remains to impose the plastic admissibility condition. This leads to the following conditions which must be simultaneously satisfied:

$$\text{zone 1: } \left|\sigma_{xx} + \frac{Q}{B}\right| \leq 2c\,; \quad \text{zones 2 and 3: } |\sigma_{xx}| \leq 2c \qquad \textbf{(2.65)}$$

The second conditions imposes $-2c \leq \sigma_{xx} \leq 2c$, while the first condition:

$$-2c - \sigma_{xx} \leq \frac{Q}{B} \leq 2c - \sigma_{xx} \qquad \textbf{(2.66)}$$

Restricting Q to positive values, only the second inequality: $\frac{Q}{B} \leq 2c - \sigma_{xx}$ is considered. Since every admissible value of σ_{xx} leads to an estimate less than the theoretical value, the best estimate is obtained by maximizing the right hand side. We deduce the following static bound:

$$Q \leq Q_+^{stat} = 4Bc \qquad \textbf{(2.67)}$$

This bound can be improved by considering a more complex SA stress field, such as the one shown in Fig. (**2.19a**), composing of 3 zones, symmetrical with respect to the central axis.

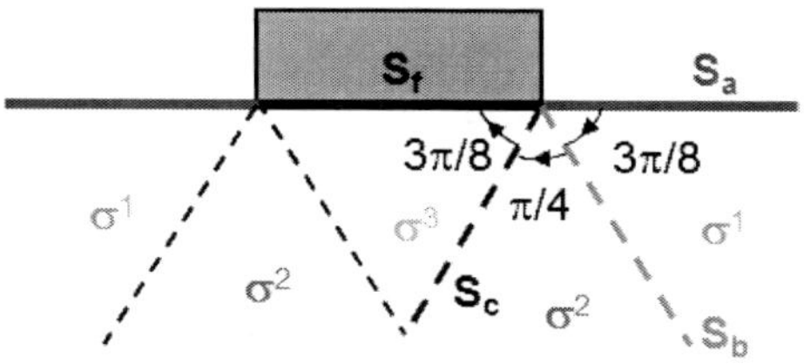

Fig. (2.19a). A more refined admissible stress field considered.

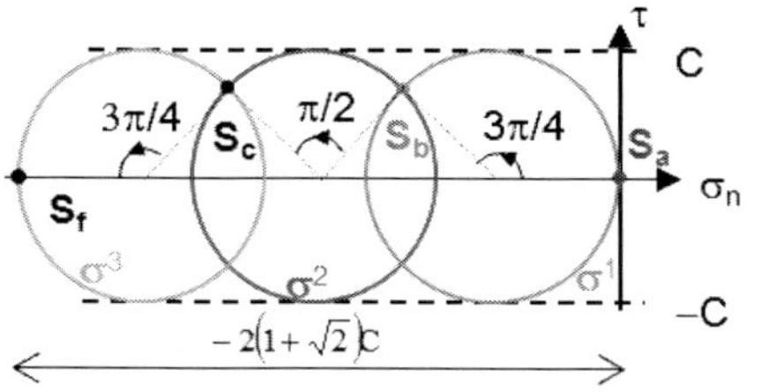

Fig. (2.19b). Mohr-circle representation of the stress field.

The stress tensors in zones 1, 2 and 3 are represented by the orange, red and green Mohr's circles in Fig. (**2.19b**). The stress vectors at the free surface S_a and at the interfaces S_b and S_c are represented by points of the same name on the Mohr-plane. Obviously this piece-wise constant stress field satisfies the zero boundary stress condition at the free surface S_a. The fact that point S_b is the intersection of circles $\boldsymbol{\sigma}^1$ and $\boldsymbol{\sigma}^2$ guarantees the continuity of the stress vector across the interface between zones 1 and 2. Similarly the stress vector defined by point S_c is continuous across the interface between zones 2 and 3. The stress field is therefore SA. Moreover, the fact that all the circles being bounded between the two horizontal lines $\tau = \pm c$ in the Mohr-plane guarantees that the strength criterion is nowhere violated. Therefore, this the stress field is also PA. This stress field leads to another estimate of the limit load given by the normal stress component of surface S_F:

$$Q \leq Q_+^{stat} = 2(1 + \sqrt{2})BC \approx 4.83Bc \tag{2.68}$$

Since a more exterior estimate is better for the static bound, the last value represents an improved estimate of the limit load compared to the previous value of $4Bc$.

In fact these results can be further improved by using the mechanism proposed by Prandtl.

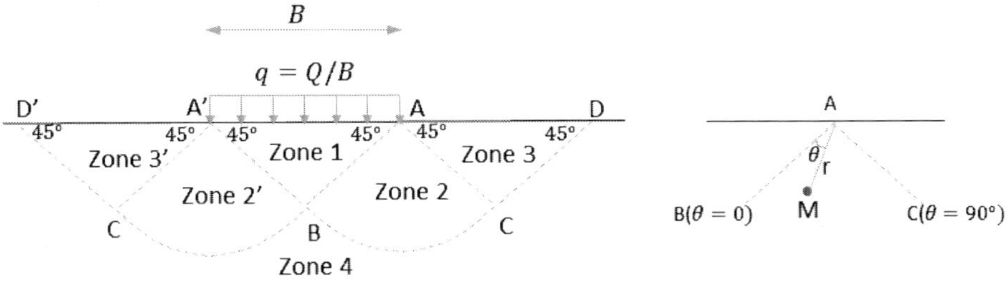

Fig. (2.20). Stress field considered in Prandtl's solution.

Let us consider a more complicated piece-wise continuous stress distribution, symmetrical relative to the vertical axis passing through the centre of the foundation as shown in (Fig. **2.20**). Writing $q = Q/B$, the stress tensors in zones 1 and 3 are constant, and are given by:

$$\boldsymbol{\sigma}^1 = \begin{bmatrix} -(q-2c) & 0 \\ 0 & -q \end{bmatrix} ; \quad \boldsymbol{\sigma}^3 = \begin{bmatrix} -2c & 0 \\ 0 & 0 \end{bmatrix} \tag{2.69}$$

Evidently they verify static equilibrium and just satisfy the strength criterion since the principal stresses differ by exactly twice the material cohesion c. The stress in zone 2 is not constant but varies from point to point. A generic point M inside zone 2 can be represented by the cylindrical coordinates (r, θ) where AB is taken to be the reference axis for the angular coordinate and the segment AC corresponding to $\theta = \pi/2$ is the right hand boundary of zone 2. The stress tensor inside this zone only depends on the angular coordinate: $\boldsymbol{\sigma}^2 = \boldsymbol{\sigma}^2(\theta)$. The mean stress $p = tr(\boldsymbol{\sigma}^2)$ varies from $-(q-c)$ at $\theta = 0$ to $-c$ at $\theta = \pi/2$ and is continuous across AB and AC. We therefore deduce that the following expressions of the mean, major and minor principal stresses in zone 2:

$$p = (\pi + 1 - 2\theta)c \,; \quad \sigma_I = p + c \,; \quad \sigma_{III} = p - c \tag{2.70}$$

Besides having a variable intensity, the stress tensor in zone 2 also rotates: the major principal stress σ_I changes direction from being vertical at AB to horizontal at AC. In fact at any generic point M, the inclination of σ_I relative to the vertical axis is exactly equal to the angular coordinate θ of that point, as defined above. The stress tensor relative to the system of coordinates (x, y) therefore writes:

$$\text{zone 2: } \boldsymbol{\sigma}^2 = c \begin{bmatrix} -(\pi + 1 - 2\theta + cos2\theta) & sin2\theta \\ sin2\theta & -(\pi + 1 - 2\theta - cos2\theta) \end{bmatrix} \tag{2.71}$$

In fact this stress field is obtained by combining the equations of static equilibrium and the yield condition, as well as the continuity of the stress vector at the boundaries AB and AC. For details, the reader can refer to chapter 3 on slip line fields.

The continuity of the stress vector at interface AB requires that: $q = (\pi + 2)c$. Since this loading is obtained by displaying a statically and plastically admissible stress field, the relation above defines another static bound:

$$Q_+^{stat} = (\pi + 2)Bc \approx 5.14Bc \qquad \textbf{(2.72)}$$

Which is better than the previous static bounds as it is more "exterior". In fact the analysis as such is not complete, since we have not defined the stress field below the curve D'C'BCD and show that the stress there verifies also the static and plastic admissibility conditions. This construction is in fact not trivial and had been done by Shield in 1954 (cited by Salençon 1984).

We now use the kinematic theorem to obtain an exterior estimate of the limit load Q_+.

2.6.1.2 *Kinematic Approach*

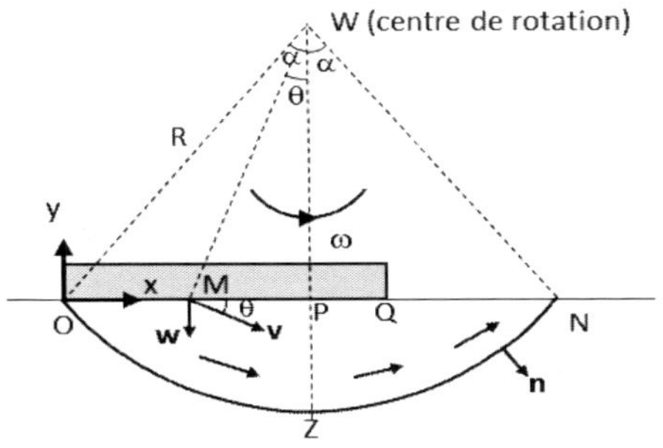

Fig. (2.21). First potential failure mechanism considered in the kinematic approach.

Consider the velocity field (*i.e.* the potential failure mechanism) shown in (Fig. **2.21**). This corresponds to a rigid body rotation around the point W of the physical subdomain OZNQP, with a sliding discontinuity across the circular interface OZN. This velocity field, defined by the parameters (R, α, ω), satisfies the kinematic boundary conditions since it is null far away from the foundation and does not lead to an infinite resisting power and is therefore admissible. As we will see shortly, due to the linearity of the functions P^e and P^r relative to the velocity field, and the linear dependence of the latter on the angular speed of rotation ω, this last parameter defining the amplitude of the velocity field does not intervene. Only the parameters (R, α) defining the geometric characteristics intervene. This is a general characteristic of the kinematic approach.

With notations defined above, the amplitude of velocity at point M is given by:

$$v = \|\boldsymbol{v}\| = \omega * \overline{WM} = \omega * \frac{\overline{WP}}{cos\theta} = \omega \frac{R\ cos\alpha}{cos\theta} \quad \textbf{(2.73)}$$

The vertical component of $\boldsymbol{v}$, denoted by w, at the same point M writes:

$$w = v\ sin\theta = \omega\ \overline{WP}\ tan\theta = \omega\overline{MP} = \omega(\overline{OP} - \overline{OM}) = \omega(R\ sin\alpha - x) \quad \textbf{(2.74)}$$

The only contribution to the external power comes from the surface force $(0, \frac{Q}{B})$ at the surface S_F. At a generic point M at a distance x from the point O, the ground velocity is $(v_x, \omega(R\ sin\alpha - x))$. We thus obtain:

$$P^e = \int_0^B \begin{Bmatrix} 0 \\ Q/B \end{Bmatrix} \cdot \begin{Bmatrix} v_x \\ \omega(R\ sin\alpha - x) \end{Bmatrix} dx = \omega Q \left(Rsin\alpha - \frac{B}{2} \right) \quad \textbf{(2.75)}$$

The resisting power developed over the failure surface OZN is given by:

$$P^r = \int_{OZN} c.v\ d\Sigma = c.\omega R. 2\alpha R = 2\omega c\alpha R^2 \quad \textbf{(2.76)}$$

Plastic admissibility condition requires that $P^e \leq P^r$ for all kinematically admissible velocity field, which means admissible values of the couple (R, α). From inspection, we find that this domain can be defined by:

$$\mathbb{D}^{v} = \left\{(R,\alpha):\ 0 < R < \infty;\ 0 < \alpha < \frac{\pi}{2};\ \ Rsin\alpha > \frac{B}{2}\right\} \quad \textbf{(2.77)}$$

The last inequality expresses the necessary condition that the external force does positive work in a failure mechanism. A few steps of computations lead to the following condition of stability:

$$stability \Longrightarrow Q \le g(R,\alpha) = \frac{2\alpha c R^2}{Rsin\alpha - \frac{B}{2}}\ \ \forall (R,\alpha) \in \mathbb{D}^{v} \quad \textbf{(2.78)}$$

To obtain the above inequality, we have taken into account the assumed positivity of the physical quantities ω and $\left(Rsin\alpha - \frac{B}{2}\right)$. The above condition requires the solution of a minimization problem:

$$Q_{+}^{kin} = \max_{(R,\alpha)\in\mathbb{D}^{v}} g(R,\alpha) \quad \textbf{(2.79)}$$

The solution is given by the pair of equations: $\dfrac{\partial g}{\partial R} = 0\,;\ \dfrac{\partial g}{\partial \alpha} = 0$. The first of these yields:

$$Rsin\alpha = B \quad \textbf{(2.80)}$$

This means that the center of rotation W lies on the vertical line passing through the right side edge of the foundation. The second equation, on account of the above result, leads to the following equation on the variable α:

$$tan\ \alpha = 2\alpha \quad \textbf{(2.81)}$$

Leading to the solution: $\alpha_{opt} = 1.166\ (\sim 67°)\ ;\ R = 1.088\,B$ and a kinematic bound:

$$Q_{+}^{kin} = g\left(\frac{B}{sin(\alpha_{opt})}, \alpha_{opt}\right) = 5.52Bc \quad \textbf{(2.82)}$$

As a further example to illustrate the use of discontinuous velocity fields, let us consider the following field defined by five zones (with a sixth immobile zone) as shown in Fig. (**2.22**).

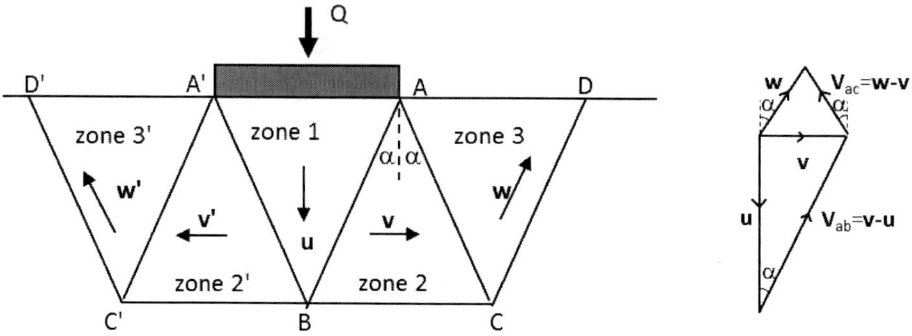

Fig. (2.22). A second velocity field to evaluate the limit load of a shallow foundation.

This field is symmetrical relative to the central vertical axis. Zone 1 moves vertically downwards with a velocity $\boldsymbol{u}$, zones 2 and 2' moves horizontally outwards with a velocity $\boldsymbol{v}$ ($\boldsymbol{v}'$ is equal and opposite to $\boldsymbol{v}$), and finally zones 3 and 3' have velocities $\boldsymbol{w}$ and $\boldsymbol{w}'$ along inclined directions parallel to the failure surfaces CD and C'D'. The velocity jump across the surfaces AB, BC, AC and CD are respectively:

$$\text{AB: } \boldsymbol{V}_{ab} = \boldsymbol{v} - \boldsymbol{u}\,; \quad \text{BC: } \boldsymbol{v}\,; \quad \text{AC: } \boldsymbol{V}_{ac} = \boldsymbol{w} - \boldsymbol{v}\,; \quad \text{CD: } \boldsymbol{w} \qquad \textbf{(2.83)}$$

For symmetry we need only consider zones 1, 2 and 3 to compute the energy contributions. We will denote the norm of the above velocities by the same name but with non-bold characters. In order that the jump of continuity be parallel to the interfaces (to keep the resisting power finite), the velocities $\boldsymbol{u}$, $\boldsymbol{v}$ and $\boldsymbol{w}$ must satisfy some geometric conditions as expressed by the hodograph in Fig. (**2.21**). These conditions allow to express the velocity jumps across all interfaces in terms of the only unknown variable u:

$$v = u.tan\alpha\, ; \quad V_{ab} = \frac{u}{cos\alpha}\, ; \quad V_{ac} = w = \frac{u}{2\,sin\alpha} \tag{2.84}$$

The external power is simply the applied force Q multiplied by the velocity u:

$$P^e = Q.u \tag{2.85}$$

The resisting power is given by:

$$P^r = \sum_i \int_{S_i} c.\,\|\boldsymbol{V}_i\| = c\sum_i S_i\|\boldsymbol{V}_i\| \tag{2.86}$$

Where $\boldsymbol{V}_i$ is the velocity jump over a failure surface S_i. The following table summarizes the surface areas and velocity jumps.

SurfaceS_i	Longueur (Valeur de S_i)	Amplitude du Saut de Vitesse $\|\boldsymbol{V}_i\|$
AB, A'B	$B/(2sin\alpha)$	$u/cos\alpha$
BC, BC'	B	$u\ tan\alpha$
AC, A'C'	$B/(2sin\alpha)$	$u/2cos\alpha$
CD, C'D'	$B/(2sin\alpha)$	$u/2cos\alpha$

This leads to:

$$P^r = 2ucB\left(\frac{1}{sin\alpha\ cos\alpha} + tan\alpha\right) \tag{2.87}$$

Now the plastic admissibility condition leads to:

$$stability \Longrightarrow Q \leq 2Bc\left(\frac{1}{sin\alpha\ cos\alpha} + tan\alpha\right)\ \forall\alpha \text{ admissible} \tag{2.88}$$

Minimization of the right hand side leads to the simple equation: $3sin^2\alpha = 1$. We then obtain easily the solution $\alpha_{opt} = sin^{-1}(1/\sqrt{3})$ (environ 35°) leading to another estimation of the limit load:

$$Q_+^{kin} = 5.65Bc \tag{2.89}$$

This kinematic bound is higher (*i.e.* "more exterior") than the previous value and therefore is less optimum. Gathering the results from the static and kinematic approaches, we can conclude that the theoretical limit load lies in the follow interval:

$$5.12Bc \leq Q_+ \leq 5.52Bc \qquad \textbf{(2.90)}$$

As in the case of static approach, this result can be further improved by a more refined mechanism as shown in Fig. (**2.23**).

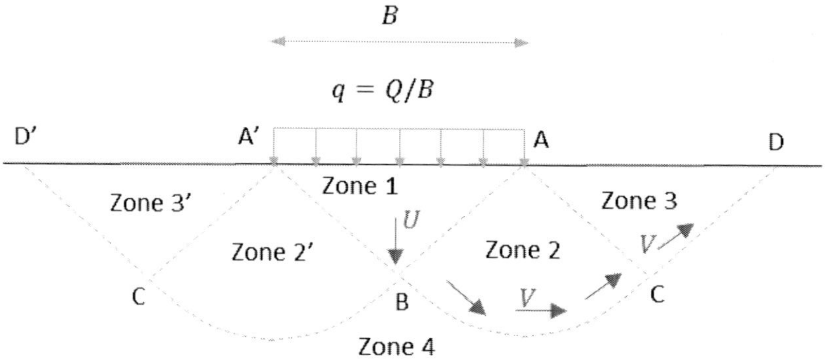

Fig. (2.23). Velocity field giving the theoretical exact solution.

The velocity fields in zones 1 and 3 are homogeneous, respectively of amplitudes U and V. In zone 2, along every line emerging from point A, the velocity is perpendicular to that line like a rigid body rotation about point A, except that the amplitude is constant and everywhere equal to V. Like the stress field presented in the static approach, the velocity field considered is symmetrical about the vertical axis; hence only the right side parts need to be described. Zone 1 descends vertically like a rigid body with a uniform velocity U. Zone 3 "slides" like a rigid body along the interface CD with a uniform velocity V. In zone 2, the velocity field $\boldsymbol{v}^2$ is very particular. Its amplitude is constant, given by $\boldsymbol{v}^2 = V$. The direction of $\boldsymbol{v}^2$ at a generic point M is always perpendicular to the radial vector AM. To avoid the resisting power increasing without bound, the volumetric strain rate must be null and the velocity jumps across interfaces AB, AC, BC and CD must be parallel to these interfaces. This leads to the conclusion that the amplitude of V must be constant and is given by:

$$V = \frac{U}{\sqrt{2}} \tag{2.91}$$

Using the cylindrical coordinates previously defined, we can write:

$$V_r(r,\theta) = 0\ ;\quad V_\theta(r,\theta) = V = \frac{U}{\sqrt{2}} \tag{2.92}$$

The above velocity field leads to a non-zero strain rate in zone 2; using classic formulae, we obtained:

$$d_{rr} = 0\ ;\quad d_{\theta\theta} = 0\ ;\quad d_{r\theta} = -\frac{U}{2\sqrt{2}r} \tag{2.93}$$

The principal strains can easily be deduced:

$$d_1 = \frac{U}{2\sqrt{2}r}\ ;\quad d_3 = -\frac{U}{2\sqrt{2}r} \tag{2.94}$$

This allows the evaluation of the π_v-function in zone 2:

$$\pi_v(\boldsymbol{d}) = c(|d_1| + |d_3|) = \frac{c\,U}{\sqrt{2}r} \tag{2.95}$$

With these ingredients we can now proceed to compute the external power and the resisting power and then the limit load. The external power represents the rate of work done by the vertical load Q following a vertical velocity U; it is simply $P^e = QU$. There are four contributions to the resisting power; they come from the strain rate field $\boldsymbol{d}$ in zone 2, and from the sliding discontinuity of the velocity field along the interfaces AB, AC, BC and CD. The first contribution can be evaluated *via* a surface integral $\int_0^{\pi/2}\int_0^{B/\sqrt{2}} \frac{c\,U}{\sqrt{2}r} r\, dr d\theta = \pi BcU/4$, where $B/\sqrt{2}$ is the length of AB and AC. The other integrals can be evaluated knowing the velocity jump and the arc lengths of the interfaces. We get after some simple calculations:

$$P^i = (\pi + 2)BcU \tag{2.96}$$

The above expressions of P^e and P^i lead immediately to the following kinematic (exterior) bound:

$$Q_{+}^{kin} = (\pi + 2)Bc \approx 5.14Bc \quad \textbf{(2.97)}$$

This is a remarkable result: the above kinematic bound is identical to the best static bound previously computed. We can therefore conclude that we have found the exact solution to our problem:

$$Q_{+} = (\pi + 2)Bc \quad \textbf{(2.98)}$$

Exact solutions of this kind are extremely rare and are only encountered for idealized problems of very simple geometry, but these solutions as the one presented above are excellent illustrations of the theoretical tools.

2.6.2. Stability of a Vertical Slope in a Cohesive-Frictional Soil Mass

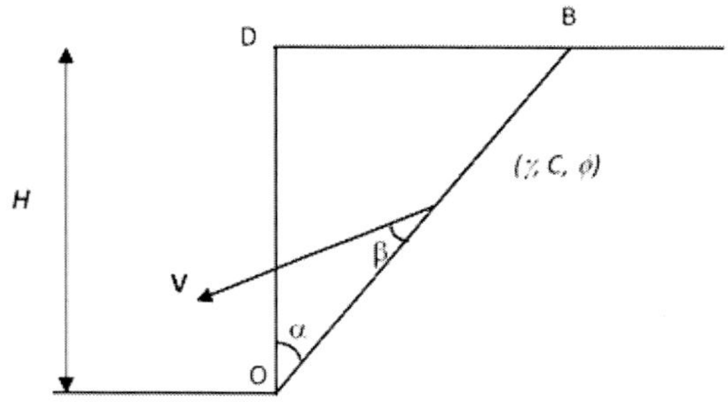

Fig. (2.24). Geometry and failure mechanism of a vertical soil slope using kinematic approach.

This example as shown in Fig. **(2.24)** serves to illustrate the treatment of the non-trivial velocity jump condition in the case of cohesive-frictional soils obeying the Mohr-Coulomb yield criterion. Consider a vertical slope of height H in a soil mass of unit weight γ, cohesion c and friction angle ϕ. The failure mechanism assumed consists of a triangular soil wedge ABD subject to a virtual velocity field, which is homogeneous, of amplitude V (a positive scalar), with an inclination of β relative to the failure surface which is itself inclined with an angle α relative to the vertical direction, as shown in Fig. (**2.24**).

Angles α and β are a priori positive. For this mechanism to be physically viable, the velocity needs to have a downward component, hence $\alpha + \beta < \pi/2$. Moreover, to avoid the resisting power increasing without bound, we must have $\beta \geq \phi$. To summarise, the domain of compatibility for the pair of parameters (α, β) is:

$$\mathbb{D}^v = \left\{(\alpha, \beta): \alpha > 0; \alpha + \beta < \frac{\pi}{2}; \beta \geq \phi\right\} \quad \textbf{(2.99)}$$

This is represented graphically in Fig. (**2.25**).

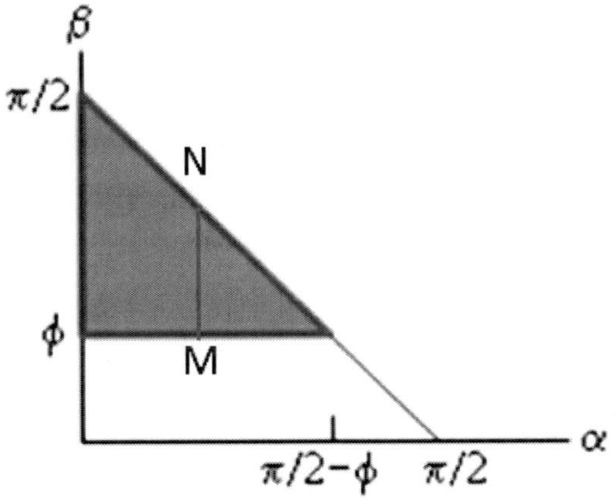

Fig. (2.25). Domain of admissible values of the parameters (α, β).

The only external applied force being gravity, we can easily show that the external power is simply given by $P^e = \gamma V\, cos(\alpha + \beta)\Omega$ where Ω is the area of the mobile soil wedge OBD. Simple calculations lead to:

$$P^e = \frac{\gamma\, H^2 V\, tan\, \alpha cos(\alpha+\beta)}{2} \quad \textbf{(2.100)}$$

Along the failure surface, we have:

$$\pi_s(\boldsymbol{n}, [\![\mathbf{v}]\!]) = C\, V \frac{\sin\beta}{\tan\phi} \quad \textbf{(2.101)}$$

leading to:

$$P^r = C\,VH \frac{\sin\beta}{\cos\alpha\ \tan\phi} \qquad \textbf{(2.102)}$$

The fundamental inequality $P^e \leq P^r$ for $\forall\ (\alpha,\beta) \in \mathbb{D}^v$ leads to:

$$\frac{\gamma H}{C} \leq \min_{(\alpha,\beta)\in\mathbb{D}^v} \frac{2\,sin\beta}{tan\,\phi\ sin\,\alpha\ cos(\alpha+\beta)} \qquad \textbf{(2.103)}$$

It is evident that the RHS is monotone increasing with respect to β. Hence the inequality $\beta \geq \phi$ can be replaced by the equality:

$$\beta = \phi \qquad \textbf{(2.104)}$$

This is a very important result because the above condition appears to be always valid for all known solutions. In other words, along an optimum failure surface in a Mohr-Coulomb type material, the velocity jump $[\![\mathbf{v}]\!]$ there must make an acute angle exactly EQUAL to the friction angle relative to the failure surface although the finiteness of the resisting power only require this angle to be EQUAL OR LARGER than the friction angle. Geometrically, the original minimisation over the whole domain $\mathbb{D}^v$ can be reduced to a minimisation along its lower boundary line.

Substitution of the last conditions leads to the following optimisation relative to the only unknown α:

$$\frac{\gamma H}{C} < \min_{0<\alpha\leq\pi/2-\phi} \frac{2\,cos\phi}{sin\,\alpha\ cos(\alpha+\phi)} \qquad \textbf{(2.105)}$$

whose only solution in the domain $0 < \alpha < \pi/2 - \phi$ is:

$$\alpha_{opt} = \frac{\pi}{4} - \frac{\phi}{2} \qquad \textbf{(2.106)}$$

leading to the kinematic bound:

$$\frac{\gamma H}{C} < \left(\frac{\gamma H}{C}\right)_+^{kin} = 4\sqrt{K_p} \qquad \textbf{(2.107)}$$

2.6.3. Failure Surfaces in the Form of Logarithmic Spirals in Cohesive-Frictional Soils

The condition $\beta = \phi$ obtained in the last problem is in fact general result. The result has an important implication. When the failure surface is not linear but curved, and that the velocity field relative to a sub-domain correspond to a rigid body rotation, a case frequently encountered in practice, then the above kinematic condition results in the failure surface being a logarithmic spiral as well will show in Fig. (**2.26**).

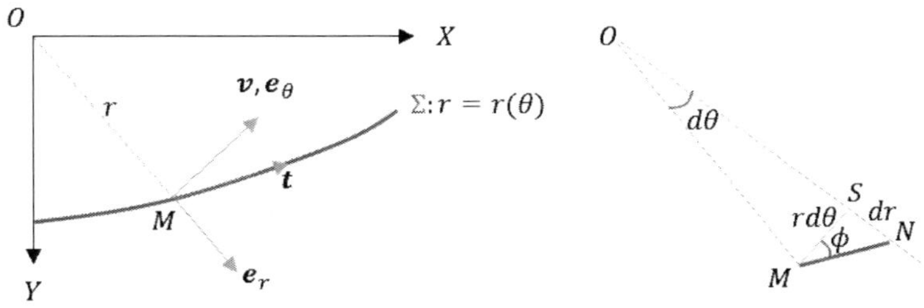

Fig. (2.26). Failure mechanism in the form of rotational movement of a rigid block and spiral logarithmic failure surface.

In the above figure, the red curve Σ corresponds to the failure surface and the soil mass above this surface is subject to a rotational rigid-body movement around point O, so that at every point M, the velocity $\boldsymbol{v}$ is orthogonal to the position vector $\overrightarrow{OM} = r\, \boldsymbol{e}_r$, or in other word is parallel to the basis vector $\boldsymbol{e}_\theta$. Over an infinitesimal portion MN of the failure surface Σ, MS is parallel to both $\boldsymbol{v}$ and $\boldsymbol{e}_\theta$ and measures $rd\theta$, NS therefore measures dr. The angle MSN is infinitesimally near to a right angle and the kinematical condition requires the angle NMS to be equal to the friction angle ϕ. We can therefore write:

$$tan\phi = \frac{dr}{rd\theta} \tag{2.108}$$

which can be integrated to give:

$$ln\left(\frac{r}{r_0}\right) = (\theta - \theta_0)tan\phi \quad \textbf{(2.109)}$$

or equivalently:

$$r = r_0 exp\big((\theta - \theta_0)tan\phi\big) \quad \textbf{(2.110)}$$

where (r_0, θ_0) is any particular point crossed by the failure surface. As one can easily recognise, the above equation defines a log-spiral. Having presented the basic working procedures in using the bound theorems to estimate the limit loads, we will show in the following paragraphs some useful theoretical results and tools for the construction of the domain K.

2.7. CONVEXITY OF THE DOMAIN OF SUPPORTABLE LOADS

We can show that the domain of supportable loads K is convex if the domain of plastically admissible stresses G is convex using either the static or the kinematic theorem.

Take first of all the point of view of the static theorem. Suppose $\boldsymbol{Q}_A$ and $\boldsymbol{Q}_B$ are both supportable hence strictly inside K. In other words we can find two stress fields $\boldsymbol{\sigma}_A$ and $\boldsymbol{\sigma}_B$ respectively SA-$\boldsymbol{Q}_A$ and SA-$\boldsymbol{Q}_B$ and that at every material point $\boldsymbol{x}$, $\boldsymbol{\sigma}_A(\boldsymbol{x}) \in G$ and $\boldsymbol{\sigma}_B(\boldsymbol{x}) \in G$, or equivalently $f(\boldsymbol{\sigma}_A(\boldsymbol{x})) \leq 0$ and $f(\boldsymbol{\sigma}_B(\boldsymbol{x})) \leq 0$. We can show that a generic point $\boldsymbol{Q}^*$ lying on the segment joining $\boldsymbol{Q}_A$ and $\boldsymbol{Q}_B$ also belongs to K. We can in fact write: $\boldsymbol{Q}^* = \lambda\boldsymbol{Q}_A + (1-\lambda)\boldsymbol{Q}_B$ with λ a positive real number in the interval $(0,1)$. On account of the linearity of the equilibrium conditions SA1, SA2 and SA3, this loading $\boldsymbol{Q}^*$ is in static equilibrium with the stress field $\boldsymbol{\sigma}^* = \lambda\boldsymbol{\sigma}_A + (1-\lambda)\boldsymbol{\sigma}_B$. The convexity of G implies that $\boldsymbol{\sigma}^*(\boldsymbol{x})$ is also inside G, since the convex yield function implies: $f(\boldsymbol{\sigma}^*) < \lambda f(\boldsymbol{\sigma}_A) + (1-\lambda)f(\boldsymbol{\sigma}_B) \leq 0$. Hence $\boldsymbol{\sigma}^*(\boldsymbol{x})$ is both SA-$\boldsymbol{Q}^*$ and PA implying that $\boldsymbol{Q}^*$ is inside K, establishing the convexity of K.

Now take the point of view of the kinematic theorem. The fact that $\boldsymbol{Q}_A$ and $\boldsymbol{Q}_B$ are both inside K implies that: $P^e(\boldsymbol{Q}_A, \boldsymbol{v}) \leq P^r(\boldsymbol{v})$ and $P^e(\boldsymbol{Q}_B, \boldsymbol{v}) \leq$

$P^r(\boldsymbol{v})$ for all kinematically admissible velocity field $\boldsymbol{v}$. On account of the linearity of P^e, we can show that the loading $\boldsymbol{Q}^*$ satisfies the condition:

$$P^e(\boldsymbol{Q}^*, \boldsymbol{v}) = \lambda P^e(\boldsymbol{Q}_A, \boldsymbol{v}) + (1-\lambda)P^e(\boldsymbol{Q}_B, \boldsymbol{v}) \leq \lambda P^r(\boldsymbol{v}) + (1-\lambda)P^r(\boldsymbol{v}) = P^r(\boldsymbol{v}) \quad \textbf{(2.111)}$$

for all kinematically admissible velocity field $\boldsymbol{v}$, hence $\boldsymbol{Q}^*$ is also inside K, establishing a second time the convexity of K.

Note that P^e and P^r are linear relative to each of their arguments, in the sense that for arbitrary real scalars λ_1 and $\lambda_2 > 0$, we have $P^e(\lambda_1 \boldsymbol{Q}, \lambda_2 \boldsymbol{v}) = \lambda_1 \lambda_2 P^e(\boldsymbol{Q}, \boldsymbol{v})$ and $P^r(\lambda_2 \boldsymbol{v}) = \lambda_2 P^r(\boldsymbol{v})$. It is essential to restrict the scalar λ_2 to positive values because the value of $P^r(\boldsymbol{v})$ undergoes non-trivial changes when we replace its argument $\boldsymbol{v}$ by $-\boldsymbol{v}$.

An important consequence of this linearity is that when we chose a particular velocity field $\boldsymbol{v}(\boldsymbol{b})$, with $\boldsymbol{b} \in \mathbb{D}^v$, to determine *via* the fundamental relation $P^e(\boldsymbol{Q}, \boldsymbol{v}(\boldsymbol{b})) = P^r(\boldsymbol{v}(\boldsymbol{b}))$ a kinematic bound $\boldsymbol{Q}(\boldsymbol{b})$, the result will not change if we multiply $\boldsymbol{v}(\boldsymbol{b})$ by any positive scalar; the relation $P^e(\boldsymbol{Q}, \lambda \boldsymbol{v}(\boldsymbol{b})) = P^r(\lambda \boldsymbol{v}(\boldsymbol{b}))$ would simply be reduced back to $P^e(\boldsymbol{Q}, \boldsymbol{v}(\boldsymbol{b})) = P^r(\boldsymbol{v}(\boldsymbol{b}))$. Therefore if $\boldsymbol{v}(\boldsymbol{b})$ is proportional to a particular scalar, say b_1 (this must be positive) then this scalar does not intervene in the optimization of $\boldsymbol{Q}(\boldsymbol{b})$ relative to the set $\mathbb{D}^v$. The parameter b_1 would simply be cancelled out in the above fundamental equality.

2.8. CONSTRUCTION OF DOMAIN *K* BY CONSIDERING RADIAL VECTORS

We have seen in the worked examples how one particular load factor Q_i can be singled out by keeping all others $\{Q_j;\ j \neq i\}$ constant. With the convexity of K, we can conceived a general methodology to determine an interior and an exterior approximate representation of K, by considering radial vectors of the form $\boldsymbol{Q} = \Lambda\boldsymbol{q}$, with $\boldsymbol{q}$ a given unit vector $\|\boldsymbol{q}\| = 1$ and $\Lambda \geq 0$, as shown in Fig. (**2.27**). We can determine by respectively the static and kinematic theorems a lower (interior) Λ^{stat} and an upper (exterior) Λ^{kin} bound of Λ, such that:

$$\Lambda^{stat} \leq \Lambda \leq \Lambda^{kin} \quad \textbf{(2.112)}$$

Consider firstly the static approach. Using equation (2.12), we can write:

$$\mathbb{F}(\Lambda^{stat}\boldsymbol{q}) = \max_{x\in\Omega} f\left(\Lambda^{stat}\boldsymbol{\sigma}(\boldsymbol{q},\boldsymbol{x})\right) = 0 \quad \textbf{(2.113)}$$

Solution of the above implicit equation gives the static bound Λ^{stat}.

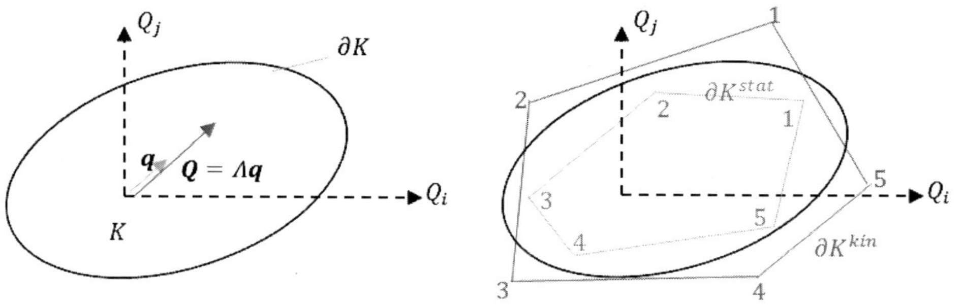

Fig. (2.27). Polygonal approximation to the domain of admissible loads.

By considering a few unit vectors $\boldsymbol{q}_1, \boldsymbol{q}_2, \boldsymbol{q}_3$ ...*etc.*, we arrive at a polygonal interior-estimate of the domain K. On account of the convexity of K, the segments joining the corners are also inside K. Fig. (**2.27**) gives a graphical illustration for the case of a pentagonal estimation using 5 different unit vectors $(\boldsymbol{q}_1, \boldsymbol{q}_2, \boldsymbol{q}_3, \boldsymbol{q}_4, \boldsymbol{q}_5)$, where the static approximation ∂K^{stat} is shown in blue lines.

For the kinematic approach, the relation (2.52) and the definition (2.54) applied to a kinematically admissible family of velocity fields $\boldsymbol{v}(\boldsymbol{b})$ allows to write:

$$\boldsymbol{b} \in \mathbb{D}^{\boldsymbol{v}} : \Lambda > \frac{P^r(\boldsymbol{v}(\boldsymbol{b}))}{P^e(\boldsymbol{q}, \boldsymbol{v}(\boldsymbol{b}))} \implies \Lambda \boldsymbol{q} \notin K \tag{2.114}$$

Note that for a physically viable failure mechanism, the external power is necessarily positive ($P^e(\boldsymbol{q}, \boldsymbol{v}(\boldsymbol{b})) > 0$). In the limiting case of equality and for any particular set of kinematically admissible parameters $\boldsymbol{b}$, this leads to a point $\Lambda \boldsymbol{q}$ outside K. The best estimate is therefore given by the smallest value among all possible elements of $\mathbb{D}^{\boldsymbol{v}}$, leading to a kinematic (exterior) bound:

$$\Lambda^{kin} = \min_{\boldsymbol{b} \in \mathbb{D}^{\boldsymbol{v}}} \frac{P^r(\boldsymbol{v}(\boldsymbol{b}))}{P^e(\boldsymbol{q}, \boldsymbol{v}(\boldsymbol{b}))} \tag{2.115}$$

We can also attempt to approximate the domain K from the exterior by a polygon. However, it cannot be guaranteed that the sides of the polygon are entirely outside the domain K. Hence the inclusion relationship $K \subset K^{kin}$ is not guaranteed for this polygonal domain.

CONCLUSION

In geotechnical analysis, the constitutive relation which is usually difficult to be determine precisely or the initial condition is difficult or even impossible to be defined, the use of stability analysis or ultimate limit state approach is commonly adopted. For most of the stability analysis, the lower and upper approaches are commonly adopted and are introduced in a rigorous manner in this chapter. These approaches are then applied to various types of problems, and the applicability and power of the lower and upper bound approaches are clearly illustrated. Analytical which are usually rigorous or good approximate solutions are available for many types of problems without the solution of difficult nonlinear hyperbolic plasticity differential equations. The concept behind these methods can be extended to general problems with the use of numerical methods. Many design figures and tables used in geotechnical engineering are also developed based on the upper and lower bound approaches. The importance of these two principles are clearly illustrated through the case studies in the present chapter.

REFERENCES

Bottero, A., Negre, R., Pastor, J., and Turgeman, S. (1980). Finite element method and limit analysis theory for soil mechanics problems, Computer Methods in Applied Mechanics and Engineering, 22, 131-149.

Chen, W. F. (1975). Limit analysis and soil plasticity, Amsterdam: Elsevier Scientific Publishing Company.

Chen, W. F., and Liu, X. L. (1990). Limit analysis in soil mechanics, Amsterdam, Netherlands, New York: Elsevier, Distributors for the U.S. and Canada, Elsevier Science Pub.

Chambon, P., and Corté, J. F. (1994). Shallow tunnels in cohesionless soil: Stability of tunnel face, J. Geotech. Engg., ASCE, 120(7), 1148-1165.

Davis, E. H., Gunn, M. J., Mair, R. J., and Seneviratne, H. N. (1980). The stability of shallow tunnels and underground openings in cohesive material, Géotechnique, 30(4), 397-416.

Leca, E., and Dormieux, L. (1990). Upper and lower bound solutions for the face stability of shallow circular tunnels in frictional material, Géotechnique, 404(4), 581-606.

Li, D.Z., Cheng, Y.M. (2013). Formulation of the upper bound axisymmetric problems as Second Order Programming, International Journal of Numerical Methods in Geotechnical engineering (under review).

Lysmer, J. (1970). Limit analysis of plane problems in soil mechanics, Journal of Soil Mechanics and Foundation, Divison, ASCE, 96(SM4), 1311-1334.

Salençon, J. (1883). Calcul à la rupture, Presses de l'Ecole Nationales des Ponts et Chaussée. Paris.

Shield, R. T. (1954). Plastic potential theory and Prandtl bearing capacity solution. J. Appl. Mech. Trans. A.S.M.E., 21, 193-194.

Sloan, S. W. (1988). Lower bound limit analysis using finite elements and linear programming, International Journal for Numerical and Analytical Methods in Geomechanics, 12(1), 61-77.

Sloan, S. W. (1989). Upper Bound Limit Analysis Using Finite-Elements and Linear-Programming, International Journal for Numerical and Analytical Methods in Geomechanics, 13(3), 263-282.

Sloan, S. W., Kleeman, P. W. (1995). Upper bound limit analysis using discontinuous velocity-fields, Computer methods in applied mechanics and Engineering, 127(1-4), 293-314.

Subrin, D., Wong, H. (2002). Stabilité du front d'un tunnel en milieu frottant: un nouveau mécanisme de rupture 3D, C.R. Mécanique 330, 513-519.

Wong, H., Subrin, D. (2006). Stabilité frontale d'un tunnel, mécanisme 3D en forme de corne et influence d la profondeur, Revue Européenne de Génie Civil. 10(4), 429-456.

Yu, H. S. (2006). Plasticity and Geotechnics. Springer USA, ISBN 0-387-33597-8.

CHAPTER 3

Slip Line, Limit Equilibrium and Limit Analysis Methods

Abstract: In this chapter, three major stability analysis methods are introduced. Each method will be discussed with many recent findings discussed. Based on the slip line method and extremum limit equilibrium method, the classical slope stability, lateral earth pressure and bearing capacity problems are unified under one formulation. A Fortran code is also provided for the lateral earth pressure analysis based on the limit analysis approach, and this code will be useful to many readers.

Keywords: Limit analysis, Limit equilibrium, Extremum principle, Slip line, Unification.

3.1. INTRODUCTION

For stability analysis, there are various methods available to the engineers, and the choice of the method will depends on the complexity of the geometry and the convenience in the solution. In this chapter, the slip line method, limit equilibrium method and limit analysis will be introduced for the lateral earth pressure, ultimate bearing capacity and slope stability problems. It is interesting to note that these three topics are usually considered separately in most of the books or research studies, and different methods of analyses have been proposed for individual problem even though they are governed by the same requirements for the ultimate conditions. Since the governing equations and boundary conditions for these problems are actually the same, Cheng and Li (2015) view that each problem can be viewed as the inverse of the other problems which will also be demonstrated in the present chapter. After the introduction of the three

Y. M. Cheng, H. Wong, C. J. Leo & C. K. Lau

basic stability analysis methods, a section on the unification of the three most important stability problems will be discussed.

The three stability methods together with the corresponding numerical solution techniques will be discussed with the use of different computer programs developed by the author. The limit equilibrium methods as discussed in this chapter are available in the program SLOPE 2000 developed by Cheng, which can be downloaded at the site http://www.cse.polyu.edu.hk/~ceymcheng/. SLOPE2000 is also one of the analysis modules in the large scale geotechnical analysis and design package GEOSUITE 1.0/2.0. For the slip line and limit analysis programs which are used in this chapter, they are more suitable for single material problem and are not yet mature enough for general conditions, hence these programs are not available for general download but can be obtained from Cheng at ceymchen@polyu.edu.hk.

3.2. SLIP-LINE METHOD

At the ultimate condition, both equilibrium and yield conditions must be satisfied. Combining the Mohr-Coulomb yield criterion (which is generally adequate for soil) and the equilibrium equations, a set of hyperbolic partial differential equations of plastic equilibrium can be developed. In order to solve the governing partial differential equation, it is more convenient to transform the governing equations to curvilinear coordinates along the directions of the failure planes for mathematical convenience. Once the equations are solved, the failure modes with the corresponding systems of stresses will be automatically determined. The slip directions or slip lines constitute a network which is called slip-line field. The governing equations can be solved with adequate boundary conditions to investigate the stresses at the ultimate condition, and the solution of the problem is commonly taken as the rigorous solution, as the solutions are either similar to those from other methods or are better. Since the governing equations are written along the slip lines, the slip line fields corresponding to the solutions are commonly considered as the failure mechanism of the governing problem. For example, the bearing capacity of footing and the lateral earth pressure behind a retaining wall are commonly analyzed by the slip line analysis, but not for the slope stability problem.

Kötter (1903) was the first to derive the slip-line equations for two-dimensional ultimate problems, while Prandtl (1920) was the first to obtain an analytical solution for footing by assuming the weight of soil to be negligible. His results were then applied by Reissner (1924) and Novotortsev (1938) to different problems on the bearing capacity of footing on weightless soil. The inclusion of soil weight in the solution of the governing partial differential equation is analytically impossible, and Sokolovskii (1965) has proposed a finite difference approximation of the slip-line equations for which the accuracy can be further improved by an iteration scheme (Cheng 2003b), and such iteration to update the coordinates of the grid points on the slip line field has been demonstrated to be important for passive pressure evaluation. Sokolovskii (1965) has solved many types of problems on the bearing capacity of footings, slopes as well as the lateral earth pressure on retaining walls. De Jong (1957) has developed a graphical procedure for the solutions which appeared to be seldom used nowadays. There are other approximate solutions for the governing differential equations which include the applications of perturbation methods (Spencer, 1962) and series expansion methods (Dembick *et al.*, 1964), but these methods are not popular and versatile enough for more complicated problems and are seldom considered now. More recent results and numerical techniques are given by Cheng and Au (2005) for bearing capacity problem and Cheng (2003b) and Cheng *et al.*, (2007b) for lateral earth pressure problem.

3.2.1. Slip-Line Method for Plane Strain Problem

Consider a general body force γ which may not align with XY axes as shown in Fig. (**3.1**). Such consideration is commonly required when there is a rotation of axes applied to a problem or when earthquake load is considered, otherwise, ξ is usually 90°. Under the rotation of axes technique as proposed by Cheng (2003b) is used, ξ may not be 90°.

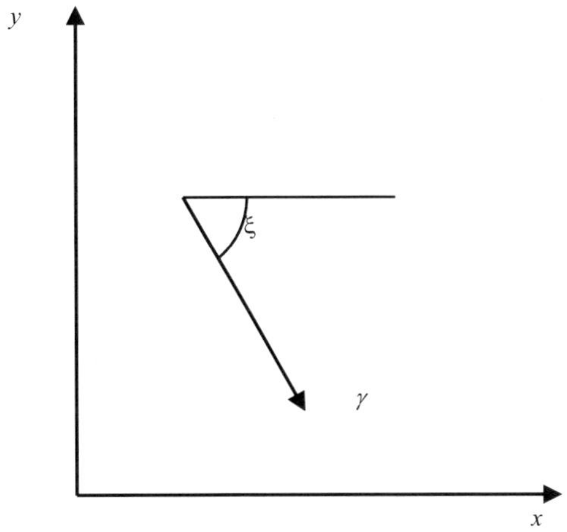

Fig. (3.1). A general two-dimensional body force system with non-vertical gravity.

Stress equilibrium gives:

$$\begin{aligned} \frac{\partial \sigma_x}{\partial x} + \frac{\partial \tau_{xy}}{\partial y} &= \gamma \cos \xi \\ \frac{\partial \tau_{xy}}{\partial x} + \frac{\partial \sigma_y}{\partial y} &= -\gamma \sin \xi \end{aligned} \tag{3.1}$$

Let R and p represent the radius and the middle of the Mohr-circle of the stresses. From (Fig. **3.2**),

$$p = (\sigma_1 + \sigma_3)/2, \qquad R = (\sigma_1 - \sigma_3)/2 \tag{3.2}$$

At yield condition, from the Mohr-Coulomb relation:

$$\sigma_x = p + R\cos 2\theta \quad \sigma_y = p - R\cos 2\theta \quad \tau_{xy} = R\sin 2\theta \tag{3.3}$$

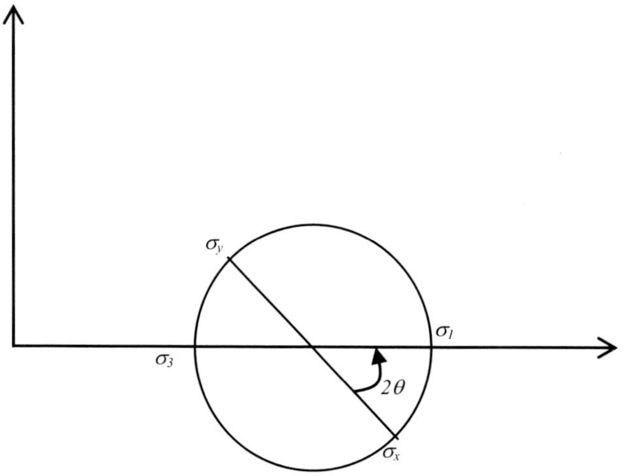

Fig.(3.2). Mohr transformation of stresses.

Put (3.3) into (3.1) gives

$$\frac{\partial p}{\partial x}(1+\sin\phi\cos 2\theta)+\frac{\partial p}{\partial y}\sin\phi\sin 2\theta+2R\left(-\frac{\partial\theta}{\partial x}\sin 2\theta+\frac{\partial\theta}{\partial y}\cos 2\theta\right)=\gamma\cos\xi \tag{3.4}$$

$$\frac{\partial p}{\partial x}\sin\phi\sin 2\theta+\frac{\partial p}{\partial y}(1-\sin\phi\cos 2\theta)+2R\left(\frac{\partial\theta}{\partial x}\cos 2\theta+\frac{\partial\theta}{\partial y}\sin 2\theta\right)=-\gamma\sin\xi \tag{3.5}$$

where ϕ is the friction angle of soil. Multiply eq. (3.4) by $\sin(\theta\pm\mu)$ and eq. (3.5) by $-\cos(\theta\pm\mu)$, where $\mu=\frac{\pi}{4}-\frac{\phi}{2}$, and then add up eq. (3.4) and (3.5) gives

$$\frac{\partial p}{\partial x}\left[\sin(\theta\pm\mu)+\sin\phi\sin(-\theta\pm\mu)\right]+\frac{\partial p}{\partial y}\left[-\cos(\theta\pm\mu)+\sin\phi\cos(-\theta\pm\mu)\right] \quad \textbf{(3.6)}$$

$$+2R\left[-\frac{\partial\theta}{\partial x}\cos(-\theta\pm\mu)+\frac{\partial\theta}{\partial y}\sin(-\theta\pm\mu)\right]=\gamma\sin(\theta+\xi\pm\mu)$$

Using the formula

$$\sin(\theta\pm\mu)=\pm\cos\left(\theta\pm\mu\mp\frac{\pi}{2}\right)=\pm\cos(\theta\mp\mu\mp\emptyset)$$

$$=\ \pm\cos(\theta\mp\mu)\cos\phi+\sin\phi\sin(\theta\mp\mu) \quad \textbf{(3.7)}$$

➔ $\sin(\theta\pm\mu)+\sin\phi\sin(-\theta\pm\mu)=\pm\cos(\theta\mp\mu)\cos\phi$ **(3.8)**

➔ $\cos(\theta\pm\mu)=\sin(\frac{\pi}{2}-\mu\mp\theta)=\sin(\frac{\pi}{2}-2\mu\mp\theta+\mu)$

$$=\sin(\phi\mp\theta+\mu)=\sin[\phi\mp(\theta\mp\mu)]$$

$$=\sin\phi\cos(\theta\mp\mu)\mp\cos\phi\sin(\theta\mp\mu) \quad \textbf{(3.9)}$$

$\therefore$ $\sin\phi\cos(-\theta\pm\mu)-\cos(\theta\pm\mu)=\pm\sin(\theta\mp\mu)\cos\phi$ **(3.10)**

$\therefore$

and $\sin(\theta+\xi\pm\mu)=\sin(\theta\mp\mu+\xi\pm2\mu)$

$$=\sin(\theta\mp\mu)\cos(\xi\pm2\mu)+\cos(\theta\mp\mu)\sin(\xi\pm2\mu) \quad \textbf{(3.11)}$$

Therefore, eq. (3.6) can be written as

$$\pm\cos(\theta\mp\mu)\cos\phi\frac{\partial p}{\partial x}\pm\sin(\theta\mp\mu)\cos\phi\frac{\partial p}{\partial y}-2R\left[\cos(\theta\mp\mu)\frac{\partial\theta}{\partial x}+\sin(\theta\mp\mu)\frac{\partial\theta}{\partial y}\right]$$

$$=\gamma[\sin(\theta \mp \mu)\cos(\xi \pm 2\mu) + \cos(\theta \mp \mu)\sin(\xi \pm 2\mu)] \quad \textbf{(3.12)}$$

Now a pair of line defined by α line and β line can be defined as the failure plane direction as shown in Fig. (**3.3**).

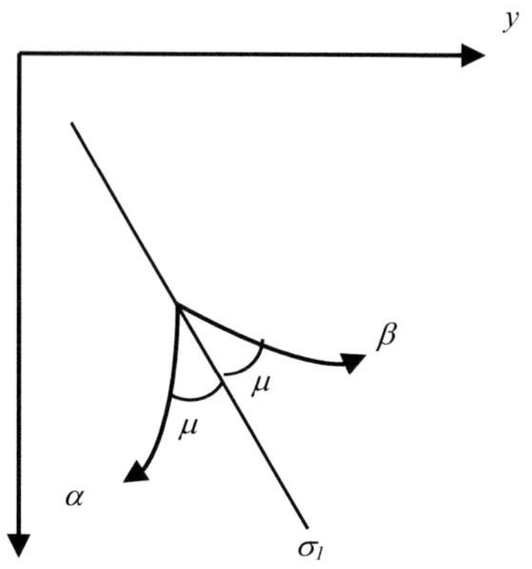

Fig. (3.3). The α and β characteristic lines.

The slopes of these two lines are:

$$\frac{dy}{dx} = \tan(\theta - \mu), \qquad \frac{dy}{dx} = \tan(\theta + \mu) \quad \textbf{(3.13)}$$

The directional derivatives along these two lines are given by

$$\frac{\partial}{\partial S_\alpha} = \cos(\theta - \mu)\frac{\partial}{\partial x} + \sin(\theta - \mu)\frac{\partial}{\partial y}$$

$$\therefore \quad \frac{\partial x}{\partial S_\alpha} = \cos(\theta - \mu), \qquad \frac{\partial y}{\partial S_\alpha} = \sin(\theta - \mu) \quad \textbf{(3.14)}$$

and $$\frac{\partial}{\partial S_\beta} = \cos(\theta + \mu)\frac{\partial}{\partial x} + \sin(\theta + \mu)\frac{\partial}{\partial y} \quad \textbf{(3.15)}$$

$$\therefore \qquad \frac{\partial x}{\partial S_\beta} = \cos(\theta + \mu), \qquad \frac{\partial y}{\partial S_\beta} = \sin(\theta + \mu) \tag{3.16}$$

where S_α, S_β are the arc length. Since $\cos\phi = \sin 2\mu$, from (3.12) and (3.15), we can transform the two equations to axes along the failure surfaces as:

$$-\sin 2\mu \frac{\partial p}{\partial S_\alpha} + 2R\frac{\partial \theta}{\partial S_\alpha} + \gamma\left[\sin(\xi + 2\mu)\frac{\partial x}{\partial S_\alpha} + \cos(\xi + 2\mu)\frac{\partial y}{\partial S_\alpha}\right] = 0 \tag{3.17}$$

$$\sin 2\mu \frac{\partial p}{\partial S_\beta} + 2R\frac{\partial \theta}{\partial S_\beta} + \gamma\left[\sin(\xi - 2\mu)\frac{\partial x}{\partial S_\beta} + \cos(\xi - 2\mu)\frac{\partial y}{\partial S_\beta}\right] = 0 \tag{3.18}$$

Eqs. (3.4) and (3.5) are now transformed to eq. (3.17) and (3.18) which are the characteristic equations along the failure surfaces. Analytical solutions to eqs. (3.17) and (3.18) can be evaluated if the unit weight of soil is 0, otherwise, numerical method has to be adopted which are complicated for complicated geometry with non-homogeneous conditions. There are three possible boundary conditions for the present problem: Cauchy type, Riemann type and mixed type. To solve the slip line equations, iterative finite difference method is commonly used. Starting from two points A and B with known solutions as shown in Fig. (**3.4**), the results at point P along the intersection of α and β lines are given by the following relations.

Along α characteristic line:

$$R = \frac{R_P + R_A}{2}\ ; \quad \theta = \frac{\theta_P + \theta_A}{2}$$
$$\frac{\partial p}{\partial S_\alpha} = p_P - p_A; \quad \frac{\partial \theta}{\partial S_\alpha} = \theta_P - \theta_A; \quad \frac{\partial y}{\partial S_\alpha} = y_P - y_A; \quad \frac{\partial x}{\partial S_\alpha} = x_P - x_A \tag{3.19}$$

Along β characteristic line:

$$R = \frac{R_P + R_B}{2} \;;\; \theta = \frac{\theta_P + \theta_B}{2} \tag{3.20}$$

$$\frac{\partial p}{\partial S_\beta} = p_P - p_B \;;\quad \frac{\partial \theta}{\partial S_\beta} = \theta_P - \theta_B \;;\quad \frac{\partial y}{\partial S_\beta} = y_P - y_B \;;\quad \frac{\partial x}{\partial S_\beta} = x_P - x_B$$

If point P in Fig. (**3.4**) is close to A and B, eqs.(3.17) and (3.18) can be rewritten as:

$$-(p_P - p_A)\sin 2\mu + (R_P + R_A)(\theta_P - \theta_A) \approx -\gamma(y_P - y_A)\sin 2\mu - \gamma(x_P - x_A)\cos 2\mu \tag{3.21}$$

$$(p_P - p_B)\sin 2\mu + (R_P + R_B)(\theta_P - \theta_B) \approx -\gamma(y_P - y_B)\sin 2\mu - \gamma(x_P - x_B)\cos 2\mu \tag{3.22}$$

where $(x_P - x_A)$ and $(x_P - x_B)$ are given by Sokolovskii (1965) as

$$x_P - x_A \approx \tan(\frac{\theta_P + \theta_A}{2} - \mu)(y_P - y_A) \approx (\frac{dx}{dy})_1(y_P - y_A) \tag{3.23}$$

$$x_P - x_B \approx \tan(\frac{\theta_P + \theta_B}{2} + \mu)(y_P - y_B) \approx (\frac{dx}{dy})_2(y_P - y_B) \tag{3.24}$$

To solve for equations (3.21) and (3.22), boundary conditions of known values of x, y, p and θ must be available on each of the slip lines. Such boundary condition can usually be specified for specific problem. For the two segments CA and CB in Fig. (**3.4**) on two corresponding α and β lines, the point P at the intersection of the slip lines passing through A and B can be initially estimated as:

$$x_P \approx x_A + x_B - x_C \;;\; y_P \approx y_A + y_B - y_C \;;\; p_P \approx p_A + p_B - p_C \;;\; \theta_P \approx \theta_A + \theta_B - \theta_C \tag{3.25}$$

In eq. (3.25), the results for point C can be estimated as the average values of the corresponding results from points A and B. After obtaining the initial estimates for x, y and p, equations (3.26) and (3.27) are used to determine the first iteration values of p_P and θ_p by substituting the initial estimate of x_P, y_P and R_P ($=p_P\sin\phi + c\cos\phi$) into equations (3.21) and (3.22) which gives:

$$\theta_P = \frac{1}{R_A + R_B + 2R_P}\{[\gamma(y_A - y_B)\sin 2\mu - \gamma(2x_P - x_A - x_B)\cos 2\mu + (p_B - p_A)\sin 2\mu + (R_P + R_A)\theta_A + (R_P + R_B)\theta_B\} \quad \textbf{(3.26)}$$

$$p_P = \frac{1}{\sin 2\mu + (\theta_P - \theta_B)\sin\phi}\{[\sin 2\mu - (\theta_P - \theta_B)\sin\phi]p_B - 2c(\theta_P - \theta_B)\cos\phi - \gamma(y_P - y_B)\sin 2\mu - \gamma(x_P + x_B)\cos 2\mu\} \quad \textbf{(3.27)}$$

$(\frac{dx}{dy})_1$ and $(\frac{dx}{dy})_2$ can be determined from eqs.(3.23) and (3.24) while x_p and y_p can be determined from eqs.(3.28) and (3.29).

$$y_P = \frac{x_B - x_A + (\frac{dx}{dy})_1 y_A - (\frac{dx}{dy})_2 y_B}{(\frac{dx}{dy})_1 - (\frac{dx}{dy})_2} \quad \textbf{(3.28)}$$

$$x_P = x_B + (\frac{dx}{dy})_2(y_P - y_B) \quad \textbf{(3.29)}$$

where x_P, y_P, θ_p and p_P obtained in the second step will be used in a refined analysis by using eqs. (3.23), (3.24), (3.28) and (3.29) to produce a more accurate solution. The differences between x_P, y_P, θ_p and p_P before and after each iteration are checked. If the changes are less than 0.1% or any small value, convergence is assumed to be achieved. If the accuracy for a specific point is achieved, then the values for other points within the soil can be calculated in a similar manner. The importance of iteration in the solution of the finite difference equations will be illustrated in the later section. After the solution of the slip line equations for the whole domain, for retaining wall problem with wall friction, the equivalent stress σ along the retaining wall can be determined from *p* by eq.(3.30) given by Sokolovskii (1965).

$$\sigma = p\frac{\sin(\Delta \pm \delta)}{\sin\Delta} \quad ; \quad \sin\Delta = \frac{\sin\delta}{\sin\phi} \quad \textbf{(3.30)}$$

The actual active or passive pressure acting along the retaining wall can be recovered from eq.(3.31) as

$$\sigma_{a,p} = \sqrt{(\sigma \cos\delta - c \cot\phi)^2 + (\sigma \sin\delta)^2} \quad \textbf{(3.31)}$$

In eq. (3.30), the positive sign refers to the passive case while the negative sign refers to the active case. After the determination of the lateral earth pressure or bearing stress, the active/passive pressure coefficients due to γ, c and q or the bearing capacity factor can be back determined then. Sokolovskii (1965) and Kumar (2001, 2001) have neglected the cohesive strength c in formulating eq. (3.31) which is included by Cheng (2003b). Cheng (2003b) has also shown that the formulation as described above can be applied to a general c-ϕ material behind a retaining wall under seismic loading condition simply by a rotation of axes approach.

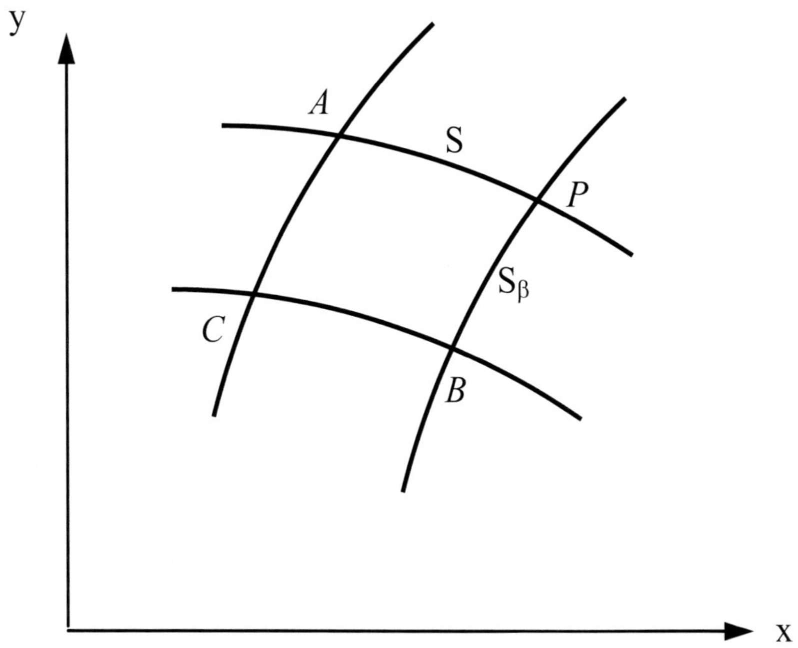

Fig. (3.4). α and β lines and finite difference solution of slip line problem.

Based on the solution of the governing partial differential equations as given by eqs. (3.12) and (3.22), Cheng has developed program SLIP for the solution of bearing capacity problem and KA/KP for solution of general lateral earthquake problem with consideration of earthquake coefficient. The programs have been validated with other programs and published results, and extensions of these programs have been done so that these programs are more versatile for more complicated cases. Typical slip line fields for bearing capacity factors N_c, N_q and N_γ are given in Figs. (**3.5** and **3.6**). Based on such known boundary condition (set q to practically 0 as the initial boundary condition for N_c and N_γ determination and 1 unit for N_q determination), the slip line fields are evaluated from right to left. For N_c and N_q determination, it is assumed that the weight of soil is negligible, and a uniform distributed load (UDL) will be obtained at the base of the footing. Based on such condition, the classical Prandtl's failure mechanism will be automatically obtained from the iterative finite difference analysis for N_c and N_q determination with unit weight of soil being 0, and the failure mechanisms are the same for both bearing capacity factors. The failure mechanism as shown in Fig. (**3.5**) is also exactly the same as the analytical solutions from eqs. (3.17) and (3.18) if the unit weight of soil is 0 (Prandtl's mechanism). On the other hand, if the effect of the weight of soil is considered alone, a triangular pressure will be obtained at the base of the foundation which is shown in Fig. (**3.6**), and N_γ is determined by the average stress of this triangular pressure. The failure mechanism for this case deviates from the classical solution in two aspects: (1) the log-spiral zone will be distorted and is no longer a true log-spiral zone; (2) the triangular zone underneath the foundation as shown in Fig. (**3.5**) will also be distorted to that as shown in Fig. (**3.6**). Interestingly, even though the classical failure mechanism as shown in Fig. (**3.5**) is different from that as shown in Fig. (**3.6**) for the effect of soil weight, the N_γ factor obtained by limit analysis using the mechanism in Fig. (**3.5**) (Chen 1975) is still close to that as obtained by slip line analysis with the failure mechanism as that in Fig. (**3.6**). As shown in Tables **3.1-3.5**, the N_γ factors obtained by the slip line method and limit analysis are relatively similar, but the differences between these factors and those by Meyerhof (1963), Hansen (1970) and Vesic (1973) are noticeable. While it is well known that N_c, N_q are the same among all methods for level ground, if the ground outside the foundation is sloping at an angle *β*, this understanding (using correction factors as compared with slip line solutions) will no longer be true which are illustrated in Tables **3.2** and **3.3**. While the iterative finite difference solutions by program SLIP agree exactly with the analytical solution by Cheng and Au (2005), minor differences of these factors with those by Meyerhof (1963), Hansen (1970) and Vesic (1973) are noticed. It appears that the correction factors by Meyerhof, Hansen and Vesic are still acceptable for practical use even though they are not the exact solutions. For the N_γ factor, the classical results by Vesic (1973) and Hansen (1970) have however over-estimated the bearing capacity and are not safe to be used. This is important for engineering design in Hong Kong, as there are many engineering problems in Hong Kong which relies on this factor on sloping ground.

Cheng and Au (2005) have further investigated the N_γ factor under earthquake condition (by an equivalent pseudo-static earthquake coefficient) which is given in Table **3.5**, while analytical solutions for the N_c and N_q factors are also provided by Cheng and Au (2005). Classically, the effects due to surcharge, cohesive strength and friction angle are directly superimposed in the bearing capacity and lateral earth pressure problems. Cheng (2002) has demonstrated that such direct superposition approach is usually good enough for engineering use except for some very special case. In program SLIP or KA/KP, the global analysis or the analysis of individual effect can be considered, but there will be no corresponding coefficients in bearing capacity or lateral earth pressure then, as every factor is coupled with other factors for such case.

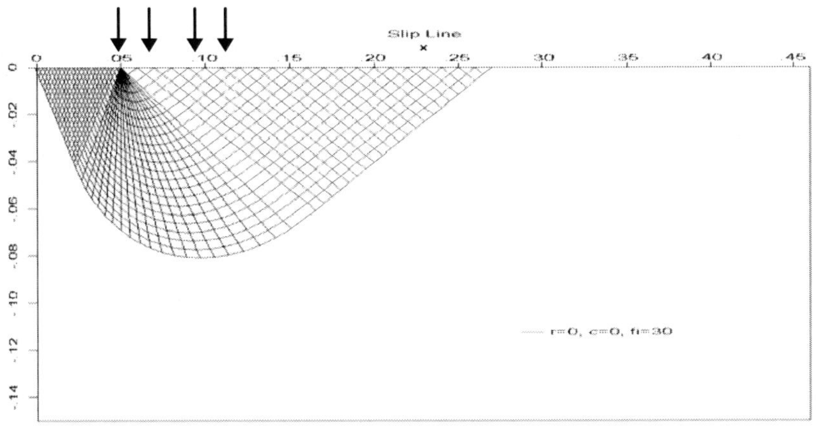

Fig. (3.5). Typical solution for slip lines fields for N_c and N_q factors (uniform foundation load).

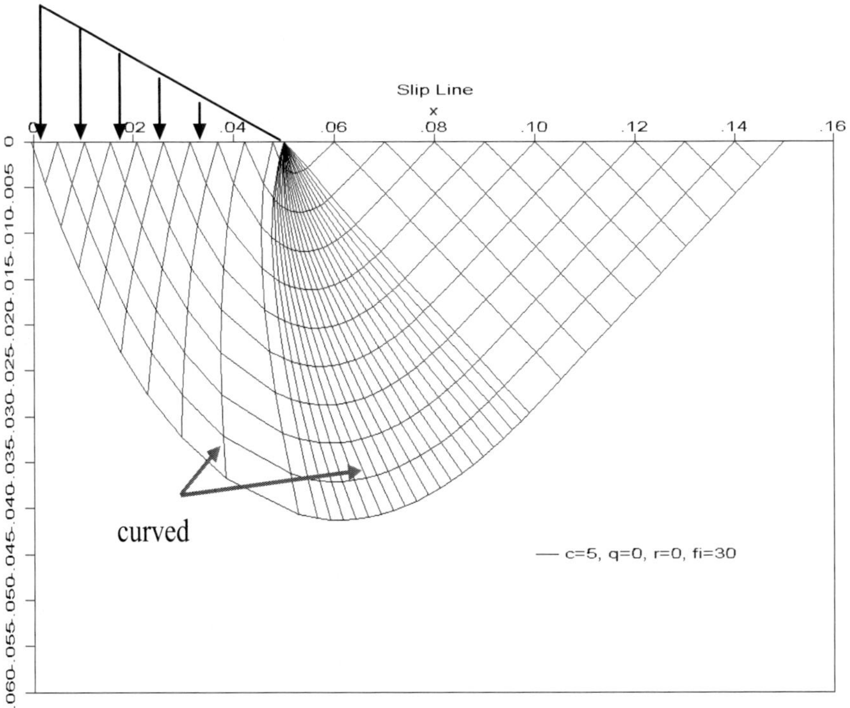

Fig. (3.6). Typical Slip lines for N_γ (triangular foundation load), ϕ=30°.

Table 3.1. Comparison of Nγ on level ground by various methods of analysis (Nc, Nq are the same among all methods).

ϕ\N_γ	SLIP	Sokolovskii	Hansen	Meyerhof	Vesic	Chen
10	0.56	0.56	0.4	0.4	1.2	0.72
20	3.16	3.16	2.9	2.9	5.4	3.45
30	15.31	15.3	15.1	15.7	22.4	15.2
40	86.6	86.5	79.4	93.6	109.3	81.79

Note: The results by Chen (1975) is based on limit analysis using associated flow rule.

Table 3.2. Comparison of N_c for sloping ground with ground slope angle β with various methods of analysis (ϕ=30°).

β\Nc	SLIP	Eq.	Vesic	Hansen
10	24.33	24.32	24.66	28.09
20	19.57	19.57	20.15	26.04
30	15.68	15.68	16.48	24
40	12.5	12.5	13.46	21.94

Table 3.3. Comparison of N_q for sloping ground with various methods of analysis (ϕ=30°).

β\Nq	SLIP	Eq.	Vesic	Hansen
10	12.94	12.94	12.49	11.6
20	7.96	7.95	7.44	6.74
30	2.74	2.74	3.29	3.35
40	5.67	5.67	5.24	4.88

Table 3.4. Comparison of N_γ for sloping ground for different methods of analysis (ϕ=35°).

β\N_γ	SLIP	Vesic	Hansen
10	22.5	32.59	21.38
20	14.9	19.43	12.42
30	9.7	8.58	6.18
35	0	4.32	3.9

Table 3.5. N_γ under different earthquake coefficient.

ϕ\k_h	0	0.05	0.1	0.15	0.2
25	6.86	6.43	6.03	5.54	5.04
30	15.31	14.53	13.22	12.99	12.15
35	35.15	33.85	32.43	30.86	29.36
40	86.6	83.95	80.8	78.32	74.86

For the solution of lateral earth pressure problem, the stress condition on the ground surface behind the retaining wall is usually prescribed (usually prescribed to be a very small value), and the slip line equations are solved to give the lateral earth pressure on the retaining wall. Typical slip line fields for vertical rough retaining wall with level back are shown in Fig. (**3.7**). Based on the consideration on the effect of cohesive strength, surcharge and unit weight of soil, three lateral earth pressure coefficients K_{ac}, K_{aq} and $K_{a\gamma}$(or K_{pc}, K_{pq} and $K_{p\gamma}$) are also obtained for either the active or passive condition.

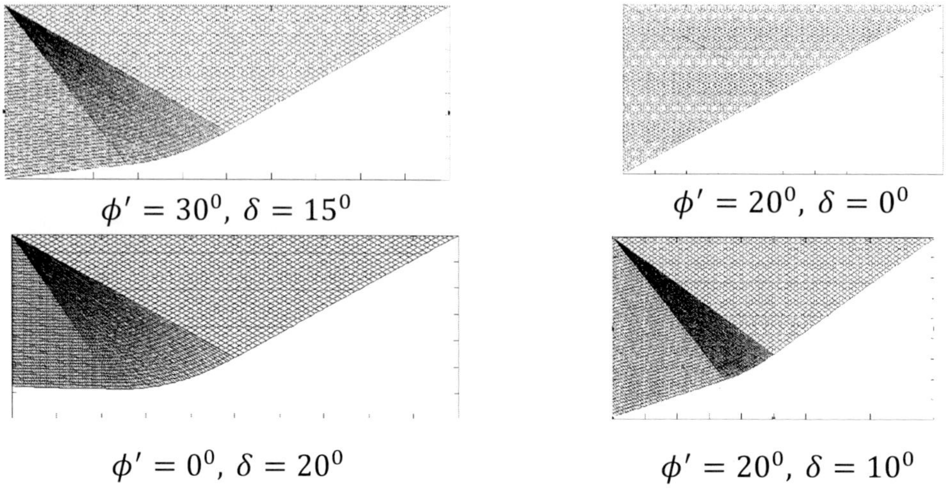

Fig. (3.7). Slip network for passive earth pressure arising from unit weight of soil from slip line method.

The importance of iteration analysis as discussed previously are considered with respect to active and passive pressures, and the results are illustrated in Tables **3.5** and **3.6** with comparisons with the results by Kerisel and Absi (1990). From these two tables by Cheng (2003b), it is observed that the use of multiple iterations results in more accurate values for the passive pressure coefficients (alternatively, a very fine grid can be used to improve the solution). For active pressure coefficients, use of multiple iterations offers no significant advantage as the effects can only be observed in the 4th or 5th significant figure. The use of multiple iterations in active and passive pressure determination have great difference from the analysis, and this phenomenon can be explained by reference to Coulomb equations for active and passive pressure. It is well known that Coulomb active pressure is close to that by slip line solution which is equivalent to active pressure determination will not be sensitive to the precise location of the failure mechanism or equivalently the use of multiple iterations. With no iteration analysis in slip line analysis, minor errors in the precise location of the slip lines have a small effect on the computation. On the other hand, Coulomb passive pressure can be a poor passive pressure determination, which is equivalent to passive pressure determination being sensitive to the precise location of the failure mechanism and hence the use of multiple iterations. For earthquake load, the classical the Mononobe-Okabe (1929) equation appears to be popular among the engineers, but Cheng (2003b) has demonstrated that this equation have over-estimated the passive pressure coefficients when the wall friction is high which is not favourable, which are illustrated in Tables **3.7** and **3.8**.

Table 3.6. Active pressure coefficients for $\delta=\phi$ and $\omega/\phi=0.8$ (δ and ω are the wall friction and the inclination of ground beyond the retaining wall with respect to horizontal direction, ε is wall inclination).

	Kerisel and Absi			**No Iteration**			**2 Iterations**		
$\varepsilon\backslash\phi$	20	30	40	20	30	40	20	30	40
80	0.743	0.634	0.533	0.754	0.644	0.541	0.754	0.644	0.541
90	0.633	0.488	0.364	0.631	0.486	0.362	0.631	0.486	0.362
100	0.517	0.357	0.233	0.524	0.361	0.235	0.524	0.361	0.235

Table 3.6. Passive pressure coefficients for δ=ϕ and ω/ϕ=0.8.

	Kerisel and Absi			No Iteration			2 Iterations		
ε\ϕ	20	30	40	20	30	40	20	30	40
80	3.900	10.800	47.000	3.800	10.122	39.318	3.939	11.072	47.985
90	4.500	13.900	69.000	4.280	12.431	54.431	4.508	13.984	69.245
100	5.100	17.700	100.000	4.871	15.571	77.575	5.243	18.086	101.837

Table 3.7. A comparison of Kpγ values for a vertical wall (β = 90°, δ = 0, ω=0) with different earthquake coefficient K.

ϕ	K	Slip Line (Cheng 2003b)	Kumar and Chitikela (2002)	Kumar (2001)	Soubra (2000)	Morrison & Ebeling (1995)	Mononobe-Okabe(1929)
30°	0	3.000	3.000	3.000	3.000	3.000	3.000
	0.1	2.821	2.817	2.821	2.820	2.892	2.821
	0.2	2.621	2.610	2.618	2.630	2.646	2.629
	0.3	2.394	2.374	2.393	2.420	2.461	2.418
	0.4	2.130	2.096	2.128	-	2.185	2.175

(Table 3.7) contd.....

	0.5	1.792	1.738	1.787	-	1.866	1.866
40°	0	4.599	4.590	4.599	4.600	-	4.599
	0.1	4.381	4.377	4.400	4.380	-	4.380
	0.2	4.146	4.138	4.148	4.150	-	4.152
	0.3	3.897	3.880	3.894	3.910	-	3.912
	0.4	3.627	3.598	3.624	-	-	3.657
	0.5	3.333	3.286	3.328	-	-	3.381

Table 3.8. A comparison of Kpγ values for a vertical wall (β = 90°, δ = ϕ, ω=0).

ϕ	K	Slip line (Cheng 2003b)	Kumar and Chitikela (2002)	Kumar (2001)	Soubra (2000)	Morrison & Ebeling (1995)	Mononobe-Okabe(1929)
30°	0	6.556	6.563	6.677	6.860	7.077	10.095
	0.1	6.084	6.083	6.187	6.350	6.661	9.020
	0.2	5.566	5.562	5.655	5.790	6.154	7.921
	0.3	4.996	4.986	5.065	5.170	5.538	6.784
	0.4	4.343	4.327	4.390	-	4.846	5.577
	0.5	3.526	3.501	3.545	-	3.923	4.208
40°	0	18.150	18.186	18.863	19.620	-	92.586
	0.1	17.106	17.131	17.736	18.460	-	83.254
	0.2	16.011	16.002	16.607	17.230	-	73.902
	0.3	14.853	14.852	15.385	15.940	-	64.522
	0.4	13.619	13.606	14.082	-	-	55.097
	0.5	12.287	12.260	12.677	-	-	45.600

3.2.2 Slip Line Analysis for Axi-Symmetric Problem

Besides the plane strain problem, Cheng and Hu (2005), Cheng *et al.*, (2007b) and Cheng *et al.*, (2008b) have also extended the slip line analysis to axi-symmetric problem for a more general case. The equilibrium equations for a toroidal element Fig. (**3.8**) can be written in cylindrical coordinates system r, θ, z as:

$$\frac{\partial \sigma_r}{\partial r} + \frac{\partial \tau_{rz}}{\partial z} + \frac{\sigma_r - \sigma_\theta}{r} = 0, \qquad \frac{\partial \tau_{rz}}{\partial r} + \frac{\partial \sigma_z}{\partial z} + \frac{\tau_{rz}}{r} = \gamma \tag{3.32}$$

Hill and Cox (2000) have assumed the condition of stable hole at great depth so that eq. (3.32) can be simplified by taking σ_z as constant. The four stress components can be expressed in terms of the mean stress σ and the inclination angle ψ which is formed by extending the major principal stress to the r axis (Fig. **3.9**) as:

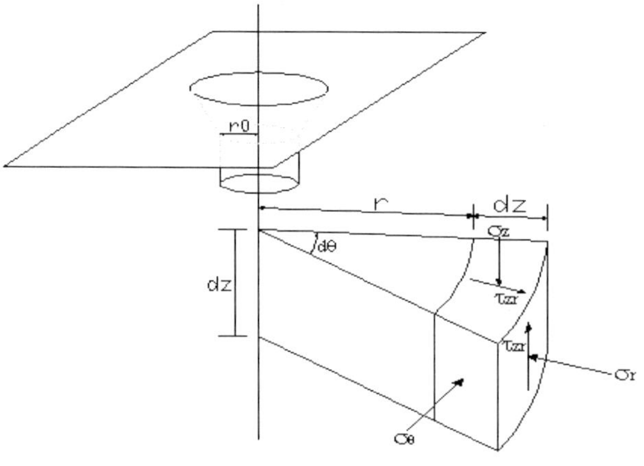

Fig. (3.8). Cylindrical coordinate system and stress components for axi-symmetric problem (z is the vertical direction).

$$\sigma_r = \sigma(1 + \sin\phi\cos 2\psi) - c\cdot\cot\phi \qquad \sigma_z = \sigma(1 - \sin\phi\cos 2\psi) - c\cdot\cot\phi \tag{3.33}$$

$$\tau_{rz} = \sigma \sin\phi \sin 2\psi \ \ \sigma_\theta = k\sigma_1 = k\sigma(1+\sin\phi) - kc \cdot \cot\phi \qquad \textbf{(3.34)}$$

where $\sigma = (\sigma_1 + \sigma_3)/2 + c \cdot \cot\phi$, c is cohesive strength and k is a ratio between σ_θ and σ_1.

k is taken to be 1.0 by most of the researchers in the past for simplicity which is known as the Haar-von Karman hypothesis. This ratio must lie somewhere between 1.0 and the active pressure coefficient k_a theoretically, and more precisely should be taken as a value between k_0 (at-rest) and 1.0 in general (depending on the mobilization of active pressure). Parter (1977) has considered both k=k_0 and k=k_a in the analysis with limit equilibrium method, where k_0 is the at-rest earth pressure condition. Cheng *et al.,* (2008b) have pointed out the problem in adopting k=k_a in general. Berezantzev (1958) has adopted the Harr-von Karman hypothesis in his simplified slip line solution of axi-symmetric active pressure equation which is actually a special case of that by Cheng *et al.,* (2008b). Jenike and Yen's approach (1962) have implicitly implied that the intermediate principal stress is equal to the minor principal stress while Hill and Cox (2000) have further considered the case of k = 1.0. Hill and Cox also mentioned that their solutions (2000) are also not applicable to those plastic regime where k is not 1.0. During shaft excavation, the radial stress will decrease while the tangential stress will increase. The assumption of intermediate principal stress equal to the minor principal stress by Jenike and Yen (1962) or Hill and Cox (2000) is not a reasonable assumption which tends to over-estimate the ultimate active pressure. Drescher (1986) has also obtained the solution using Harr-von Karman hypothesis while Cheng *et al.,* (2007b) have demonstrated that the Harr-von Karman hypothesis is a good assumption in general.

For the characteristic α and β lines, the geometry dictates the slopes of the slip lines which can be written as

$$\frac{dz}{dr} = \tan(\psi + m\mu) \quad \text{where} \quad \mu = \frac{\pi}{4} - \frac{\phi}{2} \tag{3.34}$$

and m takes the value -1 for an α line and +1 for a β line (shown in Fig. **3.9b**). Equilibrium and yield equations from eqs. (3.32), (3.33) and (3.34) give a set of hyperbolic partial differential equations that will reduce to two ordinary differential equations expressing the changes in stress along each characteristic line in terms of the changing inclination ψ and position (r, z) as

$$d\sigma + m2\sigma \tan\phi d\psi + \frac{(1-k-k\sin\phi)\sigma - c(1-k)\cot\phi}{r}dr$$
$$+m\frac{k(1+\sin\phi)\sigma + c(1-k)\cot\phi}{r}\tan\phi dz = \gamma(m\tan\phi dr + dz) \tag{3.35}$$

To simplify the analysis, the variables will be normalized with the radius r_0 which take the form

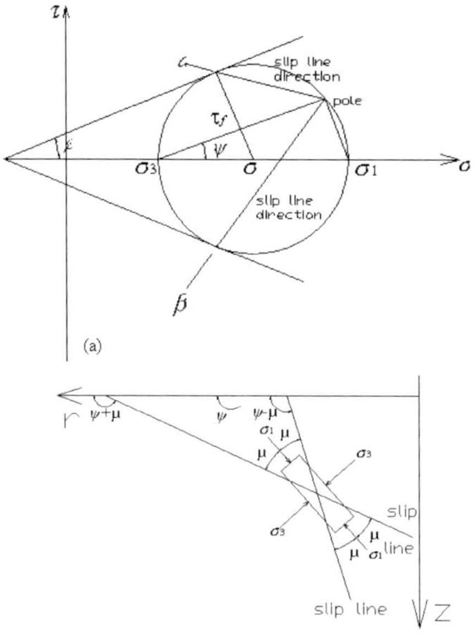

Fig. (3.9). (a) Mohr circle under failure condition. (b) Sign convention and notation.

$$\Omega = \frac{\sigma}{r_0\gamma}, R = \frac{r}{r_0}, \quad Z = \frac{z}{r_0}, \quad C = \frac{c}{r_0\gamma}, \quad Q_0 = \frac{q_0}{r_0\gamma} \tag{3.36}$$

where r_0 is the radius of the circular excavation, z is the vertical depth under consideration, γ is the unit weight of soil, q_0 is the external surcharge on ground surface (which can be inclined) behind the circular excavation. Substituting eq.(3.36) into eqs.(3.34) and (3.35), we have:

$$\frac{dZ}{dR} = \tan(\psi + m\mu) \quad \text{where} \quad \mu = \frac{\pi}{4} - \frac{\varphi}{2} \tag{3.37}$$

$$d\Omega + m2\Omega \tan\phi d\psi + \frac{(1-k-k\sin\phi)\Omega - C(1-k)\cot\phi}{R}dR$$
$$+m\frac{k(1+\sin\phi)\Omega + C(1-k)\cot\phi}{R}\tan\phi dZ = m\tan\phi dR + dZ \tag{3.38}$$

Approximate analytical solution to eq.(3.38) based on the Berezantzev's assumption on the failure mechanism (1958) has been provided by Cheng *et al.,* (2008b), while the numerical solution to eq.(3.38) will be discussed here. It should be noted that the simplified approximate analytical solution can provide a relatively good estimation of the axi-symmetric active pressure but not the failure mechanism. Houlsby and Wroth (1982) and Drescher (1983, 1986) have adopted method of characteristics for the solution of simple axi-symmetric problem, but the effects of inclined ground and wall friction are not considered. The limitations of the previous works by various researchers can be overcome by the solution of eqs. (3.37) and (3.38) which can be written in finite difference form for iterative analysis. For a point W (R_W , Z_W , Ω_W , ψ_W) as shown in Fig. (**3.10**) (a typical Riemann type problem) which is the intersection point of an α line which passed through a known "point" P (R_P, Z_P, Ω_P, ψ_P) and a β line which passed through another known "point' Q (R_Q, Z_Q, Ω_Q, ψ_Q), the finite difference solution can be derived:

$$A = (\psi_P - \psi_Q)\tan\phi + [\frac{k(\sin\phi+1)\tan\phi(Z_W - Z_Q) + (1-k-k\sin\phi)(R_W - R_Q)}{R_W + R_Q}$$

$$-\frac{k(\sin\phi+1)\tan\phi(Z_W - Z_P) - (1-k-k\sin\phi)(R_W - R_P)}{R_W + R_P}] \tag{3.39}$$

$$B = A*(\Omega_P + \Omega_Q) + \tan\phi(R_Q - R_P) - (2Z_W - Z_P - Z_Q) - 2C(1-k)[\frac{R_W - R_P}{R_W + R_P}\cot\phi +$$

$$+\frac{Z_W - Z_P}{R_W + R_P} - \frac{R_W - R_Q}{R_W + R_Q}\cot\phi + \frac{Z_W - Z_Q}{R_W + R_Q}] \tag{3.40}$$

$$C_1 = (A-2)*\Omega_P*\Omega_Q + \Omega_Q[\tan\phi(R_W - R_P) - (Z_w - Z_P) - 2C(1-k)(\frac{R_W - R_P}{R_W + R_P}\cot\phi$$

$$+\frac{Z_W - Z_P}{R_W + R_P})] - \Omega_P[\tan\phi(R_W - R_Q) + (Z_W - Z_Q) + 2C(1-k)(\frac{R_W - R_Q}{R_W + R_Q}\cot\phi - \frac{Z_W - Z_Q}{R_W + R_Q})] \tag{3.41}$$

$$D = \psi_P + \psi_Q - \frac{k(\sin\phi+1)\tan\phi(Z_W - Z_Q) + (1-k-k\sin\phi)(R_W - R_Q)}{(R_W + R_Q)\tan\phi}$$

$$-\frac{k(\sin\phi+1)\tan\phi(Z_W - Z_P) - (1-k-k\sin\phi)(R_W - R_P)}{(R_W + R_P)\tan\phi}] \tag{3.42}$$

$$E = Z_P - Z_Q + \tan\phi(2R_W - R_P - R_Q) - 2C(1-k)[\frac{(R_W - R_P)\cot\phi + Z_W - Z_P}{R_W + R_P}$$

$$+\frac{(R_W - R_Q)\cot\phi - Z_W + Z_Q}{R_W + R_Q}] \tag{3.43}$$

The finite difference equations for point W can be written as

$$R_W = \frac{Z_Q - Z_P + R_P\tan(\frac{\psi_W + \psi_P}{2} - \mu) - R_Q\tan(\frac{\psi_W + \psi_Q}{2} + \mu)}{\tan(\frac{\psi_W + \psi_P}{2} - \mu) - \tan(\frac{\psi_W + \psi_Q}{2} + \mu)} \tag{3.44}$$

$$Z_W = \frac{Z_Q + Z_P + (R_W - R_P)\tan(\frac{\psi_W + \psi_P}{2} - \mu) + (R_W - R_Q)\tan(\frac{\psi_W + \psi_Q}{2} + \mu)}{2} \tag{3.45}$$

$$\Omega_W = \frac{B}{2A+4} + \frac{\sqrt{B^2 - 4(A+2)C_1}}{2|A+2|} \tag{3.46}$$

$$\psi_W = \frac{D}{2} + \frac{(\Omega_P - \Omega_Q)(A-2) + 2E}{2(2\Omega_W + \Omega_P + \Omega_Q)\tan\phi} \tag{3.47}$$

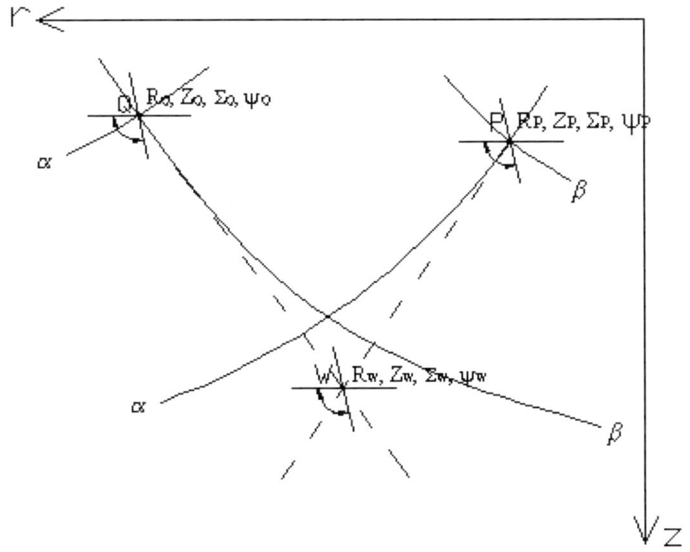

Fig. (3.10). Computation scheme of new point W from known points P and Q.

Refer to Fig. (**3.10**) where results for points P and Q are known and results for W are to be determined. R_W and Z_W can be determined from eqs. (3.44) and (3.45). Ω_w and ψ_w can then be calculated from eqs. (3.46) and (3.47) by using an initial trial of $2.33\psi_W = (\psi_P + \psi_Q)/2$. Since the characteristics lines are curved in general, the solution can be improved by updating ψ_W from eq. (3.47) and repeating the iteration analysis until convergence is achieved (within a specified tolerance of 0.1% by Cheng *et al.*, 2007b).

For the general problem as shown in (Fig. **3.11**), the ground line KS makes an angle α with the horizontal direction. Proceeding from the known boundary condition (external surcharge q_0) on the free surface KS (Fig. **3.11**), the entire stress field within KLMT can be determined. Firstly, the boundary KS is subdivided with to a set of equally spaced points, and a fine grid is suggested for analysis. The solution then marches in towards the wall boundary. The whole calculation iterates until an acceptable accuracy (0.1% or any other value) is achieved. A very coarse mesh applied to a circular excavation for calculating active pressure case is illustrated in Fig. (**3.11**) which is used to illustrate the marching scheme used in the computer program WELLKA developed by Cheng *et al.,* (2007b), and a typical slip line field is shown in Fig. (**3.12**).

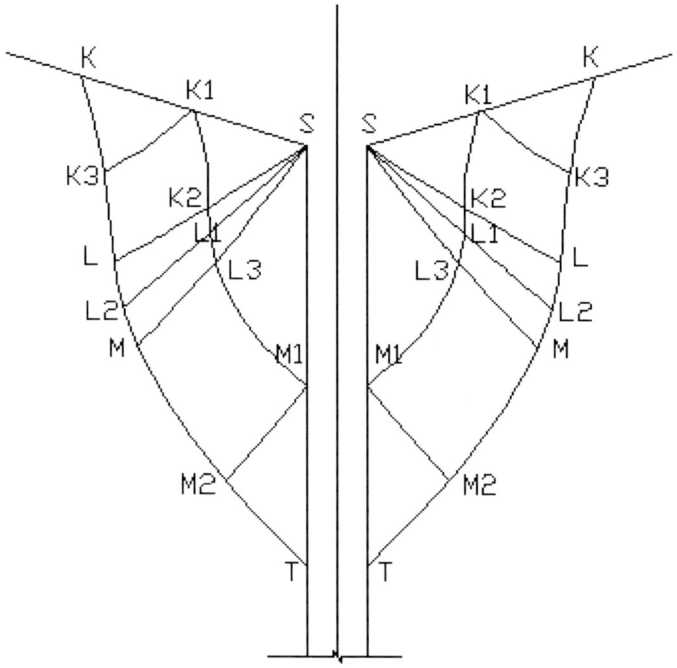

Fig. (3.11). Typical stress characteristic mesh for axi-symmetric problem.

3.2.2.1 Steps for solution:

(1) Active zone — this is a typical Cauchy type boundary condition and the boundary conditions at S, K1 and K are known. K2 is found from S and K1, K3 is found from K1 and K; L is found from K3 and K2. S, K2 and L become the known boundary conditions for the fan zone.

(2) Fan zone — Node S can be viewed as a degenerate β characteristic with unique R and Z, but varying ψ and Ω. The first two terms of eq.(3.35) show that the stress increases exponentially with rotation $\Delta\psi$ of the β line from K2 to L1 in the fan zone. For example, L1 is found from S and K2. Similarly, L2 is found from L1 and L. S, L3 and M now become the known boundary conditions for the passive zone.

(3) Passive zone—Adjacent to the side of the circular wall, R is known and the value of ψ can be determined from the wall condition. The mobilization of the wall friction is assumed to be known. The solutions for Ω and Z at M1 can therefore be determined from the boundary condition at L3. M2 can then be found from the results at M1 and M. Finally, Ω and Z can be found at T using M2 together with the known values of R and ψ at T. The active pressure at excavation wall can then be determined according to following formulae:

$$\Delta = \arcsin\frac{\sin\delta}{\sin\phi} \text{ and } P = \Omega\frac{\sin(\Delta-\delta)}{\sin\Delta} \quad \textbf{(3.48)}$$

where δ is the wall friction angle wall, Ω is the mean stress at the wall. The dimensionless active earth pressure P_a and the actual active earth pressure p_a are given by

$$P_a = \sqrt{(P\cos\delta - C\cot\phi)^2 + (P\sin\delta)^2} \text{ and } p_a = r_0\gamma P_a \quad \textbf{(3.49)}$$

Based on the above procedures, the effect of friction angle, wall friction and cohesive strength, backfill angle as well as external surcharge can all be determined for general problems which is a major extension of the works by Houlsby and Wroth (1982).

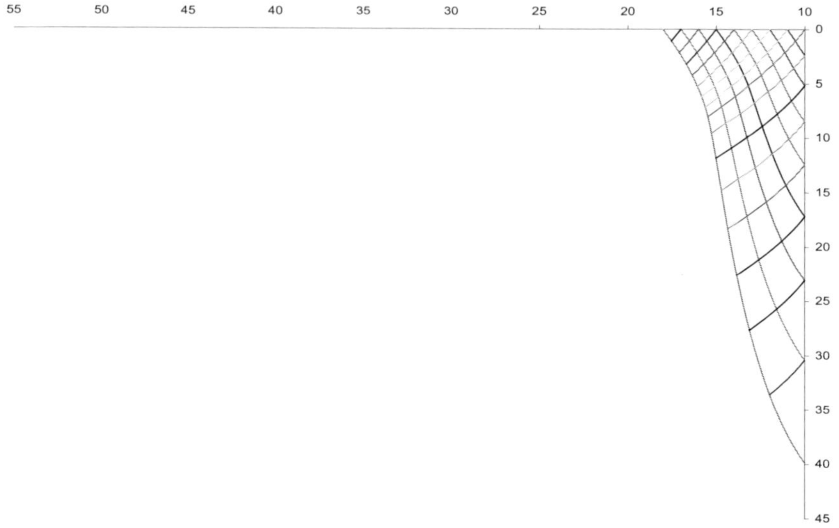

Fig. (3.12). Typical slip lines field for ϕ=40° and k=1.

It is found that the failure surface from iterative finite difference solution has an interesting shape different from the corresponding plane strain problem. The failure surface is similar to plane strain at top (R is greater than z and is close to plane strain condition), but arch action will has noticeable effect at greater depth. Even more interesting is that the angle of ground inclination α behind the wall affects only the active pressure coefficient at the top of the excavation. If the depth is great enough, the arch action (circular effect) will mask out this effect so that the ground inclination angle will only play a very small role on the active pressure. This new result has demonstrated that the arch action is more important than the ground inclination angle when the depth is great. The active pressure coefficient due to self-weight of soil and surcharge is found to decrease with depth. The active pressure due to self-weight of soil may increase indefinitely with depth (at a reduced rate with depth) or attains a maximum value which depends on the lateral stress coefficient k and the friction angle of soil. The actual active pressure due to surcharge will decrease with depth, which means that the circular effect (arch action) has masked the effect of surcharge if the depth is great. The behavior of the active pressure due to self-weight of soil and external surcharge under axi-symmetric condition is drastically different from the corresponding plane strain results. For the cohesive strength of soil, its effect on the active pressure increases with depth and attains a maximum value if the depth is great. The maximum value can be attained by any value of lateral stress coefficient k which is also different from the effect of self-weight of soil. Cheng *et al.,* (2008b) have also demonstrated that the plane strain relation $k_{ac}=[k_{aq}-1]\cot\phi$ is also generally applicable for axi-symmetric problem. Based on the results by Cheng *et al.,* (2008b), it can be concluded that the arch action is actually due to the effects of the intermediate stress plays which is absent from the plane strain condition.

3.2.3. Discussion on Slip Line Analysis

One of the limitations of the slip-line method is the neglect of the stress-strain relationship of the soil, and only the equilibrium and yield conditions are used in the formulation of the stability/ultimate condition. Such idealization may however be acceptable and even necessary for many types of geotechnical problems, as the initial stresses and the constitutive models are usually not known with good accuracy. For the case of plane deformations, the characteristic equations with the given stress boundary conditions have the same number of equations as the unknown stresses so that the problem is statically determinate. If the boundary conditions involve stress and rates of displacement, then the constitutive relation of the soil must be specified in order to obtain a solution.

In general, in a slip-line solution, only part of the soil mass is assumed to be in the state of plastic equilibrium. The solution consists of constructing a slip-line field in the region, which satisfies all the stress boundary conditions that directly concern the region, as well as the equilibrium and yield conditions at every point inside the region. The stress field so obtained has been termed as partial stress field. The stress distribution outside this partial stress field region is not defined or determined. For the solution to be valid, there should exist an associated stress distribution in the non-yielded region, which is in equilibrium with the partial stress field and nowhere violates the yield criterion. The extended stress field defines the stress distribution over the whole body concerned. Even if such an extended stress field can be specified, there is no guarantee that this extended stress solution will be correct.

If the partial stress distribution in the slip-line solution is acceptable as a lower-bound solution, the stress–strain relations should be used to determine whether the give stress and displacement states correspond. If the solution is the correct one, the associated displacement mode will be compatible with a continuous distortion satisfying the displacement boundary conditions and everywhere the rate of plastic work will be positive. Provided that this is the case, then, the solutions so obtained is also an *upper* bound to the collapse value according to the upper-bound rules of limit analysis, and hence is identical with the correct value.

It must be kept in mind that a partial stress field obtained from the slip-line equations is not necessarily the correct solution nor is it known when it is an upper-bound or a lower-bound solution. If a compatible displacement or velocity field can be associated with the specified partial stress field through a given stress–strain relation, the slip-line solution is an upper-bound solution. If, in addition, the partial stress distribution in the plastic zone can be extended to the entire body, satisfying the equilibrium equations, the yield criterion and the stress boundary conditions, the slip-line solution will also a lower bound and is hence the correct solution.

3.3 INTRODUCTION TO LIMIT EQUILIBRIUM METHOD

Classically, the limit equilibrium method (LEM) is commonly adopted by the engineers for the simplified analysis of many geotechnical problems, which include the lateral earth pressure, bearing capacity and slope stability problem. This is due to the fact that limit equilibrium method employs only force and moment equilibrium without tedious mathematics, and such approach is well received for normal engineering use. More importantly, the use of slip line or other similar methods (variational principle or others) are usually limited to problems with regular geometry and single material. In fact, many computer programs have been developed for the solution of various stability problems by LEM, but no general purpose program based on the slip line method has been developed so far, due to the inherent limitations of the method as discussed previously. Interestingly, for many well-known standard problems where the rigorous solutions are known, LEM will be able to give good approximate solutions with ease. In views of the versatile application of the LEM, it is not surprising that this is the most popular method to the engineers. LEM is more commonly adopted in lateral earth pressure and slope stability problem. The well-known Coulomb active and passive earth pressures as well as the trial wedge method are actually highly simplified limit equilibrium method for lateral earth pressure problem. Under such condition, the factor of safety of the system is actually taken to be 1.0 in the lateral earth pressure determination. On the other hand, when LEM is used for slope stability problem, the factor of safety is commonly determined. Other than this requirement, there is no major difference between LEM for lateral earth pressure and slope stability problem, and Cheng *et al.,* (2013a, 2013b) has demonstrated that by varying the lateral pressure until critical factor of safety is 1.0 in slope stability methods, the lateral earth pressure as determined will be similar to that based on the Coulomb's method. The major difference between LEM for slope stability and lateral earth pressure problems are the requirement on the internal force distribution. LEM for lateral earth pressure analysis usually neglects this assumption and the whole soil mass is considered as a rigid body but neglect the moment equilibrium requirement, while refined internal force distribution is usually allowed in slope stability analysis so that moment equilibrium can also be satisfied.

3.3.1. Definition of the Factor of Safety for Slope Stability Analysis

There are different definitions to the term factor of safety. In general, the factor of safety for slope stability analysis is commonly defined as the ratio of the ultimate shear strength divided by the mobilized shear stress at failure state. The precise definition of factor of safety is however governed by the force or moment equilibrium equation as:

1. Moment equilibrium: the factor of safety F_m is defined with respect to moment as:

$$F_m = \frac{M_r}{M_d} \tag{3.50}$$

where M_r is the sum of the resisting moments and M_d is the sum of the driving moment. For a circular failure surface, the centre of the circle is usually taken as the moment point for convenience, but other point of rotation can actually be used if necessary (very uncommon). For a non-circular failure surface, any arbitrary point can be used for the moment consideration, and as long as the system truly obey force and moment equilibrium, then the factor of safety will not depends on the location of the moment point. Interestingly, a moment high above the soil mass can usually result in better convergence in iteration analysis. For Bishop or similar method which do not satisfy horizontal force equilibrium, the factor of safety will depends on the choice of the moment point as 'true' moment equilibrium require force equilibrium. Actually, the use of the moment equilibrium equation alone without enforcing force equilibrium cannot guarantee 'true' moment equilibrium.

2. *Force equilibrium*: the factor of safety F_f is defined with respect to force as:

$$F_f = \frac{F_r}{F_d} \tag{3.51}$$

where F_r is the sum of the resisting forces and F_d is the sum of the driving forces.

For "simplified methods" which can only fulfill either the force or moment equilibrium, these two definitions will be slightly different in the values and meaning, however, most of the design codes do not have a clear requirement on the differentiation of these two factors of safety, and a single factor of safety with no clear about the precise definition is actually specified in many design codes. In this respect, a slope may actually possess more than one factors of safety. It should also be pointed out that for soil nail or similar, the loading can be specified as active or passive in some computer programs, and the corresponding forces or moment from the nails are either deducted from the bottom of eqs.(3.50) and (3.51) or added to the top of these two equations. Such operation will result in different factor of safety, particularly when the nail load is large.

3.3.2. Formulation of Limit Equilibrium Methods

Slope stability problem by classical limit equilibrium method is actually a statically indeterminate problem, and there are different assumptions and methods of analysis available to the engineers. Most of the limit equilibrium methods are based on the techniques of slices which can be vertical (most common due to the vertical weight being the most critical factor in most cases), horizontal or inclined (more suitable when there are horizontal/inclined layers with soil parameters critically different from the other soil layers). Various assumptions and methods have been proposed in the past by bishop (1955), janbu (1957), lowe and karafiath (1960), morgenstern and price (1965), spencer (1967), janbu (1973) and many others. The two-dimensional slice methods have been well summarized by fredlund and krahn (1984), nash (1987), morgenstern (1992), duncan (1996) and cheng and lau (2014). The common features of the methods of slices have been summarized by zhu *et al.,* (2003) as:

(A) The sliding mass defined by the failure surface is divided into a finite number of slices. The slices are usually cut vertically, but horizontal as well as inclined cut have also been used by various researchers. In general, the differences between different methods of cutting are not major, and vertical cut is preferred by most engineers at present as gravity is the major source of instability.

(B) There is only a single factor of safety which is applied throughout the whole failure mass.

(C) Assumptions regarding inter-slice forces/thrust line are used to render the problem determinate.

(D) The factor of safety is determined from force or/and moment equilibrium equations based on the assumptions.

The classical limit equilibrium analysis considers only the ultimate limit state of the system without any consideration on the development of strain, and this feature is similar to that by the slip line method or limit analysis. Based upon the static equilibrium conditions and the concept of limit equilibrium, the number of equations and unknown variables of LEM are summarized in Tables **3.9** and **3.10**.

Table 3.9 Summary of known Equations (n=number of slice).

Equations	Condition
n	Moment equilibrium for each slice
2n	Force equilibrium in X and Y directions for each slice
n	Mohr-Coulomb failure criterion
4n	Total number of equations

Table 3.10 Summary of Unknown variables.

Unknowns	Description
1	Factor of safety
n	Normal force at base of slice
n	Location of normal force at base of slice
n	Shear force at base of slice
n-1	Interslice horizontal force
n-1	Interslice tangential force
n-1	Location of interslice force (thrust line)
6n-2	Total number of unknowns

From the above analysis, it is clear that the classical limit equilibrium problem is statically indeterminate in the order of 6n-2-4n = 2n-2. Hence, additional (2n-2) assumptions have to be introduced in order to solve the problem. The locations of the base normal forces are usually assumed to be at the middle of the slice (n unknowns are eliminated), and this is usually a reasonable assumption unless the width of the slice is wide. In this respect, a limited width of slice (or adequate number of slices) is necessary for a good analysis. This assumption will reduce the number of unknowns to n-2. The most common additional assumptions are either the location of the interslice normal forces or the angle of inclination of the internal forces (equivalent the relation between the interslice normal and shear forces). Such assumption will provide additional n-1 equations so that the problem will now become over-specified by 1. There is not any classical LEM method which can avoid this limitation and can satisfy all the global and local force/moment equilibrium automatically under all cases. At present, there are more than 20 different types of slope stability analysis methods developed by different researchers with different assumptions. The extremum principle by Cheng *et al.,* (2010) which is practically equivalent to the lower bound method is an advancement as compared with the classical LEM in that all the local and global force/moment equilibrium can be satisfied without any assumption on the interslice force function, but the amount of computations required for the extremum determination is much more than that based on the classical LEM methods.

By considering the overall moment equilibrium and overall force equilibrium separately, two different expressions are obtained for the factor of safety, F_m and F_f with a given internal force distribution function. Classically, slope stability analysis is commonly formulated in terms of forces instead of stresses for convenience, and we will keep this approach in this chapter.

With respect to Fig. (**3.13**):
At the base of slice—base normal stress= σ , base shear stress= τ , pore pressure= u
Failure criterion: ultimate shear strength $s = c' + (\sigma - u)\tan\phi'$
Mobilized shear strength $\tau = s/F$, where F is the factor of safety. For the base normal and shear forces P and T,

Now $P = \sigma l$ $\quad T = \tau l$ $\quad$ so $T = \frac{1}{F}(c'l + (P - ul)\tan\phi')$ **(3.52)**

Resolve vertically: $P\cos\alpha + T\sin\alpha = W - (X_R - X_L)$ **(3.53)**

Rearranging and substituting for T gives

$$P = [W - (X_R - X_L) - \frac{1}{F}(c'l\sin\alpha - ul\tan\phi'\sin\alpha)]/m_\alpha \quad \textbf{(3.54)}$$

where $m_\alpha = \cos\alpha(1 + \tan\alpha\frac{\tan\phi'}{F})$

Resolve horizontally:

$$T\cos\alpha - P\sin\alpha + E_R - E_L = 0 \quad \textbf{(3.55)}$$

Rearranging and substituting for T gives

$$E_R - E_L = P\sin\alpha - \frac{1}{F}[c'l + (P - ul)\tan\phi']\cos\alpha \quad \textbf{(3.56)}$$

Overall MOMENT equilibrium (about O):

$$\Sigma Wd = \Sigma TR + \Sigma Pj \quad \textbf{(3.57)}$$

Rearranging and substituting for T gives

$$F_m = \frac{\Sigma[c'l + (P - ul)\tan\phi']R}{\Sigma(Wd - Pf)} \quad \textbf{(3.58)}$$

For circular slip surfaces $f = 0$; $d = R\sin\alpha$; $R = \text{constant}$

so $$F_m = \frac{\Sigma[c'l + (P - ul)\tan\phi']}{\Sigma W\sin\alpha} \quad \textbf{(3.59)}$$

Overall FORCE equilibrium

In the absence of surface loading

$$\Sigma(E_R - E_L) = 0 \quad \textbf{(3.60)}$$

$$\Sigma(X_R - X_L) = 0 \quad \textbf{(3.61)}$$

From eq. (3.60),

$$\Sigma(E_R - E_L) = \Sigma P\sin\alpha - \Sigma\frac{1}{F_f}(c'l + (P - ul)\tan\phi')\cos\alpha = 0$$

So

$$F_f = \frac{\Sigma(c'l + (P - ul)\tan\phi')\cos\alpha}{\Sigma P\sin\alpha} \quad \textbf{(3.62)}$$

In order to solve for F_m and F_f, P must be evaluated, and this requires the evaluation of X_R, P_L the interslice shear forces. As the problem is indeterminate an assumption must be made. Some common assumptions are:

$X_R - X_L = 0$ Bishop (1955)

$\frac{X}{E}$ = constant=λ Spencer (1967)

$\frac{X}{E} = \lambda f(x)$ Morgenstern and Price (1965) **(3.63)**

Bishop (1955) showed that F_m is much less sensitive to the assumption about interslice forces than F_f. As the expression of P contains the interslice shear forces X_R and X_L, assumptions are required to render the problem determinate for solution. It is these assumptions and whether overall/local force equilibrium, moment equilibrium or both are considered that distinguish the different methods of limit equilibrium analysis in literature.

3.3.3. Interslice Force Function

The interslice shear force X is assumed to be related to the interslice normal force E by the relation $X = \lambda f(x)E$, and $\lambda f(x)$ can be considered as the inclination of the internal force. With reference to (Fig. **3.14**), L is the overall horizontal length of the slip surface while x is the horizontal distance of the section under consideration with respect to the left exit end of the slip surface. x_i is dimensionless and is define as x/L so that $0.\leq x \leq 1.0$. The angle of inclination of the internal force is taken as θ_i, and $\lambda_i f(x_i)$ is defined as $\tan\theta_i$. λ for a slip surface is then defined as the maximum of λ_i of all sections. For $f(x)$, there is actually not theoretical basis to determine it for a general case under the classical formulation, and there are 7 types of $f(x)$ commonly used form are implemented in SLOPE2000:

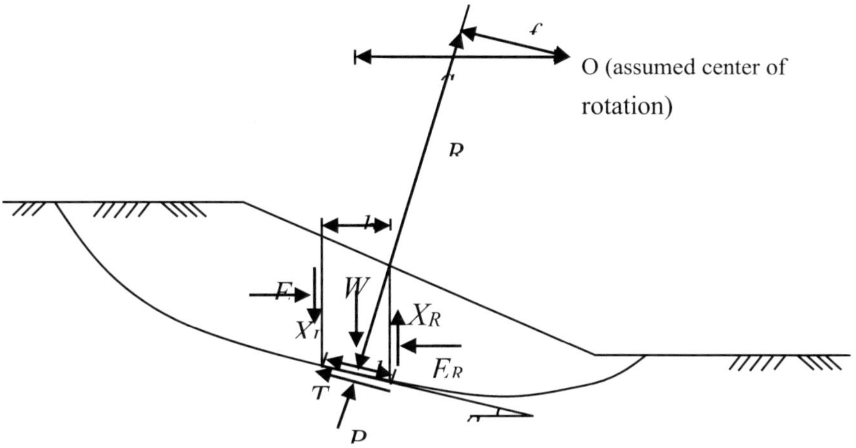

Fig. (3.13). Internal forces for limit equilibrium analysis.

Type 1: *f(x)* = 1.0. This case is equivalent to the Spencer method and is commonly adopted by many engineers. It will be demonstrated later that *f(x)*=1.0, which is the most popular option to the engineers, may not be realistic and good enough in some special cases, though it may be acceptable for normal engineering problems. This option is popoular simply because it is simple and there is no particular good choice for *f(x)* for general case.

Type 2: *f(x)* = sin(x). This is a relatively popular alternative to *f(x)*=1.0. This function is adopted simply purely because of its simplicity, but some engineers also report that convergence is usually better with this function than assuming *f(x)* = 1.0.

Type 3: *f(x)* = trapezoidal shape as shown in (Fig. **3.15**). Type 3 *f(x)* can reduce to type 1 as a special case, but this option is seldom adopted in practice due to the difficulty in defining the parameters for this function. In fact, the author has never used this option for real projects.

Type 4: *f(x)* = error function or the Fredlund-Wilson-Fan force function (1986) which is defined by *f(x)* = $\Psi \exp(-0.5c^n\eta^n)$, where Ψ, c and n have to be defined by the users. η is a normalized dimensional factor which has a value of -0.5 at left exit end (x=0) and 0.5 at right exit end of the failure surface (x=1). η varies linearly with the x-ordinates of the failure surface. The factor Ψ appears originally in the work by Fan *et al.,* (1986) is now absorbed into the parameter λ so that this parameter can be removed in actual computation. This error function is actually based on the curve fitting solution from an elastic finite element stress analysis by Fan *et al.,* (1986). Since the stress state in the LEM is the ultimate condition instead of an elastic stress state, the suitability of this interslice force function cannot be justified in general. It is also difficult to define the suitable parameters for a general problem with soil nails, water table and external loads, but it appears that this function is also easy to converge in the stability calculation.

Type 5: Corps of Engineers interslice force function. *λf(x)* is assumed to be constant and is equal to the slope defined by the two extreme ends of the failure surface. Moment equilibrium is not considered if this function is adopted.

Type 6: Lowe-Karafiath interslice force function (1960). *λf(x)* is assumed to be the average of the slope of the ground profile and the failure surface at the section under consideration. Similar to the Corps of Engineers method, moment equilibrium cannot be achieved by this function as *λf(x)* is prescribed to be known.

Type 7: *f(x)* is defined as the tangent of the base slope angle at the current section, and this assumption is used in the Load Factor method in China. Similar to the previous two cases, this method cannot satisfy moment equilibrium.

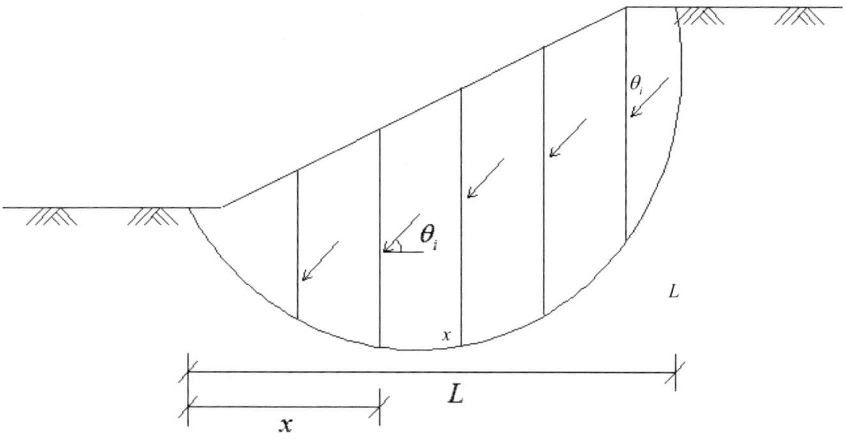

Fig. (3.14). Definition of tanθ for defining λ and *f(x)* for limit equilibrium analysis.

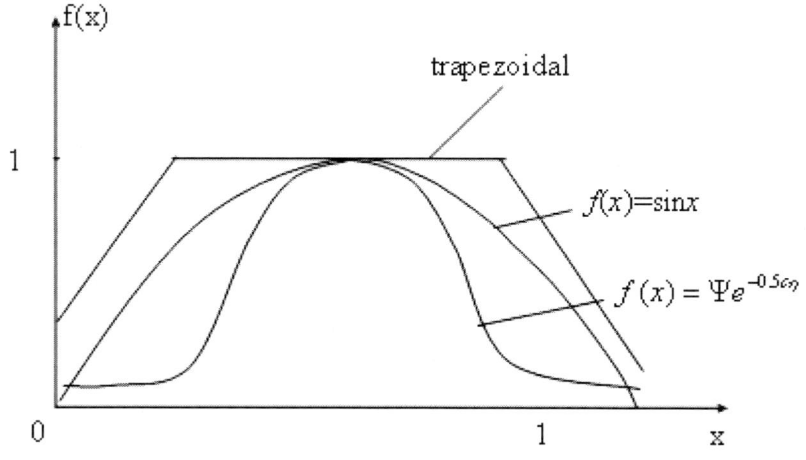

Fig. (3.15). Different types of f(x) commonly implemented into computer programs.

For the above first four types of interslice force functions, they are commonly adopted in the Morgenstern-Price and GLE methods (Sarma method by some researchers, but appear to be not rare), and both the moment (globally) and force equilibrium (locally and globally) can be satisfied simultaneously. The back-computed thrust line may however lies outside the soil mass which means that local moment equilibrium is actually not achieved. An interslice force function should satisfy all local and global equilibrium, but there is no simple way or theoretical background in defining this function except for the extremum principle by Cheng *et al.,* (2010).

For type 5 to type 7 interslice force functions, only force equilibrium is ultilized in the formulation. The factors of safety from these methods are however usually close to those by the "rigorous" methods, and are usually better than the results by the Janbu simplified method (1957). In fact, the Janbu simplified method (1957) is given by the case of $\lambda=0$ for the Corps of Engineers method, Lowe-Karafiath method (1960) and the Load factor method, and results from the Janbu simplified analysis (1957) can also be taken as the first approximation trial in the Morgenstern-Price analysis (1965) if iteration method is used.

Based on a Mohr circle transformation analysis, Chen and Morgenstern (1983) have established that $\lambda f(x)$ at the two ends of a slip surface should be equal to the ground slope angle. Other than this requirement, there is no simple way to establish $f(x)$ for a general problem. Although there is no simple way to define $f(x)$, Morgenstern (1992) has pointed out that for normal problems, factor of safety from different methods of analyses are similar so that the assumptions on the internal force distributions are not major issues for practical use, except for some isolated cases. In views of the difficulty in prescribing a suitable $f(x)$ for general case, most engineers will choose $f(x)$ = 1.0 for simplicity which appears to be satisfactory for most cases. Cheng *et al.,* (2010) have however established the upper and lower bounds of the factor of safety and the corresponding *f(x)* based on the extremum principle. The true factor of safety of any system must lie in between these two limits. Cheng *et al.,* (2010) have also demonstrated that the results from extermum principle can satisfy the requirement as stipulated by Chen and Morgenstern (1983).

Besides the interslice force assumption, the locations of the interslice internal force (thrust line) or the base normal forces can also be taken as the basic assumption in formulating the limit equilibrium method. Janbu rigorous method (1973) appears to be appealing in that the local moment equilibrium is used explicitly while only global instead of local moment equilibrium is enforced with the interslice force function (hence the back-computed line of thrust can lie outside the soil mass which is not possible). As suggested by Janbu (1973), the line of thrust ratio is usually taken as 1/3 of the interslice height, which is basically compatible with the classical lateral earth pressure distribution. It should be noted that the equilibrium of the last slice is actually not checked in the Janbu rigorous method (1973), as the problem is over specified by 1 by nature. The compliance of local moment equilibrium from the Janbu rigorous method (1973) is hence not strictly rigorous. A practical limitation of this method is the relatively poor convergence in the iteration analysis, particularly when the failure surface is highly irregular or there are external loads. This is due to the fact that the line of thrust ratio is pre-determined, and a small change in the line of thrust can give rise to a large change in the internal force, hence convergence is usually poorer. This method is hence seldom implemented into commercial program, but is available as an option in SLOPE2000 with some minor revisions to improve the convergence. If the method is modified by assuming $ht/h=\lambda f(x)$, where ht=height of line of thrust above slice base and h=length of the vertical interslice, the convergence of this method may be be improved for suitable *f(x)*. There is however difficulty in defining *f(x)* for the line of thrust for general case, hence this approach is practically not used. Cheng *et al.,* (2013b) has developed another version of Janbu rigorous method with better convergence which is implemented in the program SLOPE 2000.

With reference to Fig. (**3.16**), eqs. (3.52) to (3.55) which are the Mohr-Coulomb relation and vertical force equilibrium will hold for every method of analysis, including the Janbu rigorous method.

Resolve parallel to base of slice:

$$T+(E_R-E_L)\cos\alpha=(W-(X_R-X_L))\sin\alpha$$

$$E_R-E_L=(W-(X_R-X_L))\tan\alpha-\frac{1}{F}(c'l+(P-ul)\tan\phi')\sec\alpha \quad \textbf{(3.64)}$$

Take moments about centre of base of slice (for thin slice):

$$E_R b\tan\alpha_t-X_R b-(E_R-E_L)h_t=0$$

or $$X_R=E_R\tan\alpha_t-(E_R-E_L)\frac{h_t}{b} \quad \textbf{(3.65)}$$

From eq. (3.60)

$$\Sigma(E_R-E_L)=\Sigma(W-(X_R-X_L))\tan\alpha-\frac{1}{F_f}\Sigma(c'l+(P-ul)\tan\phi')\sec\alpha=0 \quad \textbf{(3.66)}$$

Hence

$$F_f=\frac{\Sigma(c'l+(P-ul)\tan\phi')\sec\alpha}{\Sigma(W-(X_R-X_L))\tan\alpha} \quad \textbf{(3.67)}$$

The solution is reached iteratively. First it is assumed that $X_R-X_L=0$. Then values of E and X are calculated using eq. (3.64) and (3.65) above, and the values of shear force X lag by one iteration as compared with E. As moment equilibrium is satisfied by eq. (3.65), hence $F_f=F_m$. In Janbu's original formulation (1973), P is eliminated and the following expression for F is obtained:

$$F = \frac{\sum[c'b + (W - (X_R - X_L) - ub)\tan\phi']/n_\alpha}{\sum(W - (X_R - X_L))\tan\alpha} \quad \text{in which } n_\alpha = \cos\alpha . m_\alpha \tag{3.68}$$

Factor of safety is sometimes sensitive to line of thrust. *X* obtained should be checked with E to ensure the correct F.O.S. If violated, the thrust line should be adjusted. This approach is however not possible if an automatic search for global minimum is progressing.

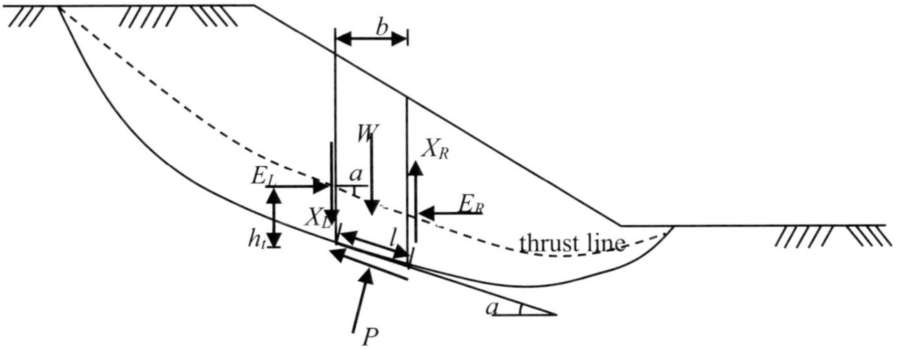

Fig. (3.16). Line of thrust in the Janbu rigorous formulation.

For the determination of the factor of safety using the interslice force function, λ and *F* are the two unknowns which can be determined by different methods. In most of the commercial programs, the factor of safety is obtained by iteration method with an initial trial factor of safety (usually 1.0) which is effective for most cases. The use of the iteration method is actually equivalent to expressing the complicated factor of safety polynomial in a functional form as (fixed point theorem):

$$F = f(F) \tag{3.69}$$

Chen and Morgenstern (1983) and Zhu *et al.*, (2001) have proposed the gradient type Newton-Rhapson technique in the evaluation of the factor of safety *F* and λ. The gradient type methods are more complicated in the formulation but are fast in solution. Chen and Morgenstern (1983) have also provided the expressions for the derivatives of the moment and shear required in the Newton-Rhapson analysis. Zhu *et al.,* (2001) admitted that the initial trials of F and λ can greatly affect the efficiency of the computation. In some cases, poor initial trials can even lead to divergence in analysis. Zhu *et al.,* (2001) proposed a variant of the gradient type method which appears to work fine for smooth failure surface, but this method can also breaks down and diverges. The author's experience is that for non-smooth or deep seated failure surface, it is not easy to estimate a good initial trial value, and Zhu's proposal may not work well for these cases even though the answer exist (Cheng *et al.,* 2008c).

The interslice forces at any interface are the same for the slice to the left and to the right of that interface, an overall equation can hence be assembled in a way similar to that in the stiffness method which will result in a matrix equation (Cheng, 2003a) or equivalently a polynomial equation. The complete solution of all the real positive factors of safety from the matrix can be determined by the double QR method by Cheng (2003a) (double QR method computes all the positive roots while single QR method cannot), which is a useful numerical method to calculate all the positive roots associated with the Hessenberg matrix arising from force equilibrium. It should be noted that imaginary numbers or negative numbers may also satisfy the factor of safety polynomial, so the double QR method instead of the classical QR method is necessary to determine the real positive factors of safety. As an alternative, Cheng (2003a) and Cheng and Zhu (2004) have proposed that the factor of safety based on the force equilibrium is determined directly from the polynomial arising from the force equilibrium with a given λ, and these approaches can avoid the problems that may be encountered using the Newton-Rhapson method or iteration method. These approaches can still work under difficult problems while Chen and Morgenstern's (1983) or Zhu's (2001) methods may failo to give the answer. To solve for the factor of safety, the following steps can be used:

(1) λ is increased gradually. From slice 1 to n, based on λ and *f(x)*, the factors of safety using force equilibrium can be determined from the polynomial by the Gauss-Newton method with a line search step selection or the double QR method (direct computation with the expense of time). The internal forces *E,X,P* and *T* can be then be determined directly without using any iteration analysis.

(2) Based on the internal forces as determined in step 1 for a positive and physically acceptable factor of safety (extremely small value may occasionally appear but should be removed), moment equilibrium is then back checked (net moment of the system). If the moment equilibrium is not satisfied with that specific factor of safety, increase λ and repeat step 1.

(3) In the actual implementation, the signs of the unbalanced system moments associated with different λ are monitored. If the sign of the unbalanced moment changes between λ and $\lambda+\delta\lambda$, then the result will be refined by bisection method until the unbalanced moment is small enough.

There are many cases where iteration analysis (as adopted in many commercial programs) may fail to converge, but the factors of safety actually exist and can be determined (Cheng *et al.,* 2008c). Using the polynomial form by Cheng and Zhu (2004) or the matrix form and the double QR method by Cheng (2003a), it is possible to determine the factor of safety without iteration analysis. Zheng *et al.,* (2009) have also suggested procedures to solve for the factor of safety which is similar to the procedure as suggested above. The advantage of the methods by Cheng is that the factor of safety and the internal forces with respect to force equilibrium are obtained directly without any iteration analysis. Cheng (2003a) has also demonstrated that there can be at most n possible factor of safety (including negative value and imaginary number) from the double QR analysis for a failure mass with n slices, which is actually the famous theorem for an nth order polynomial with n roots.

Since all the possible factors of safety for each λ are examined, this method is the ultimate method in the determination of the factor of safety. If other methods of analysis fail to determine the factor of safety, this method may still work. On the other hand, if no physically acceptable solution is found from the double QR method, the problem under consideration has no solution by nature of the specified *f(x)*. Failure to convergence is thus equivalence to the specification of an unsuitable *f(x)*.

3.3.4. Discussion on the Interslice Force Function

From the author's experience on the application of LEM in many complicated and large scale slope stability problems in Hong Kong, China, and Europe, some of the important questions about fundamental problems of the LEM as raised by the engineers include:

1. The meaning of failure to converge during stability calculation – this is particularly important for slopes with external loads and soil reinforcement which may create local stress concentration so that the problem is difficult to be defined by a simple interslice force function as 1.0.

2. *f*(*x*) can be important for some special problems but there is no way to define it.

3. For cases where *f*(*x*) is important, there will be a wide range of results based on different *f*(*x*), and the choice of a sutiable result is important to the engineers.

These questions are important to both researchers and engineers for certain difficult problems, but there are very few previous studies devoted to these three questions. These questions also review the fundamental problem of the basic formulation of the classical LEM methods.

Failure to converge with the "rigorous" methods is well known to many engineers, in particular, for complicated problems with heavy external loads or soil reinforcement. From the author's experience, Janbu rigorous is less satisfactory while Sarma's method appears to be the best in terms of convergence. Cheng *et al.,* (2008c) have carried out a detailed study on the fundamental problem in convergence and have concluded that there are two reasons for failure to converge with the rigorous methods. First, the iteration method that is commonly used to determine the factor of safety may fail to converge because the interslice shear force is commonly assumed to be zero in the first step of the iterative analysis for simplicity. Cheng (2003a) has developed the double QR method which can evaluate the factor of safety and internal forces directly from the solution of a Hessenberg matrix. Based on this method, failure to converge in stability calculations due to the search path (first reason) can be totally eliminated. There are, however, many cases for which the double QR method determines that a physically acceptable answer *does not exist* for a given $f(x)$ (usually for non-smooth slip surface with soil nails), which means that no meaningful factor of safety will be available unless $f(x)$ can be varied. Actually, some engineers have already questioned the meaning of "no factor of safety" for a given slip surface, as such a concept does not appear in the corresponding structural engineering discipline. So far, there is little previous study on this type of failure to converge, and no rigorous method can guarantee convergence for the general case and provide complete satisfaction of the equilibrium requirements.

Since a critical slip surface may not necessarily converge according to the existing "rigorous" methods of analysis (Cheng *et al.,* 2008c), there is always a chance that the critical slip surface may be missed during optimization analysis. Cheng *et al.,* (2008c) have found a practical case where failure to converge by using commercial program can leads to an error of over 30%, though this is not common in general. It is also interesting to note that it has never been proven that a slip surface that fails to converge in stability analysis is not a critical slip surface, but all commercial programs will simply neglect those slip surfaces that fail to converge, as nothing can be done if no answer can be found. Although this problem may not be critical in practice except for very limited case, a failure surface with no FOS is still physically surprising. A system without factor of safety is not the real behavior, and it is just a human deficiency in making the wrong assumption in $f(x)$ or the stability calculation, and this situation never appears in other engineering disciplines. Factor of safety always exist for a problem, but it is possible that we are not able to determine by classical computation simply because of the use of wrong assumption or mathematical technique, and this is supported by the study by Cheng *et al.*, (2008c) that many smooth slip surfaces can still fail to converge using the popular Spencer's method.

Cheng has solved the above three problems raised by engineers using an approach which is equivalent to the variational principle (Cheng *et al.*, 2010, Cheng *et al.*, 2011). The upper and lower bounds to the factor of safety for any problem can be evaluated within a reasonable time for any complicated problem by using modern heuristic optimization method. The present approach can also be viewed (has also been proved by Cheng *et al.*, 2013a) as a lower bound approach, based on which $f(x)$ can be determined, and failure to converge will be virtually eliminated in the stability analysis.

Baker and Garber (1978), Baker (1980) and Revilla and Castillo (1977) have applied calculus of variation to determine the factor of safety of a slope which does not require any assumption on the internal force distribution. The variational formulation by Baker (1980) was criticized by De Jong (1980, 1981) as the stationary value may have an indefinite character rather than a minimum. The global minimum is not necessarily given by the gradient of the function as zero if the global minimum lies at the boundary of the solution domain (Cheng 2003). This conclusion was also supported by Castilo and Luenco (1980, 1982) which was based on a series of counter-examples. Baker (2003) later incorporated some additional physical restrictions into the basic limiting equilibrium framework so as to guarantee that the slope stability problem has a well-defined minimum solution. Although the variational principle requires very few assumptions with no convergence problems during the solution, it is difficult to adopt when the geometry or the ground/loading conditions are complicated. This approach is also never adopted in any commercial program or used in any practical problem.

Pan (1980) and cheng *et al.,* (2008a) have stated that slope stability problem is actually a dual optimization problem which appears to be not well known outside china. on one hand, the soil mass should redistribute the internal forces and utilize its strength to resist the failure, which will result in a maximum factor of safety for any given slip surface. this is called the maximum extremum principle, and the factor of safety will then be compatible with that for reinforced concrete/stell design, and such value should be considered as accurate instead of approximate. on the other hand, the slip surface with the minimum factor of safety is the most possible slip surface, which is called the minimum extremum principle. the maximum and minimum extremum principles are actually equivalent to the lower and upper bound methods. mathematically, the solution from the use of variational principles is an extremum of a function, and this is also equal to the global maximum/minimum of the function. the "present proposal" can hence be viewed as a form of the discretized variational principle.

The maximum extremum principle is not new in engineering, and at the ultimate limit state of a reinforced concrete beam, the compressive zone of the concrete beam will start to yield and the yield zone will propagate until a failure mechanism is formed. the ultimate limit state design of a reinforced concrete beam under application of a moment is hence equivalent to the maximum extremum principle. for any prescribed failure surface, the maximum "strength" of the system will be mobilized when a continuous yield zone is formed which is similar to a concrete beam. pan's extremum principle (1980) can provide a practical guideline for slope stability analysis, and it is equivalent to the calculus of variation method used by baker and garber (1978), baker (1980) and revilla and castillo (1977). this dual extremum principle is proved by chen (1998) based on the lower and upper bound analyses, and it is further elaborated upon with applications to rock slope problems by chen *et al.*, (2001). the maximum extremum is actually the lower bound solution, and the present approach is actually a lower bound approach as well as a variational principle approach.

To determine the bounds to the factor of safety and $f(x)$, the slope as shown in Fig. (**3.17**) can be considered. For a failure surface with n slices, there will be n-1 interfaces and hence n-1 $f(x_i)$. All $f(x)$ must lie within the range of 0 to 1.0, while the mobilization factor λ and the objective function FOS based on Morgenstern-Price's method (1965) will be determined for each set of $f(x_i)$. The maximum and minimum factors of safety of a prescribed failure surface satisfying force and moment equilibrium will then be given by the various possible $f(x_i)$ satisfying eq.(3.70) and the physical constraints (3.71).

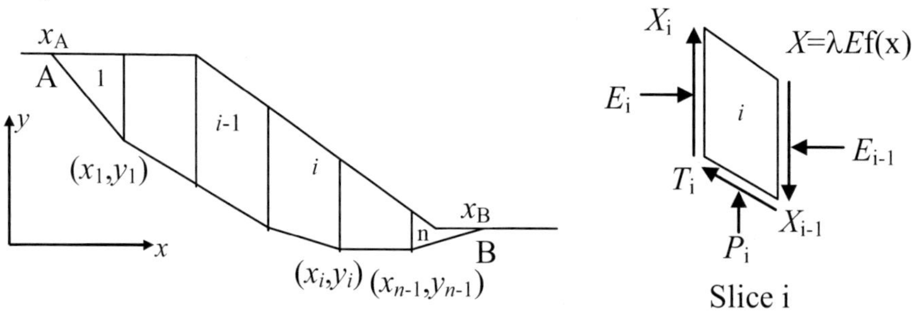

Fig. (3.17). Slip surface and slices.

Maximize (or minimize) FOS subject to $0 \leq f(x_i) \leq 1.0$ for all i **(3.70)**

In carrying out the optimization analysis as given by eq.(3.70), the constraints from the Mohr-Coulomb relation along the interfaces between slices as given by eq.(3.71) should be considered.

$$X \leq E \tan(\phi') + c'L \quad \textbf{(3.71)}$$

where L is the vertical length of the interface between slices. The constraint given by eq.(3.71) should also satisfy the requirement that the thrust line of the internal forces lies within the soil mass. Eq.(3.71) can have a major impact on the FOS in some cases, which will be illustrated by numerical examples in the following section. Since other than the $f(x)$, the Morgenstern-Price's method (1965) is totally governed by force and moment equilibrium, the maximum and minimum factors of safety found from varying $f(x)$ with the constraints (3.71) and the acceptability of the thrust line will provide the upper and lower bounds to the true factor of safety of the slope that are useful for some difficult problems.
The difficulties of the present global optimization analysis are:

1. The number of control variables for the present non-polynomial (NP) type optimization problem is not small.

2. No solution may be obtained for many combinations of $f(x_i)$. This can be due to two possible reasons: (1) no physically acceptable FOS can be found, or (2) eq.(3.71) or thrust line location is violated. It is found that violation of eq.(3.71) can be a major problem, and the objective function can be highly discontinuous over the solution domain. This is an important and difficult problem for global optimization analysis.

3. There are multiple local minima for the objective function, and the objective function is not necessarily a convex function.

The present approach relies on the use of a stable and robust global optimization method which can be effective for the present highly discontinuous problem. Various optimization methods for slope stability analysis have been used by different researchers: the simplex method by Nguyen (1985) and Chen and Shao (1983), alternating variable methods by Celestino and Duncan (1981), the conjugate-gradient method by Arai and Tagyo (1985), and the Monte Carlo method by Greco (1996) and Malkawi *et al.,* (2001). These methods are applicable for simple problems but can be trapped by the presence of a local minimum for general cases. The simulated annealing method, particle swarm optimization (PSO), harmony search (HM), ant-colony, Tabu search, and fish swarm methods aree first adopted by Cheng (2003a), Cheng (2007), Cheng *et al.,* (2015b), Cheng and Li (2007), Cheng *et al.,* (2007a, 2007b, 2008d, 2008e), Cheng and Lau (2014), while genetic algorithms have been adopted by Zolfaghari *et al.,* (2005) and Cheng *et al.,* (2007a), and the leap-frog algorithm was adopted by Bolton *et al.,* (2003). Cheng *et al.,* (2012) have also proposed the use of coupled optimization method which will be more suitable for complicated difficult problem. Cheng *et al.,* (2007a) have carried out detailed comparisons between six major types of heuristic global optimization methods, and the sensitivity of the analysis with different optimization parameters has been investigated.

Since the objective function is highly discontinuous by nature, the factor of safety is obtained by the double QR method by Cheng (2003a). The simulated annealing method which is more stable but less efficient is used to determine the extrema (or the maximum factor of safety) with any given slip surface according to eq.(3.70). To evaluate the global minimum factor of safety of a slope, a global optimization analysis (inner loop) should be carried out for the factor of safety, which is an outer loop of the global optimization analysis. To ensure that "false" failure-to-converge due to iteration analysis (Cheng *et al.,* 2010) is not encountered so as to reduce the discontinuity of the objective function, the factor of safety is determined by the more time-consuming but robust double QR method. Since discontinuity of the objective function is improved, the more efficient harmony search method (or similar) can be used for locating the critical failure surface. The complete process is computationally intensive, but the use of modern global optimization processes can make this process practical on a personal computer within an acceptable computation time.

The author has adopted and modified the simulated annealing method (Cheng 2003a, 2007) complying with eqs.(3.70) and (3.71) to evaluate the two extrema of the factor of safety, and the method is coded into a general purpose program SLOPE 2000 by Cheng (2003a). To determine the maximum and minimum extrema during the simulated annealing analysis, a tolerance of 0.0001 is used to control the optimization search. The author has adopted 15 slices in the computation, hence there are 14 $f(x_i)$ unknowns in the analysis, and the number of trials required to evaluate the two extrema ranges from 25000 to 32000, which is controlled by the tolerance during the optimization search. Based on this study, it was found that about 30-80% of the trials can converge when $f(x)$ is varied, and those trials that fail to converge are controlled by either eq.(3.71) or the thrust line location, or no physically acceptable answer can be found from the double QR analysis. The number of trials which fail to comply with eq.(3.71) is about 3-5 times that where no physically acceptable answer can be found by the double QR method, so the compliance with eq.(3.71) (and the thrust line location), which has been neglected in the Morgenstern-Price's method (1965), is actually important if an arbitrary $f(x)$ is defined.

For the slope as shown in Fig. (**3.18**) with the soil parameters given in Table **3.11**, the various factors of safety are given in Tables **3.12a** and **3.12b**. If only the lower nail is applied, the differences between the two extrema as compared with Spencer's result (1967) are about 5.9% and 4.4% when eq. (3.71) is used or not used respectively, and such a result is not negligible. The corresponding results/differences when the two soil nails are present are 10.5% and 4.1%. It can be observed that when the soil nail or external load is applied, the choice of $f(x)$ has a noticeable influence on the results, and the compliance with eq. (3.71) is a critical issue which must be considered in the determination of the extrema.

Table 3.11. Soil parameters for Fig. (3.18).

Soil	Unit weight (kN/m^3)	Saturated unit weight (kN/m^3)	c' (kPa)	ϕ' (°)
Top	18	20	5	36
Second layer	15	17	3	30

Table 3.12a: Factors of safety from lower bound approach and Spencer's analysis for (Fig. 3.18).

Case	Max. FOS no eq.3.71	Max FOS with eq.3.71	Min. FOS no eq.(3.71)	Min. FOS with eq.(3.71)	Spencer
Bottom nail	1.856	1.841	1.750	1.763	1.790
2 nails	2.661	2.600	2.398	2.498	2.515

Table 3.12b. λ from lower bound approach and Spencer's analysis for Fig. (3.18).

Case	Max. FOS no eq.3.71	Max FOS with eq.3.71	Min. FOS no eq.(3.71)	Min. FOS with eq.(3.71)	Spencer
Bottom nail	1.149	0.944	0.924	1.902	0.488
2 nails	1.435	1.281	1.149	2,011	0.547

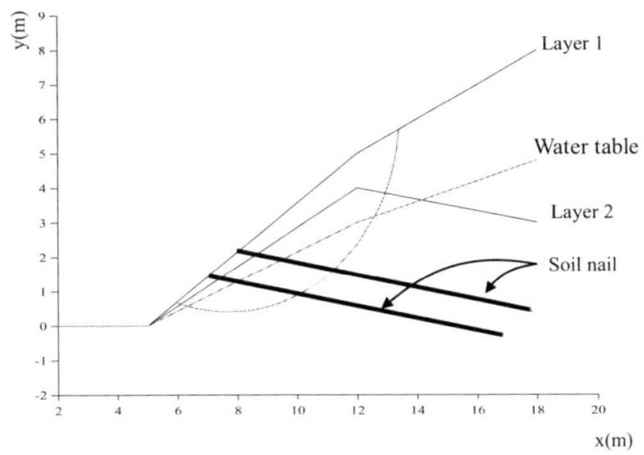

Fig. (3.18). A problem with 2 soil nails.

Consider the slope with a steep failure surface (near to the right hadn side) as shown in (Fig. **3.19**). The factors of safety are 1.542, 1.570, 1.526, and 1.550 based on Bishop's method, Janbu's simplified method (without the correction factor), the Swedish method, and Sarma's method. The extrema are 1.602 and 1.547 if eq.(3.71) is not enforced and are 1.564 and 1.559 if eq.(3.71) is considered. No physically acceptable result can be found for Spencer's method (1967) using the double QR method, and "failure to converge" is the fundamental problem in assuming $f(x)$=1.0 as there is no answer by nature for this problem using the Spencer's method (1967).

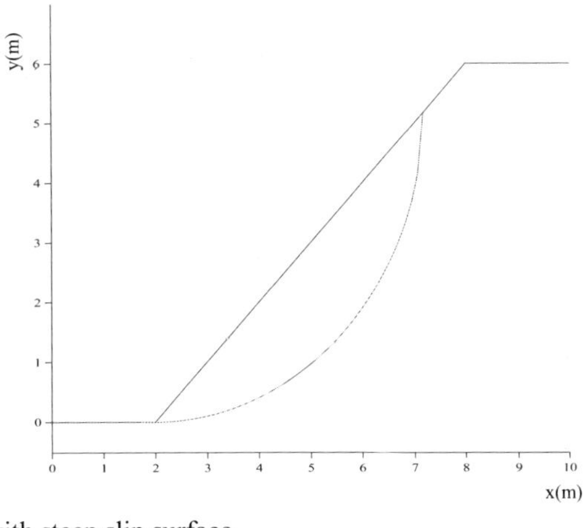

Fig. (3.19). A slope with steep slip surface.

For the two extrema from the extermum principle, the author views that the maximum extremum should be taken as the factor of safety for the prescribed failure surface, as the internal forces within the soil mass should redistribute until the maximum resistance capacity of the soil mass is fully mobilized (definititon of ultimate condition). Beyond that limit, the soil mass will start to fail. The "extremum proposal" also possesses an advantage in that it is independent of the choice of $f(x)$ or the unknown variables to define the problem. It is well known that there are also cases where $f(x)$ may have a noticeable influence on the factor of safety. There is no clear guideline on the acceptance of a suitable FOS for different $f(x)$. The use of the maximum extremum can avoid this dilemma, which is an advantage over the classical definition. The present approach is a typical lower bound approach, as statically admissible forces associated with a prescribed $f(x)$ are considered, and this understanding will be further illustrated in later section. The selection of the maximum FOS is hence justified from the lower bound theorem.

The extremum principle as discussed above can be further demonstrated by a simple bearing capacity problem for clay soil, where the bearing capacity factor N_c is well known to be $\pi+2$ and is given by Prandtl (1920), Sokolovskii (1965) and Chen (1975). So far, there is not any reported success in using the classical "rigorous" slope stability method to evaluate this factor, and an approximate solution by Rocscience (2006). Consider a smooth shallow foundation on clay as shown in Fig. (**3.20**) where the failure mechanism can be obtained analytically. The stresses within the failure mass can be determined by the iterative finite difference solution of the slip line equations as illustrated by Cheng and Au (2005), and a bearing capacity factor Nc of 5.14 (or $\pi+2$) is determined by the author from iterative finite difference analysis. Consider the vertical plane AB as shown in (Fig. **3.20**), the normal and shear stresses on this plane are determined from slip line analysis, and the integration of the normal and shear stresses along AB will give the total normal and shear forces E and X on this surface. The inclination of X/E for different dimensionless distance x are hence obtained. Denote the largest ratio of X/E as λ, *f(x)* is determined by the ratio between X/E and λ, and the corresponding *f(x)* is shown in Fig. (**3.21**) while λ is obtained as 0.388. For the left half of triangle CDO and the right half of triangle GFO in Fig. (**3.20**), the principal stresses are along the horizontal and vertical directions respectively so that *f(x)* is equal 0. Within zone ODF, there is a rapid change of the principal stresses directions and *f(x)* will no longer be zero. From D to O in (Fig. **3.21**), *f(x)* increases rapidly to 1.0 at x=0.5. For this problem, the interslice force function is determinate from slip line analysis, but this kind of function appears to be never used in the past. If the cohesive strength and unit weight of the clay are 20 kPa and zero respectively, based on an applied pressure q_u of 102.83 kPa (20N_c), the factor of safety of this system should be 1.0. A factor of safety 0.941 is however given by the Spencer's method (*f(x)*=1.0) (Rocscience 2006) which is clearly not satisfactory, and clearly the result is a lower bound to the rigorous solution. This result is not good as the simple interslice force function 1.0 is far from the actual *f(x)* as shown in (Fig. **3.21**). Based on the slip line solution, a *f(x)* close to that as shown in Fig. (**3.21**) is obtained with a factor of safety 0.995 and λ equals to 0.391 using 15 slices in the LEM analysis, and the result is very close to the theoretical factor of safety 1.0. This example has clearly demonstrated that a correct *f(x)* can be important in some cases, and the applicability of the extremum principle is not limited by the assumption of *f(x)*, as there is no need to make this assumption explicitly. Furthermore, the result has also demonstrated that the maximum of the factor of safety as adopted from the extremum principle is correct, as other *f(x)* will gives a smaller factor of safety different from the rigorous solution.

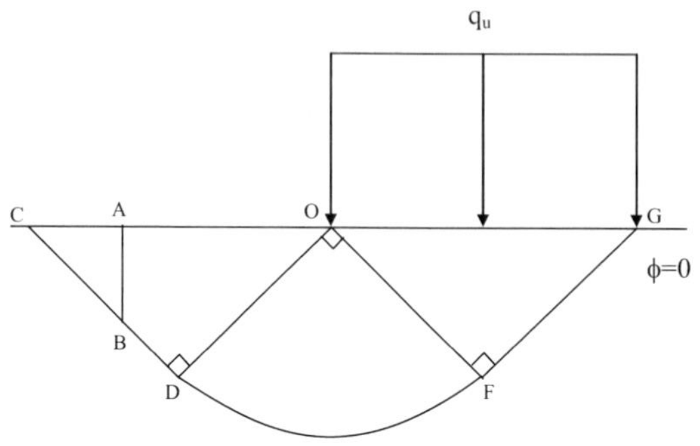

Fig. (3.20). A simple smooth footing on clay.

The interslice force function as shown in Fig. (**3.21**) can be derived from slip line analysis, but such a shape cannot be described by the commonly used interslice force function, and it also appears to be never adopted in the past. This example has also illustrated that *f(x)* can be important for some cases, and the maximum extremum which does not require *f(x)* explicitly can overcome this limitation.

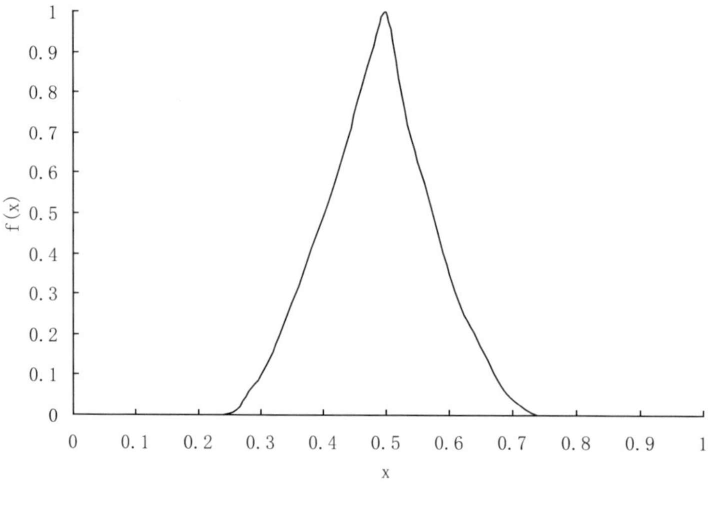

Fig. (3.21). *f(x)* for the simple footing on clay as shown in Fig. (**17**) based on slip line solution

While an arbitrary choice of $f(x)$ may be in conflict with eq.(3.71) or no solution can be evaluated, the extremum principlealways provides results that are physically consistent and acceptable. The present approach possesses the following advantages: (1) it avoids the assumption of $f(x)$/thrust line or other similar relation, (2) it exhibits virtually no failure to converge for arbitrary problem, (3) consistent and acceptable internal forces complying with force and moment equilibrium are always determined, (4) it provides the bounds to the true factor of safety, (5) it determines $f(x)$ during the evaluation of the FOS, and (6) the problem of the variational principle as discussed by De Jong (1980, 1981) and Castilo and Luenco (1980, 1982) is automatically eliminated by using the global optimization analysis.

Currently, there are more than 20 slope stability analysis methods available (but only very few are used for practice), or equivalently, more than 20 sets of FOS under different assumptions for a problem. Although the differences between these factors of safety are usually small, there are cases where $f(x)$ can be important, and it is not easy to select a suitable FOS for these difficult cases. The author views that the maximum extremum FOS can be taken as the FOS of the slope under consideration. Under this case, $f(x)$ can be determined from the present approach. The question of $f(x)$ (or equivalently, the thrust line in Janbu's rigorous method, 1973) and convergence can then be considered as settled under the present formulation. Based on the use of mixed optimization methods, which can be considered as a discretized variational principle, the "extremum principle" can also be a practical tool for application to complicated problems that cannot be solved by the classical variational principle.

3.4. UNIFICATION OF BEARING CAPACITY, LATERAL EARTH PRESSURE AND SLOPE STABILITY PROBLEMS

Classically, lateral earth pressure, bearing capacity and slope stability problems are considered separately, though these stability problems are all related to the ultimate limit state of a system. The author views that these stability problems are governed by similar ultimate requirements and should be equivalent in terms of the failure mechanism, and these problems can be considered as different view of the ultimate limit state of the system. In this section, based on the method of characteristics, the author will determine the bearing capacity factors N_c, N_q and N_γ, for shallow foundation and the lateral earth pressure coefficients K_{ac}, K_{aq} and $K_{p\gamma}$ for retaining wall. Through the method of characteristics, the bearing capacity and lateral earth pressure coefficients are demonstrated to be closely related, and these two problems can be viewed as the inverse of the other problem. The author will also apply the extremum principle to demonstrate the equivalence between the lateral active pressure, bearing capacity and slope stability problems. Overall, these three problems can be considered as equivalent in the basic principles, and the classification of a problem is simply a matter of convenience.

For bearing capacity and lateral earth pressure problems, the use of slip line method is more common, as the geometry under consideration is usually more regular in nature. For slope stability problem where the geometry is usually irregular with complicated soil reinforcements and external loads, the use of the slip line method is practically not possible. Engineers usually adopt the limit equilibrium method (an approximate slip line form) with different assumptions on the internal force distribution (*f(x)*) for the solution of the problems. Cheng *et al.*, (2010) and Cheng *et al.*, (2011) have developed the extremum principle and have demonstrated the equivalence of the maximum extremum from the limit equilibrium method and the classical plasticity solution by a simple footing on clay. Cheng *et al.*, (2010) treat *f(x)* as a variable to be determined, and complete equilibrium is enforced during the search for the maximum extremum of a system. Cheng *et al.*, (2010) have also pointed out that as long as a *f(x)* is prescribed, the limit equilibrium solution will be a lower bound to the ultimate limit state which is the lower bound theorem. Under the ultimate condition where the strength of a system is fully mobilized, *f(x)* is actually determinate by this requirement which is a boundary condition which has not been used in the past.

If the self-weight of soil is neglected (*i.e.* γ=0), equations (3.17) and (3.18) can be simplified as follows:

$$-\sin 2\mu \frac{\partial p}{\partial S_\alpha} + 2R \frac{\partial \theta}{\partial S_\alpha} = 0 \qquad \textbf{(3.72)}$$

$$\sin 2\mu \frac{\partial p}{\partial S_\beta} + 2R \frac{\partial \theta}{\partial S_\beta} = 0 \qquad \textbf{(3.73)}$$

Integrates the above equations along the α and β lines gives:

$$\ln \bar{p} - 2\theta \cot 2\mu = C_\alpha \Rightarrow \ln \bar{p} - 2\theta \tan\varphi = C_\alpha \text{(along α line)} \qquad \textbf{(3.74)}$$

$$\ln \bar{p} + 2\theta \cot 2\mu = C_\beta \Rightarrow \ln \bar{p} + 2\theta \tan\varphi = C_\beta \text{(along β line)} \qquad \textbf{(3.75)}$$

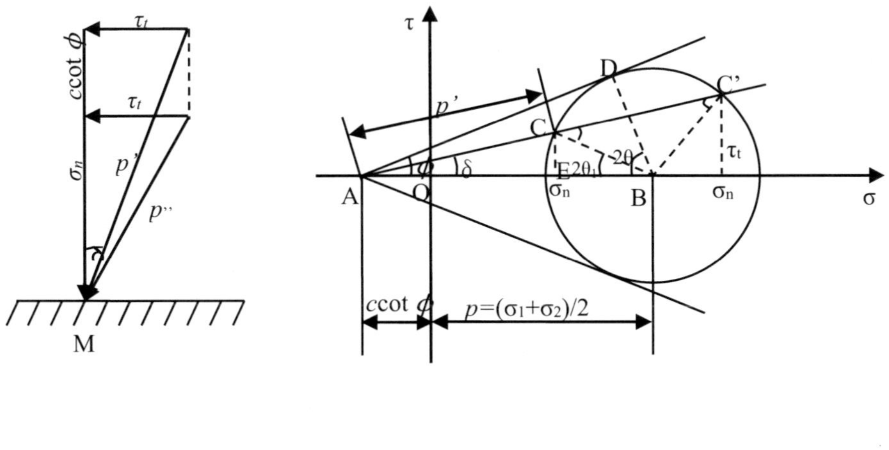

Fig. (3.22). The stress state of point M on a rough surface.

where $\overline{p} = p + c\cot\phi$, c is cohesive strength of soil and C_α and C_β are constants. The sign convention is that compressive stress bein g positive, while the angles θ and others are taken as positive if m e- asured in a counter-clockwise direction. Eqs. (3.74) and (3.75) will be used to derive the expressions for Nq, Nc, Kaq and Kac in the following sections.

For a more general condition, the friction along the interface between the soil and the footing or the retaining wall is considered. If a pressure *p'* act on one point M on the soil surface and the angle between *p'* and the normal line to boundary is δ (*i.e.* friction angle, Fig. (**3.22a**)), *p'* can be described as:

$$p' = \sqrt{(\sigma_n + c\cot\phi)^2 + \tau_t^2} \tag{3.76}$$

where σ_n and τ_t are normal and shear stress acting on the point M respectively. The stress state at the point M can be represented by the Mohr circle as shown in (Fig. **3.22b**). AD is the Mohr-Coulomb line and the angle between AD and horizontal line AB is ϕ. AC represents stress *p'*, and the angle between AC and horizontal line is δ (∠CAB), and δ is taken as positive in Fig. (**3.22a**) in a clockwise direction, which corresponds to an active pressure condition. The line AC intersects the Mohr circle at point C, and the abscissa and ordinate of point C are σ_n and τ_t respectively. Line OB denotes the characteristic stress *p*, ∠ABD = 2θ, ∠ABC = 2θ_1, and angle θ_1 is inclination angle between the major principal stress and boundary. It can easily be seen from the geometrical relationship in the Fig. (**3.22b**) that:

$$\tau_t = p'\sin\delta \tag{3.77}$$

$$\overline{\sigma_n} = \sigma_n + c\cot\phi = p'\cos\delta \tag{3.78}$$

Therefore, eq.(3.77) divided by eq.(3.78) gives,

$$\frac{\tau_t}{\sigma_n + c\cot\phi} = \frac{\sin\delta}{\cos\delta} = \tan\delta \tag{3.79}$$

For convenience, δ in eq. (3.79) is a boundary condition which is defined in terms of τ_f, σ_n, c and φ instead of purely τ_f, σ_n for simplicity. This boundary condition is usually the footing or the back of a retaining wall (surface AO in (Fig. **3.23**)) where σ_n is constant (γ is neglected here). The true friction can be transformed to the apparent friction δ easily. According to the geometrical relationship in (Fig. **3.32b**),

$$\overline{\sigma_n} = \sigma_n + c\cot\varphi = \overline{AB} - \overline{BE} = \overline{AB} - \overline{BC}\cos 2\theta_1 = \overline{AB} - \overline{BD}\cos 2\theta_1 \tag{3.80}$$

$$\tau_t = \overline{CE} = \overline{BC}\cos 2\theta_1 = \overline{BD}\sin 2\theta_1 \tag{3.81}$$

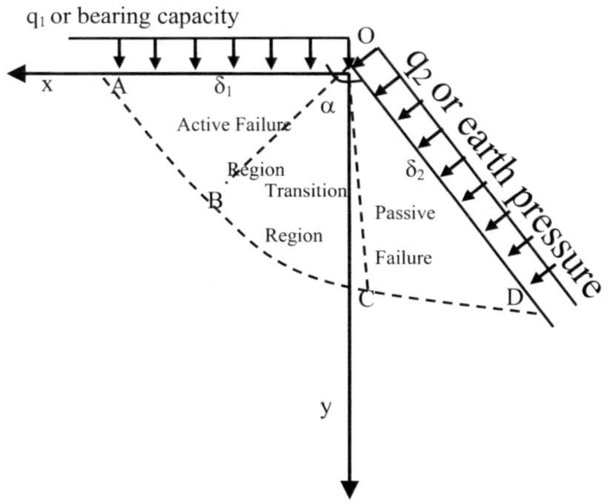

Fig. (3.23). Unified model of bearing capacity and lateral earth pressure problem.

From triangle ABD, it is known that,

$$\overline{AB} = p + c\cot\phi \quad (3.82)$$

$$\overline{BD} = \overline{AB}\sin\phi = (p + c\cot\phi)\sin\phi \quad (3.83)$$

Substitute the above equations into eqs. (3.80) and (3.81),

$$\overline{\sigma_n} = \sigma_n + c\cot\phi = (p + c\cot\phi) - (p + c\cot\phi)\sin\phi\cos 2\theta_1 \quad (3.84)$$

$$\tau_t = (p + c\cot\phi)\sin\phi\sin 2\theta_1 \quad (3.85)$$

So, eq. (3.85) divided by eq. (3.84) gives:

$$\frac{\tau_t}{\sigma_n + c\cot\phi} = \frac{(p + c\cot\phi)\sin\phi\sin 2\theta_1}{(p + c\cot\phi) - (p + c\cot\phi)\sin\phi\cos 2\theta_1} = \frac{\sin\phi\sin 2\theta_1}{1 - \sin\phi\cos 2\theta_1} \quad (3.86)$$

Solve eq. (3.79) and (3.86) simultaneously,

$$\frac{\sin\delta}{\cos\delta} = \frac{\sin\phi\sin 2\theta_1}{1 - \sin\phi\cos 2\theta_1} \quad (3.87)$$

Rearrange eq. (3.87) give,

$$\frac{\sin\delta}{\sin\phi} = \sin(2\theta_1 + \delta) \quad (3.88)$$

The general solution to eq. (3.88) is,

$$2\theta_1 + \delta = 2m\pi + \Delta \quad (3.89)$$

and

$$2\theta_1 + \delta = (2m + 1)\pi - \Delta \quad (3.90)$$

where $\Delta = \arcsin\frac{\sin\delta}{\sin\phi}$. Combine the above two solutions,

$$2\theta_1 + \delta = 2m\pi + k\Delta + (1-k)\frac{\pi}{2} \quad \textbf{(3.91)}$$

or

$$\theta_1 = m\pi + \frac{1}{2}(k\Delta - \delta) + (1-k)\frac{\pi}{4} \quad \textbf{(3.92)}$$

where k=±1; m is an integer and generally m=0 or m=±1. According to the triangle ABC in (Fig. **3.22b**),

$$p + c\cot\phi = p'\frac{\sin\Delta}{\sin(\Delta - k\delta)} = \frac{(\sigma_n + c\cot\phi)}{\cos\delta}\frac{\sin\Delta}{\sin(\Delta - k\delta)} \quad \textbf{(3.93)}$$

Based on eqs. (3.92) and (3.93), the pressure and friction angle on the soil boundary can be converted into the characteristic pressure p and θ_1. For the angle θ between the major principal stress σ_1 and boundary, if the boundary is horizontal, $\theta=\theta_1$. If the boundary inclines to the horizon at α, $\theta=\theta_1+\alpha$. It is noted here that the sign (±) of k is defined based on the direction of displacement of the boundary when the soil attains the limit equilibrium state. When the boundary of soil move in the direction of p', k=-1; when the boundary of soil move in the opposite direction of p', k=1. In other words, if the soil is in active failure state, k=-1; otherwise, the soil is in the passive failure state and k=1.

For the case of a smooth soil surface ($\delta=0$) with tangential stress (shear stress) $\tau_f= 0$, the characteristic pressure on the boundary OA can be simplified as,

$$p = \frac{(\sigma_n + c\cot\phi)}{1-\sin\phi} - \cot\phi = \frac{\sigma_n + c\cos\phi}{1-k\sin\phi} \quad \textbf{(3.94)}$$

Equation (3.94) is obtained because of the relation

$$\lim_{\delta\to 0}\frac{\sin\Delta}{\sin(\Delta-\delta)} = \frac{1}{1-\sin\phi}.$$

The ultimate limit state of a bearing capacity and lateral earth pressure problem is shown in Fig. (**3.23**). For the ultimate bearing capacity problem, q_2 is the surcharge on the ground surface, and q_1 is an unknown to be determined. By contrast, when active lateral earth pressure problem is considered, q_1 is the surcharge acting behind the retaining wall, and q_2 is the lateral earth pressure to be determined. Except for the known and unknown variables and the direction in solving the problem (from left to right or from right to left), the governing equations and conditions for the bearing capacity and lateral earth pressure problems are actually similar. Friction angle is considered in the present study, *i.e.* the friction angle at the interface OA is δ_1; and the friction angle at the interface OD is δ_2. The total slipping wedge can be divided into three limit equilibrium regions, *i.e.* active failure region OAB, transition region OBC and passive failure region OCD.

For a plane strain case, the ultimate bearing capacity of shallow foundation can be determined from the 'superposition' approach which has been shown to be an approximate but good assumption for normal problems by Michalowski (1997) and Cheng (2002). For a strip footing loaded vertically in the plane of symmetry, the ultimate bearing capacity pressure q_u is given by the bearing capacity factors N_γ, N_c and N_q as:

$$q_u = q_{u\gamma} + q_{uc} + q_{uq} = \frac{1}{2}\gamma B N_r + cN_c + qN_q \tag{3.95}$$

Similarly, the total active earth pressure is considered a combination of the effects due to the weight of the soil ($p_{a\gamma}$), the cohesive strength of soil (p_{ac}) and the surcharge loading (p_{aq}). The lateral active earth pressure coefficients $K_{a\gamma}$, K_{ac}, *and* K_{aq} and the total active earth pressure can be expressed as,

$$p_a = p_{a\gamma} + p_{ac} + p_{aq} = \gamma h K_{ar} - cK_{ac} + qK_{aq} \tag{3.96}$$

To derive N_q from the bearing capacity model, assume c=0 and γ=0 and a uniform surcharge q_2 is applied along interface OD. The boundary conditions along the slope surface OA (active boundary) are:

$$p = p'\frac{\sin\Delta}{\sin(\Delta - k\delta)} = \frac{q_{uq}}{\cos\delta_1}\frac{\sin\Delta_1}{\sin(\Delta_1+\delta_1)} \tag{3.97}$$

$$\theta = \theta_1 = \pi - \frac{1}{2}(\Delta_1+\delta_1) + \frac{\pi}{2} \tag{3.98}$$

The boundary conditions along the slope surface OD (passive boundary) are:

$$p = \frac{q_2}{\cos\delta_2}\frac{\sin\Delta_2}{\sin(\Delta_2-\delta_2)} \tag{3.99}$$

$$\theta = \theta_1 + \alpha = \pi + \frac{1}{2}(\Delta_2-\delta_2) + \alpha \tag{3.100}$$

According to equations (3.74) or (3.75), substitute the above boundary conditions into the equation (3.75) respectively and the following relation can be obtained,

$$\begin{aligned}
&\ln\left(\frac{q_{uq}}{\cos\delta_1}\frac{\sin\Delta_1}{\sin(\Delta_1+\delta_1)}\right) + 2\left(\pi + \frac{\pi}{2} - \frac{1}{2}(\Delta_1+\delta_1)\right)\tan\phi = \\
&\ln\left(\frac{q_2}{\cos\delta_2}\frac{\sin\Delta_2}{\sin(\Delta_2-\delta_2)}\right) + 2\left(\pi + \frac{1}{2}(\Delta_2-\delta_2) + \alpha\right)\tan\phi
\end{aligned} \tag{3.101}$$

From the above equation, N_q can be derived,

$$N_q = \frac{\cos\delta_1}{\cos\delta_2}\frac{\sin(\Delta_1+\delta_1)\sin\Delta_2}{\sin(\Delta_2-\delta_2)\sin\Delta_1}\exp\left[(2\alpha - \pi + \Delta_1 + \delta_1 + \Delta_2 - \delta_2)\tan\phi\right] \tag{3.102}$$

Similar procedure can be used to derive N_c. Consider the where q_2=0 and γ=0, the boundary conditions at the base of the footing become:

$$p+c\cot\phi=\frac{q_{uc}+c\cot\phi}{\cos\delta_1}\frac{\sin\Delta_1}{\sin(\Delta_1+\delta_1)} \quad \textbf{(3.103)}$$

θ is unchanged and is same as eq. (3.98). The boundary conditions along the slope surface OD become:

$$p+c\cot\phi=\frac{c\cot\phi}{\cos\delta_2}\frac{\sin\Delta_2}{\sin(\Delta_2-\delta_2)} \quad \textbf{(3.104)}$$

θ is unchanged and is same as eq. (3.100). Similarly to the previous process in deriving N_q, substitute the above boundary conditions into eq. (3.75) gives N_c as:

$$N_c=\left\{\frac{\cos\delta_1}{\cos\delta_2}\frac{\sin(\Delta_1+\delta_1)\sin\Delta_2}{\sin(\Delta_2-\delta_2)\sin\Delta_1}\exp\left[(2\alpha-\pi+\Delta_1+\delta_1+\Delta_2-\delta_2)\tan\phi\right]-1\right\}\cot\phi \quad \textbf{(3.105)}$$

For the lateral earth pressure problem, K_{aq} is derived by assuming c=0 and γ=0. The boundary conditions on the surface OA are:

$$p=p'\frac{\sin\Delta}{\sin(\Delta-k\delta)}=\frac{q_1}{\cos\delta_1}\frac{\sin\Delta_1}{\sin(\Delta_1+\delta_1)} \quad \textbf{(3.106)}$$

θ is the same as eq. (3.98). The boundary conditions on the interface OD are:

$$p=\frac{p_{aq}}{\cos\delta_2}\frac{\sin\Delta_2}{\sin(\Delta_2-\delta_2)} \quad \textbf{(3.107)}$$

θ is the same as eq. (3.100). Substitute the above boundary conditions into equation (3.75), and K_{aq} can be derived as:

From triangle ABD, it is known that,

$$\overline{AB} = p + c\cot\phi \tag{3.82}$$

$$\overline{BD} = \overline{AB}\sin\phi = (p + c\cot\phi)\sin\phi \tag{3.83}$$

Substitute the above equations into eqs. (3.80) and (3.81),

$$\overline{\sigma_n} = \sigma_n + c\cot\phi == (p + c\cot\phi) - (p + c\cot\phi)\sin\phi\cos 2\theta_1 \tag{3.84}$$

$$\tau_t == (p + c\cot\phi)\sin\phi\sin 2\theta_1 \tag{3.85}$$

So, eq. (3.85) divided by eq. (3.84) gives:

$$\frac{\tau_t}{\sigma_n + c\cot\phi} = \frac{(p + c\cot\phi)\sin\phi\sin 2\theta_1}{(p + c\cot\phi) - (p + c\cot\phi)\sin\phi\cos 2\theta_1} = \frac{\sin\phi\sin 2\theta_1}{1 - \sin\phi\cos 2\theta_1} \tag{3.86}$$

Solve eq. (3.79) and (3.86) simultaneously,

$$\frac{\sin\delta}{\cos\delta} = \frac{\sin\phi\sin 2\theta_1}{1 - \sin\phi\cos 2\theta_1} \tag{3.87}$$

Rearrange eq. (3.87) give,

$$\frac{\sin\delta}{\sin\phi} = \sin(2\theta_1 + \delta) \tag{3.88}$$

The general solution to eq. (3.88) is,

$$2\theta_1 + \delta = 2m\pi + \Delta \tag{3.89}$$

and

$$2\theta_1 + \delta = (2m+1)\pi - \Delta \tag{3.90}$$

where $\Delta = \arcsin\frac{\sin\delta}{\sin\phi}$. Combine the above two solutions,

Eq. (3.113) can also be viewed from another point of view. The active pressure along OD induced from the uniform surcharge q_1 along AO will be negative for the cohesive strength. If q_1 is chosen such that the lateral earth pressure on OD is exactly 0, this condition will correspond to the ultimate bearing capacity induced by cohesive strength. Using this concept, eq. (3.113) is also determined from eqs. (3.105), (3.108) and (3.111). The bearing capacity factors N_c and N_q are hence demonstrated to be related to the active lateral earth pressure coefficients K_{ac} and K_{aq} by eqs. (3.112) and (3.113). For the terms N_γ and $K_{a\gamma}$ related to the self-weight of soil, analytical expression is not possible and numerical computation will be adopted for the comparisons which will be explained in the later section.

To verify the above results, the unification of the active lateral earth pressure and bearing capacity problems are further assessed by the use numerical computation. Cheng (2003b) has developed an iterative finite difference program called KA for the slip line analysis of lateral earth pressure. For the numerical study in this section, another program ABC which is coded by Martin (2004) for bearing capacity problem based on the method of characteristic is used for comparison. First of all, consider the case without friction on the interface. As shown in (Table **3.13**), for N_q and K_{aq}, the values using the formula or the finite difference programs are very close to each other, and the relation between N_q and K_{aq} is consistent with eq. (3.112). The numerical results have further supported the validity of the previous results and the unification of the lateral earth pressure and bearing capacity problem. Table **3.13** shows that the results from ABC and KA are very close to the solutions from the analytical equations, and the relation between K_{ac} and N_c as given in Table **3.14** meet the requirement from eq. (3.113). The authors have however noticed a requirement on the iterative finite difference analysis in the present study. Cheng (2003b) has found that the active pressure coefficients are not sensitive to the precise location of the slip line, and no iteration in the finite difference scheme is required for practical purpose (passive pressure coefficient is very sensitive and at least two iterations are required for a good accuracy). When lateral earth pressure program KA is used for bearing capacity problem, the transition zone is much larger and the precision of analysis becomes important. For the present study, iteration in program KA is required and convergence is deemed to be satisfied when the change of the coordinates of the slip line is less than 0.1%. In general, several iterations will be required for the analysis, and the number of iterations increases with the friction angle which is as expected. With the use of iteration analysis, the numerical results from KA (active pressure program) agree well with the analytical expressions for bearing capacity problem as given in this paper. For shallow foundations with sloping ground or active lateral earth pressure problem, the inverse relation between the above two classical soil mechanical problems also holds well which is illustrated by the results as shown in (Tables **3.15** and **3.16**).

Table 3.13. Verification of the relationship between K_{aq} and N_q on level ground (smooth), and K_{aq} x N_q is very close to 1.0 for all cases.

ϕ	N_q		K_{aq}	
(°)	**Eq. (3.102)**	**ABC**	**Eq. (3.108)**	**KA**
10	2.471	2.471	0.405	0.405
20	6.399	6.34	0.156	0.156
30	18.401	18.407	0.054	0.054
40	64.195	64.256	0.016	0.016
50	319.057	319.92	0.003	0.003

Note: c=0, γ=0, α=180°.

Table 3.14. Verification of the relationship between K_{ac} and N_c on level ground (smooth), and $N_c = K_{ac} / K_{aq}$ is satisfied.

φ (°)	N_c		K_{ac}	
	Eq. (3.105)	**ABC**	**Eq. (3.111)**	**KA**
10	8.345	8.345	-3.377	-3.382
20	14.835	14.836	-2.318	-2.329
30	30.14	30.149	-1.638	-1.648
40	75.313	75.385	-1.173	-1.178
50	266.882	267.606	-0.836	-0.838

Note: γ=0, q=0, α=180°.

Table 3.15. Verification of the relationship between K_{aq} and N_q on sloping ground (smooth condition), and K_{aq} x N_q is very close to 1.0 for all cases.

α (°)	N_q	K_{aq}	
	Eq. (3.102)	**Eq. (3.108)**	**KA**
170	15.042	0.066	0.066
160	12.297	0.081	0.081
150	10.052	0.099	0.100
140	8.218	0.122	0.122
130	6.718	0.149	0.149

Note: c=0, ϕ=30°, γ=0.

Table 3.16. Verification of the relationship between K_{ac} and N_c on sloping ground (smooth), and $N_c = K_{ac} / K_{aq}$ is satisfied.

α	N_c	K_{ac}	
(°)	**Eq. (3.105)**	**Eq. (3.111)**	**KA**
170	24.322	-1.617	-1.626
160	19.567	-1.591	-1.599
150	15.679	-1.560	-1.566
140	12.501	-1.521	-1.526
130	9.903	-1.474	-1.477

Note: ϕ=30°, q=0, γ=0.

For the more general cases with friction on the interface between soil and wall/footing, the inverse relation between the bearing capacity and active earth pressure problems is also confirmed which is shown in Tables **3.17-3.20**. Through the above results, it is clear that each problem is the inverse of the other, and the choice of the problem depends on the way the problem is being viewed and solved. In other words, the variables of one problem are just the variables of the corresponding image problem.

Table 3.17. Verification of the relationship between K_{aq} and N_q on level ground (rough), and K_{aq} x N_q is very close to 1.0 for all cases.

δ_1	δ_2	N_q	K_{aq}	
(°)	(°)	**Eq. (3.102)**	**Eq. (3.108)**	**KA**
0	5	19.285	0.0519	0.0518
0	10	20.093	0.0498	0.0497

(Table 3.17) contd.....

0	15	20.834	0.0480	0.0479
0	20	21.505	0.0465	0.0464
0	25	22.083	0.0453	0.0452
0	30	22.456	0.0445	0.0444
5	0	21.167	0.0472	0.0472
10	0	23.838	0.0419	0.0419
15	0	26.299	0.0380	0.0379
20	0	28.416	0.0352	0.0351
25	0	30.020	0.0333	0.0332
30	0	30.829	0.0324	0.0322
5	5	22.184	0.0451	0.045
10	10	26.030	0.0384	0.0383
15	15	29.776	0.0336	0.0335
20	20	33.209	0.0301	0.0300
25	25	36.027	0.0277	0.0276
30	30	37.622	0.0266	0.0263

Note: c=0, ϕ=30°, γ=0.

Table 3.18. Verification of the relationship between K_{ac} and N_c on level ground (rough), and $N_c = K_{ac} / K_{aq}$ is satisfied.

δ_1 (°)	δ_2 (°)	N_c	K_{ac}	
		Eq. (3.105)	**Eq. (3.111)**	**KA**
0	5	31.671	-1.642	-1.652

(Table 3.18) contd.....

0	10	33.07	-1.646	-1.656
0	15	34.353	-1.649	-1.66
0	20	35.516	-1.652	-1.663
0	25	36.516	-1.654	-1.666
0	30	37.162	-1.655	-1.672
5	0	34.931	-1.650	-1.660
10	0	39.557	-1.659	-1.670
15	0	43.819	-1.666	-1.677
20	0	47.485	-1.671	-1.682
25	0	50.265	-1.674	-1.685
30	0	51.666	-1.676	-1.687
5	5	36.692	-1.654	-1.664
10	10	43.353	-1.666	-1.676
15	15	49.841	-1.674	-1.683
20	20	55.787	-1.680	-1.692
25	25	60.668	-1.684	-1.697
30	30	63.432	-1.686	-1.702

Note: ϕ=30°, q=0, γ=0.

Table 3.19. Verification of the relationship between K_{aq} and N_q on sloping ground (rough), and K_{aq} x N_q is very close to 1.0 for all cases.

α	N_q	K_{aq}	
(°)	**Eq. (3.102)**	**Eq. (3.108)**	**KA**
170	24.341	0.0411	0.041
160	19.898	0.0503	0.0502

(Table 3.19) contd.....

150	16.266	0.0615	0.0614
140	13.297	0.0752	0.0751
130	10.87	0.0920	0.091

Note: c=0, ϕ=30°, γ=0.

Table 3.20. Verification of the relationship between K_{ac} and N_c on sloping ground (rough), and $N_c = K_{ac} / K_{aq}$ is satisfied.

α	N_c	K_{ac}	
(°)	**Eq. (3.105)**	**Eq. (3.111)**	**KA**
170	40.428	-1.661	-1.672
160	32.733	-1.645	-1.656
150	26.442	-1.626	-1.636
140	21.299	-1.602	-1.611
130	17.096	-1.573	-1.58

Note: ϕ=30°, q=0, γ=0.

For N_γ, Sokolovskii (1965) and Cheng and Au (2005) and others determined it by solving equations (3.17) and (3.18) from right to left. For the ultimate pressure at the bottom of the footing on a level ground, the slip line program SLIP by Cheng and Au (2005) starts at with a uniform distributed load (very small and tends to zero) along OD (which is horizontal now) and the construction of the slip-line field begins from the right hand side to the left hand side. The ultimate bearing stress along OA due to the weight of soil and friction angle of the soil will be a triangular pressure along OA, which are given by Sokolovskii (1965), Booker and Zheng (2000) and Cheng and Au (2005). However, the image passive pressure problem can be determined by applying a small uniform distributed load (tends to zero) on line OA as shown in Fig. (**3.24**) and solving the slip line equations from left to right. The passive pressure along line OD from passive pressure determination is virtually the same as the bearing stress on line OA from bearing capacity determination from the program output from SLIP and KP! This approach is conceptually the reverse of the classical method of characteristic for solving the bearing capacity problem, but it is not explicitly considered in the past.

A typical slip line system for the lateral pressure problem is shown in (Fig. **3.24**), which is actually the same as the slip lines obtained by Cheng and Au (2005) for bearing capacity problem. Just from Fig. (**3.24**) alone, it is actually not possible to tell whether it is a bearing capacity problem or a lateral earth pressure problem (even though the ways of solution are the inverse). The mean pressure acting on OA is the ultimate bearing pressure due to the self-weight of soil and N_γ is then determined accordingly. The present results are actually equal to that from the Prandtl's mechanism, and the results are compared with the classical solution by Sokolovskii (1966), Hansen (1970), Vesic (1973), Meyerhof (1963) and Chen (1975). Based on the slip line program ABC developed by Martin (2004) for bearing capacity determination and the slip line lateral earth pressure program KP developed by Cheng (2003b), the equivalent N_γ for level ground and sloping ground are given in (Tables **3.21** and **3.22**). Since N_γ from SLIP is defined by the average value of the bearing stress along OA from bearing capacity determination (after Sokolovskii 1966) from the Prandtl's mechanism, numerically, it is the same as the passive pressure $K_{p\gamma}$ from KP, or

$$N_\gamma = K_{p\gamma} \quad \textbf{(3.114)}$$

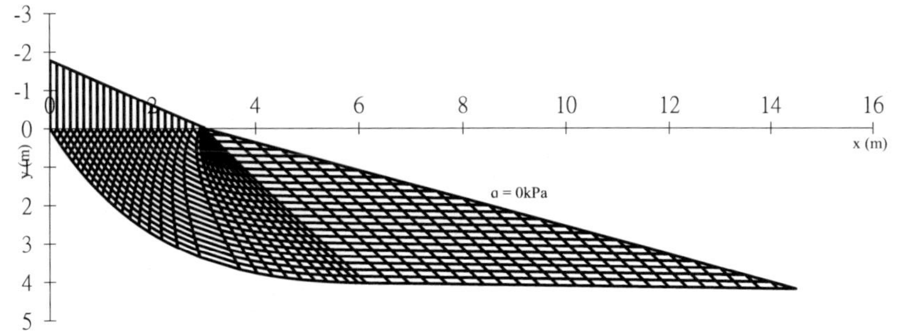

Fig. (3.24). Typical slip line pattern for lateral earth pressure problem with triangular surcharge.

N_γ can be considered as the corresponding passive earth pressure coefficient, and the classical bearing capacity equation $\gamma BN_\gamma/2$ can be considered as equivalent to $K_p\gamma H^2/2$. The classical definition of bearing capacity $\gamma BN_\gamma/2$ is a hence better definition as compared with using γBN_γ, as it actually reflects the fact that bearing capacity and passive pressure problems are just image problem of each other.

Table 3.21. Comparison of N_γ on level ground by various methods of analysis (Prandtl mechanism).

$\phi\backslash N_\gamma$	ABC	Sokolovskii	Hansen (1970)	Meyerhof (1963)	Vesic (1973)	Chen (1975)	KP
10	0.56	0.56	0.4	0.4	1.2	0.72	0.56
20	3.16	3.16	2.9	2.9	5.4	3.45	3.158
30	15.31	15.3	15.1	15.7	22.4	15.2	15.30

Table 3.22. Comparison of N_γ for different methods of analysis (ϕ=35°) for sloping ground (correction factors by Vesic 1973 and Hansen 1970 are used).

$\alpha\backslash N_\gamma$	Vesic (1973)	Hansen	KP	ABC	SLIP
180	48.03	33.93	35.20	35.18	35.15
170	32.59	21.38	22.49	22.5	22.49

(Table 3.22) contd.....

160	19.43	12.42	14.89	14.9	14.9
150	8.58	6.18	9.69	9.7	9.69
145	4.32	3.9	0	0	0

From (Tables **3.21** and **3.22**), it is clear that N_γ can also be determined from the lateral earth pressure program KP by the author. It is interesting to note that when α equals 145°, $\boldsymbol{N_\gamma}$ should equals to 0 when c=0. This result is correctly reflected in the active pressure program KA or program ABC but is not correct in the Vesic's (1973) and Hansen's (1970) bearing capacity factors for purely cohesionless soil.

Cheng *et al.,* (2010) have demonstrated that the extremum principle provides a good approximation to the ultimate condition, and the results by extremum principle are very close to (usually equal to) the slip line solutions in general. In fact, Cheng *et al.,* (2013a) have demonstrated that the extremum principle is equivalent to the solution of partial differential equations, and the slope stability problem and bearing capacity problem can be viewed as equivalent problem under the extremum condition (Cheng *et al.,* 2013b).

To determine N_c factor from slope stability analysis, a surcharge is applied underneath a foundation with a value equal to cN_c without any external surcharge while the self-weight is maintained at zero. For factor N_q, the self-weight and cohesive strength are maintained at zero while a surcharge of 1 unit is applied outside the foundation and the foundation load is maintained at N_q. If the slip surface based on the classical Prandtl's or Hill's mechanism (Chen 1975) is used, the factors of safety as determined for different ϕ are exactly <u>1.0</u> if the extremum principle is used and *f(x)* is kept on varying until the maximum factor of safety is found (Cheng 2013a). For the term N_r, a triangular pressure is applied underneath the foundation, and the maximum pressure is adjusted until the critical factor of safety from extremum limit equilibrium method is 1.0. The critical slip surface is allowed to be automatically determined from the extremum principle in this analysis, and the resulting critical slip surface with a factor of safety 1.0 from the extremum principle will be very close to that by slip line analysis. The bearing capacity factors N_r obtained from different approaches are given in (Table **3.23**). For example, for ϕ=10°, the slip line system using a very fine grid for the case of N_r is given in (Fig. **3.25**). The failure zone at the right of Fig. (**3.25**) is a typical triangular zone underneath the footing with a curved narrow transition zone, and this slip surface is far from the Hill's or Prandtl's mechanism as adopted by Chen (1975). The slip surface from the extremum principle is close to that as shown in (Fig. **3.25**), and the N_r from extremum principle limit equilibrium slope stability analysis is also very close to that by the slip line method. For N_c and N_q, if the slip surfaces are not specified but is searched from the extremum principle in the way as suggested by Cheng *et al.,* (2007a, 2010), the critical slip surfaces from extremum limit equilibrium solution are exactly equal to that from the slip line solution as shown by Cheng *et al.,* (2013b).

Table 3.23. Bearing capacity N_r from different methods (solution from Slip is based on Cheng and Au (2005) using slip line solution, solution from Sokolovskii is also based on slip line solution, solution from Chen is based on limit analysis, present solution (automatic) is based on extremum principle with search for critical slip surface).

ϕ (°)	**Program SLIP**	**Sokolovksii (1965)**	**Chen (1975)**	**Present (Automatic search)**
10	0.56	0.56	0.72	0.57

(Table 3.23) contd.....

20	3.16	3.16	3.45	3.18
30	15.31	15.3	15.2	15.3
40	86.8	86.5	81.79	86.6

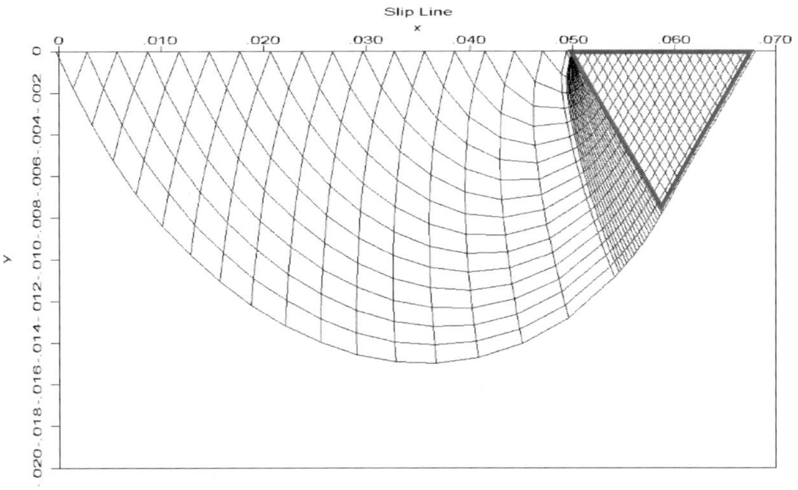

Fig. (3.25). Slip line for the bearing capacity problem when ϕ=10°.

It is also interesting to note that the lower bound principle has been clearly illustrated in the present analysis. For the case of factor N_c using the Prandtl's mechanism, the authors have analyzed the intermediate results during the extremum analysis. *f(x)* is kept on changing during the analysis, with an initial value of 1.0 for all x_i during the simulated annealing optimization analysis. With *f(x)*=1.0, the factor of safety is actually 0.925 which is far from 1.0. As *f(x)* is changed, the factor of safety converges towards 1.0 during the simulated annealing analysis, but no factor of safety greater than 1.0 can be obtained. The same results are also obtained for the case of N_q as well. The results as shown in Fig. (**3.26**) are actually good illustration of the lower bound principle. It can hence be concluded that the LEM is practically a lower bound method.

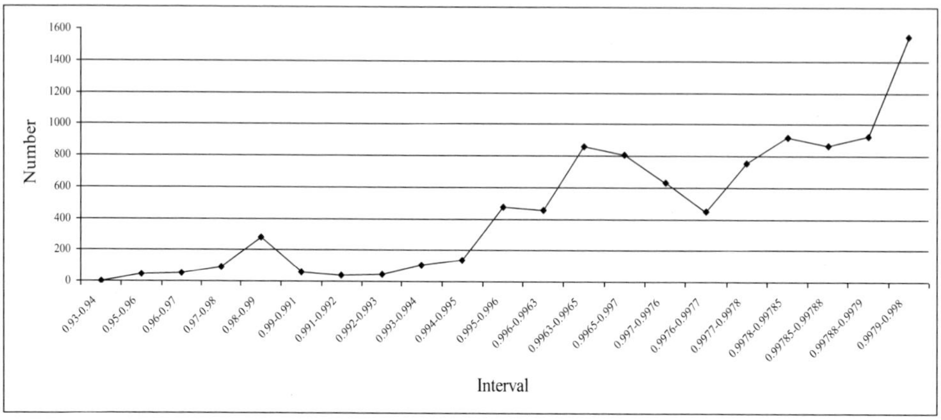

Fig. (3.26). Distribution of acceptable factor of safety during simulated annealing analysis for N_c (same results for N_q) with ϕ=30° using *f(x)* as the variable.

3.4.1. Discussion on Unification of Stability Analysis Methods

The author has applied the method of characteristics to study two classical and important geotechnical problems: active lateral earth pressure problem and ultimate bearing capacity problem. The active lateral earth pressure and bearing capacity problems are demonstrated to be equivalent, except for the ease of mathematically manipulation. Based on the slip line theory, the coefficients N_q, N_c, K_{aq} and K_{ac} are derived and the relations among them are established. Two finite difference programs ABC and KA are further used to confirm the validity of the results as derived in the present study. Other than the ways in solving eqs. (3.17) and (3.18), there are no practical difference between the two classical geotechnical problems, and the two problems can actually be viewed as equivalent problems under the ultimate condition. It is demonstrated that classical bearing capacity and active lateral earth pressure problems are actually the same, as they are controlled by both the yield and equilibrium equations. Based on the present study, for a normal shallow foundation problem, it can be viewed as a lateral earth pressure problem and vice versa. The surcharge (ultimate bearing capacity) behind an imaginary retaining wall which generate a net upward stresses outside the foundation (or equivalently the retaining wall) will be the ultimate limit state of the system. A bearing capacity problem can hence be viewed as a lateral earth pressure problem in this manner.

The author has also demonstrated the lower bound principle in Fig. (**3.26**) using the extremum analysis. Based on the minimum of the extremum from limit equilibrium method, the author has also demonstrated that there are no practical differences between the slope stability problem and bearing capacity problems, provided that the extremum from the slope stability analysis is used in the comparisons. Based on the present study, it can be concluded that the three geotechnical problems are practically the same problem – the ultimate condition of the system where the maximum resistance of the system is fully mobilized. In this respect, the three geotechnical problems can be considered as equivalent under the ultimate condition. If the three methods are not compared on the same ground, for example, an inappropaiate *f(x)* in slope stability analysis or simplified wedge analysis without the consideration of the internal stresses, then the results from the simplified methods for the three basic geotechnical stability problems will not be the same.

Sokolovskii (1965) has hinted that the three problems are equivalent in nature, but has not actually demonstrated the equivalence of the problems. The classical plasticity slip line formulation is actually a lower bound problem where the maximum resistance of a system is fully mobilized. For classical slope stability problem where a *f(x)* is prescribed, the solution is a lower bound to the true collapse state which is clearly illustrated in Fig. (**3.26**) and is also discussed by Cheng *et al.,* (2010) and Cheng *et al.,* (2013a). By varying *f(x)* until the maximum resistance of the system is fully mobilized, the corresponding solution is practically equivalent to the plasticity slip line solution. The three classical geotechnical problems can then be unified, and such unification is not surprising because the fundamental principles behind the three problems are exactly the same: equilibrium and yield. The classification of a problem is hence just a matter of convenience instead of the difference in the nature of the problem.

3.5. LIMIT ANALYSIS METHOD

Another popular stability formulation is the limit analysis which adopts an idealized stress-strain relation, and the soil is assumed to be a rigid perfectly plastic material with an associated flow rule (nonassociated flow rule can also be adopted). Without carrying out a tedious step-by-step elasto-plastic analysis, limit analysis can provide good (or even exact) solutions to many simple problems. Limit analysis is based on the bound theorems of classical plasticity theory (drucker *et al.*, 1951; drucker and prager, 1952). The general procedure of limit analysis is to assume a kinematically admissible failure mechanism for an upper bound solution (most commonly adopted) or a statically admissible stress field for a lower bound solution (less commonly adopted), and the objective function will be optimized (maximize for lower bound analysis and minimize for upper bound analysis) with respect to the control variables. In the early stage of limit analysis development, direct algebraic method or analytical method is used to obtain the solutions for problems with simple geometry and soil profile (chen, 1975). Since closed form solutions for most practical complicated nonhomogeneous problems are not available, later attention has been shifted to employing the slice techniques in traditional limit equilibrium to the upper bound limit analysis (michalowski, 1997; donald and chen, 1997).

Limit analysis are based on two theorems: (a) the lower bound theorem, which states that any statically admissible stress field not violating the yield criterion will provide a lower bound estimate of the true ultimate load; and (b) the upper bound theorem, which states that when the energy dissipated by any kinematically admissible velocity field is equated with the energy dissipated by the external loads, then the external loads will be the upper bounds on the true ultimate load (drucker and prager, 1952).

A statically admissible stress field is one that satisfies the equilibrium equations, stress boundary conditions, and yield criterion. A kinematically admissible velocity field is one that satisfies strain and velocity compatibility equations, velocity boundary conditions, and the flow rule. When combined, the two theorems will provide a rigorous bound on the true ultimate load. Application of the lower bound theorem usually proceeds as: (a) first, a statically admissible stress field is constructed which is often a discontinuous field. (b) these unknown stresses are then adjusted so that the load on the soil is maximized but the soil remains unyielded. The resulting load becomes the lower bound estimate for the actual collapse load.

Stress fields used in lower bound approaches are often constructed without a clear relation to the real stress fields. Thus, the lower bound solutions for practical geotechnical problems are often difficult to find. Collapse mechanisms used in the upper bound calculations, however, have a distinct physical interpretation associated with actual failure patterns and thus have been extensively used in practice.

3.5.1. Lower Bound Approach

The application of the classical analytical limit analysis was usually limited to simple problems. Numerical methods therefore have been used to compute the lower and upper bound solutions for more complicated problems. The first lower bound formulation based on the finite element method was proposed by lysmer (1970) for plain strain problems. The approach used the concept of finite element discretization and linear programming. The soil mass is subdivided into simple three-node triangular elements where the nodal normal and shear stresses were taken as the unknown variables. The stresses were assumed to vary linearly within an element, while stress discontinuities were permitted to occur at the interface between adjacent triangles. The statically admissible stress field was defined by the constraints of the equilibrium equations, stress boundary conditions, and linearized yield criterion. Each nonlinear yield criterion was approximated by a set of linear constraints on the stresses that lie inside the parent yield surface, thus ensuring that the solutions were strictly lower bounds. The ultimate load was maximized subjected to a set of linear constraints on the stresses using linear programming method. Other investigations have worked on similar algorithms (anderheggen and knopfel, 1972; bottero *et al.*, 1980). The major disadvantage of these formulations was the linearization of the yield criterion which generated a large system of linear equations, and required excessive computational times, especially if the traditional simplex or revised simplex algorithms were used (sloan, 1988a).

Efficient analyses for solving the lower bound problem by the finite element method and linear programming method have been developed recently (bottero *et al.*, 1980; sloan 1988 and 1989). The key concept of these analyses was the introduction of an active set algorithm (sloan 1989) to solve the linear programming problem where the constraint matrix was sparse. Sloan (1989) has shown that the active set algorithm was ideally suited to the numerical lower bound formulation and could solve a large-scale linear programming problem efficiently. A second problem associated with the numerical lower bound solutions occurred when dealing with statically admissible conditions for an infinite-half space. Assdi and sloan (1990) have solved this problem by adopting the concept of infinite elements, and hence obtained rigorous lower bound solutions for general problems.

Lyamin and sloan (1997) proposed a new lower bound formulation which used linear stress finite elements, incorporating nonlinear yield conditions, and exploiting the underlying convexity of the corresponding optimization problem. They showed that the lower bound solution could be obtained efficiently by solving the system of nonlinear equations that define the kuhn-tucker optimality conditions directly. Recently, zhang (1999) presented a lower bound limit analysis in conjunction with another numerical method - the rigid finite element method (rfem) to assess the stability of slopes. This formulation satisfies both static and kinematical admissibility of a discretized soil mass without requiring any assumption. Nonlinear programming method has to be adopted to search for the critical solution surface.

3.5.2. Upper Bound Approach

Upper bound theorem is generally carried out as follows: (a) firstly, a kinematically admissible velocity field is constructed. No separations or overlaps should occur anywhere in the soil mass. (b) secondly, two rates are then calculated: the rate of internal energy dissipation along the slip surface and discontinuities that separate the various velocity regions, and the rate of work done by all the external forces, including gravity forces, surface tractions and pore water pressures. (c) thirdly, the above two rates are set to be equal. The resulting equation, called energy-work balance equation, is solved for the applied load on the soil mass. This load would be equal to or greater than the true ultimate load.

The first application of upper bound limit analysis to slope stability problem was by drucker and prager (1952) for the critical height of a slope. A failure plane was assumed, and analyses were performed for isotropic and homogeneous slopes with various angles. In the case of a vertical slope, it was found that the critical height obtained by the upper bound theorem was identical with that obtained by the limit equilibrium method. Similar studies have been done by chen and giger (1971) and chen (1975). However, their attention was mainly limited to a rigid body sliding along a circular or log-spiral slip surface which could be minimized using an analytical technique.

Karel (1977a, 1977b) has developed an energy method for soil stability analysis. The failure mechanisms used in the method included: (a) a rigid zone with a planar or a log-spiral transition layer, (b) a soft zone confined by plane or log-spiral surfaces, and (c) a composed failure mechanism consisting of rigid and soft zones. The internal dissipation of energy occurred along the transition layer for rigid zone, within zone and along transition layer for the soft zone. However, no numerical technique was proposed to determine the least upper bound of the factor of safety. Izbicki (1981) presented an upper bound approach to slope stability analysis. A translational failure mechanism which was confined by a circular slip surface in the form of rigid blocks similar to the traditional slice method was used. The factor of safety was determined by an energy balance equation and the equilibrium conditions of the field of force associated with the assumed kinematically admissible failure mechanism. Similar to the works by karel (1977a, 1977b), no numerical technique was provided to search for the least upper bound of the factor of safety.

Michalowski (1995) presented an upper bound (kinematical) approach of limit analysis in which the factor of safety for slopes is associated with a failure mechanism in the form of rigid blocks analogous to the vertical slices used in traditional limit equilibrium methods. A convenient way to include pore water pressure has also been presented and implemented in the analysis of both translational and rotational slope collapse. The strength of the soil between blocks was assumed explicitly, and it was taken as zero or its maximum value set by the mohr-coulomb yield criterion. Donald and chen (1997) proposed another upper bound approach to evaluate the stability of slopes based on a multi-wedge failure mechanism. The sliding mass was divided into a small number of discrete blocks, with linear interfaces between the blocks and with either linear or curved bases to individual blocks. The factor of safety was iteratively calculated by equating the work done by external loads and body forces to the energy dissipated along the bases and interfaces of the blocks.

There are works related to limit analysis by the finite element method which will discussed in details in chapter 5. Many researchers (mroz and drescher, 1969; collins, 1974; chen, 1975; michalowski, 1989; drescher and detournay, 1993; donald and chen, 1997; yu *et al.*, 1998) pointed out that an upper bound limit analysis solution may be regarded as a special limit equilibrium solution but not vice versa. The equivalence of the two approaches plays a key role in the derivations of the limit load or factor of safety for materials following the non-associated flow rule.

Classically, algebraic expressions for the upper bound method are determined for the simple problems for optimization analysis. Assuming a log-spiral failure mechanism for failure surface a shown in Fig. (**3.27**), the work done by the weight of the soil is equated to the energy dissipated along the failure surface based on the upper bound approach by chen (1975) using an associated flow rule, and the height of the slope can be expressed:

$$H = \frac{c'}{\gamma} f(\phi', \alpha, \beta, \theta_h, \theta_o) \quad \textbf{(3.115)}$$

Where

$$f = \frac{\sin\beta\{\exp[2(\theta_h - \theta_o)\tan\phi] - 1\}}{2\sin(\beta - \alpha)\tan\phi(f_1 - f_2 - f_3)}\{\sin(\theta_h + \alpha)\exp[(\theta_h - \theta_o)\tan\phi] - \sin(\theta_o + \alpha)\}$$

$$f_1 = \frac{1}{3(1 + 9\tan^2\phi)}\{(3\tan\phi\cos\theta_h + \sin\theta_h)\exp[3(\theta_h - \theta_o)\tan\phi] - (3\tan\phi\cos\theta_o + \sin\theta_o)\}$$

$$f_2(\theta_h, \theta_o) = \frac{1}{6}\frac{L}{r_o}\left(2\cos\theta_o - \frac{L}{r_o}\cos\alpha\right)\sin(\theta_o + \alpha)$$

$$f_3(\theta_h, \theta_o) = \frac{1}{6}\exp[(\theta_h - \theta_o)\tan\phi]\left[\sin(\theta_h - \theta_o) - \frac{L}{r_o}\sin(\theta_h + \alpha)\right] *$$

$$\left\{\cos\theta_o - \frac{L}{r_o}\cos\alpha + \cos\theta_h\exp[(\theta_h - \theta_o)\tan\phi]\right\}$$

The critical height of the slope is obtained by minimizing eq. (3.115) (numerically) with respect to θ_0 and θ_h which has been obtained by chen (1975). Chen has also found that failure surface a is the most critical log-spiral failure surface unless β is small. When β and ϕ' are small, a deep seated failure shown by failure surface b in Fig. (**3.27**) may be more critical. The basic solution as given by eq. (3.115) can however be modified slightly for this case by joining points d and f. The virtual mass dcf can be deducted from the global mass OEF. The critical result of f(ϕ',α,β) as given by eq. (3.115) can be expressed as a dimensionless stability number n_s which is given eq. (3.116) as a replacement of eq.(3.115), and β' is angle dfc in Fig. (**3.27**). In general, the stability numbers by chen (1975) are very close to that by taylor (1948). Cheng has also determined similar normalized stability table as given in Table **3.24** based on the bishop's method by using program slope2000, and again, the results are also very similar to those based on the limit analysis by chem (1975).

$$f\left(\theta_h,\theta_o,\beta'\right)=\frac{\sin\beta'\left\{\exp\left[2\left(\theta_h-\theta_o\right)\tan\phi\right]-1\right\}}{2\sin\left(\beta'-\alpha\right)\tan\alpha\left(f_1-f_2-f_3-f_4\right)}\left\{\sin\left(\theta_h+\alpha\right)\exp\left[\left(\theta_h-\theta_o\right)\tan\phi\right]-\sin\left(\theta_o+\alpha\right)\right\} \quad \textbf{(3.116)}$$

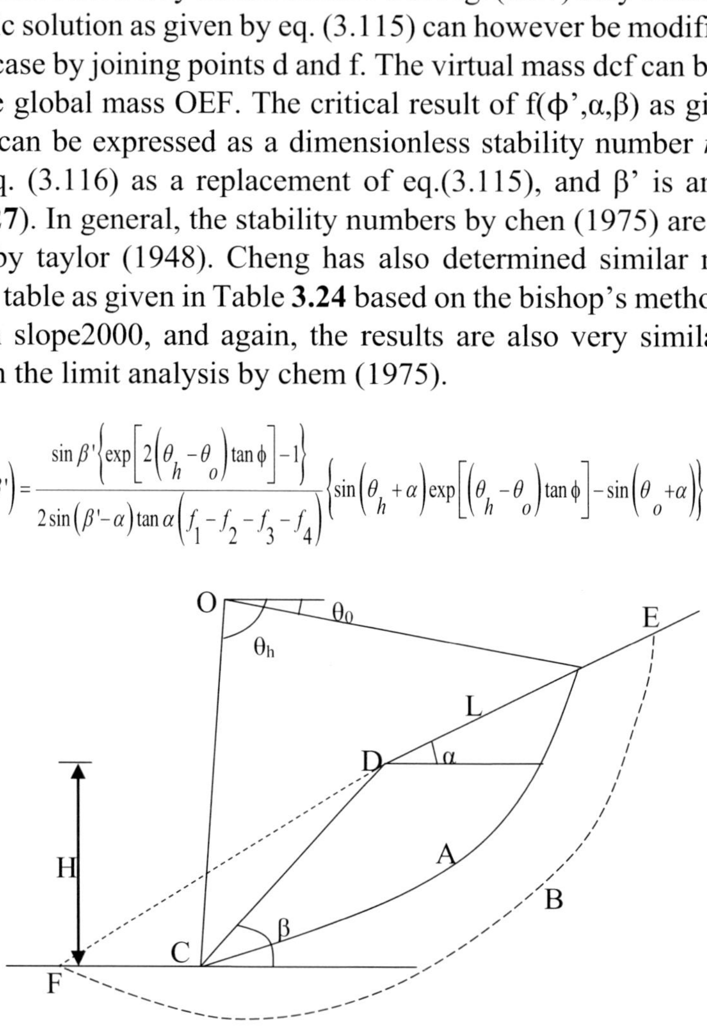

Fig. (3.27). Critical log-spiral failure surface by limit analysis for a simple homogeneous slope.

Table 3.24. Stability Chart using 2D Bishop simplified analysis by SLOPE2000 (* means below toe failure).

ϕ (°)	β\α (°)	70	65	60	55	50	45	40	35	30	25	20	15
0	0	4.80	5.03	5.25	5.46	5.67	5.87	5.41*	5.43*	5.45*	5.43*	5.45*	5.46*
5	0	5.41	5.73	6.09	6.46	6.85	7.29	7.79	8.37	9.08*	9.97*	11.43*	14.38*
	5	5.30	5.63	5.96	6.32	6.70	7.11	7.59	8.14	8.77*	9.60*	10.96*	13.69*
10	0	6.05	6.52	7.09	7.71	8.40	9.21	10.22	11.54	13.45	16.62	23.14	45.57
	5	5.95	6.44	6.99	7.58	8.26	9.05	10.06	11.36	13.24	16.33	22.78	45.00
	10	5.84	6.33	6.86	7.41	8.07	8.83	9.78	11.11	12.84	15.79	21.90	42.86
15	0	6.94	7.58	8.37	9.36	10.50	11.94	13.90	16.79	21.69	32.14	69.23	-
	5	6.77	7.50	8.30	9.25	10.36	11.80	13.74	16.59	21.48	31.86	68.97	
	10	6.67	7.38	8.18	9.09	10.20	11.61	13.51	16.33	21.13	31.36	68.18	
	15	6.53	7.22	8.01	8.89	9.96	11.25	13.14	15.85	20.45	30.25	68.18	
20	0	7.97	9.01	10.14	11.61	13.51	16.07	20.00	26.67	41.38	94.74	-	
	5	8.04	8.91	10.04	11.50	13.38	15.93	19.82	26.55	41.10	94.74		
	10	7.69	8.82	9.92	11.35	13.22	15.76	19.61	26.28	40.72	93.75		
	15	7.60	8.66	9.75	11.16	12.97	15.49	19.25	25.79	40.18	92.78		
	20	7.59	8.44	9.55	10.91	12.66	15.06	18.71	25.00	38.54	88.24		
25	0	9.42	11.01	12.57	14.80	18.04	22.93	31.47	50.28	120.00	-		
	5	9.28	10.91	12.46	14.69	17.91	22.78	31.03	50.00	120.00			
	10	9.50	10.84	12.33	14.57	17.73	22.56	31.03	49.72	119.21			
	15	9.00	10.60	12.16	14.35	17.51	22.28	30.72	49.32	118.42			
	20	8.97	10.51	12.00	14.12	17.22	21.95	30.25	48.65	116.88			
	25	8.81	10.17	11.73	13.74	16.74	21.25	29.27	46.75	111.80			
30	0	11.89	13.79	16.07	19.65	25.53	35.64	58.63	144.00	-			
	5	11.54	13.74	16.00	19.52	25.35	35.64	58.44	144.00				
	10	11.43	13.75	15.83	19.35	25.17	35.43	58.25	144.00				
	15	11.04	13.53	15.65	19.19	25.00	35.16	57.32	142.86				
	20	10.71	13.31	15.49	18.95	24.66	34.75	57.32	142.29				
	25	10.81	12.93	15.23	18.60	24.19	34.16	56.60	140.63				

(Table 3.24) contd.....

	30	10.43	12.11	14.86	18.09	23.44	32.97	54.22	134.33				
35	0	14.83	18.00	21.25	27.48	39.30	65.93	166.67	-				
	5	14.94	17.82	21.18	27.40	39.30	65.69	166.67					
	10	14.25	17.65	21.08	27.27	39.13	65.45	165.90					
	15	14.04	17.54	20.93	27.03	38.79	65.45	165.14					
	20	13.85	17.65	20.69	26.87	38.63	64.98	165.14					
	25	13.19	16.32	20.55	26.55	38.30	64.29	165.14					
	30	12.82	15.76	20.18	26.01	37.50	63.16	163.64					
	35	12.54	15.67	19.35	25.14	36.14	61.02	155.17					
40	0	20.07	24.03	30.10	42.35	72.29	185.57	-					
	5	19.13	23.68	29.41	42.06	72.00	185.57						
	10	19.82	23.72	29.27	42.35	71.43	185.57						
	15	18.95	23.53	30.28	41.86	71.43	183.67						
	20	17.61	23.38	29.32	41.47	71.43	183.67						
	25	16.93	23.23	28.85	41.10	70.87	183.67						
	30	16.36	22.70	28.57	40.72	70.04	181.82						
	35	16.04	21.05	28.13	40.00	68.97	180.00						
	40	15.72	20.00	27.69	38.46	65.93	171.43						

For the problem in (Fig. **3.27**), the result can also be expressed analytically if swedish method is used which will be shown here.

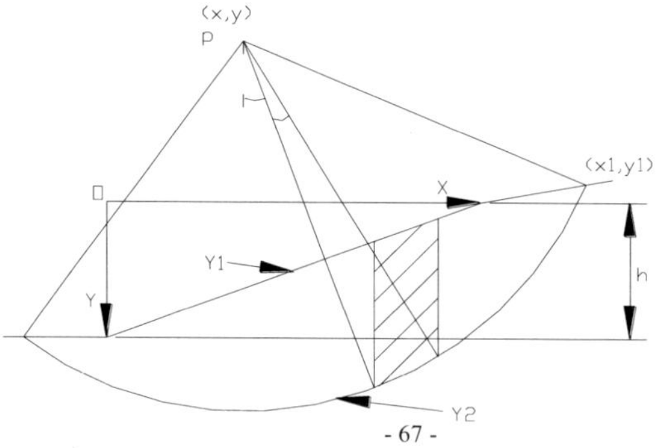

Fig. (3.28). Swedish method for problem in (Fig. **3.27**).

According to (Fig. **3.28**), the factor of safety by the Swedish Method with homogeneous soil mass and no water pressure is given by Cheng as:

$$FS = \frac{c \cdot L + \tan\phi \sum W_i \cos\alpha_i}{\sum W_i \sin\alpha_i} = \frac{\gamma \tan\phi [N] + cL}{\gamma [T]} \quad \textbf{(3.117)}$$

where L is the total length of the circular arc surface and W_i is the weight of slice number i, $[N]=(\Sigma W_i \cos\alpha_i)/\gamma$; $[T]=(\Sigma W_i \sin\alpha_i)/\gamma$. Analytically, it is given as

$$Y_1 = \begin{cases} -(X - h/\tan\beta)\tan\alpha & (X \geq h/\tan\beta) \\ h - X\tan\beta & (0 \leq X \leq h/\tan\beta) \\ H & (X \leq 0) \end{cases} \quad \textbf{(3.118)}$$

$$[N] = \int_{x-\sqrt{r^2-(h-y)^2}}^{x+\sqrt{r^2-(y_1-y)^2}} \left[y + \sqrt{r^2-(X-x)^2} - y_1 \right] \frac{\sqrt{r^2-(X-x)^2}}{r} dX - \int_{x-\sqrt{r^2-(h-y)^2}}^{0} [h - y_1] \frac{\sqrt{r^2-(X-x)^2}}{r} dX$$

$$- \int_{0}^{\frac{h}{\tan\beta}} [h - X\tan\beta - y_1] \frac{\sqrt{r^2-(X-x)^2}}{r} dX - \int_{\frac{h}{\tan\beta}}^{x_1} \left[-\left(X - \frac{h}{\tan\beta} \right) \tan\alpha - y_1 \right] \frac{\sqrt{r^2-(X-x)^2}}{r} dX \quad \textbf{(3.119)}$$

$$[T] = \int_{x-\sqrt{r^2-(h-y)^2}}^{x+\sqrt{r^2-(y_1-y)^2}} \left[y + \sqrt{r^2-(X-x)^2} - y_1 \right] \frac{X-x}{r} dX - \int_{x-\sqrt{r^2-(h-y)^2}}^{0} [h - y_1] \frac{X-x}{r} dX$$

$$-\int_{0}^{\frac{h}{\tan\beta}}[h-X\tan\beta-y_1]\frac{X-x}{r}dX-\int_{\frac{h}{\tan\beta}}^{x_1}\left[-\left(X-\frac{h}{\tan\beta}\right)\tan\alpha-y_1\right]\frac{X-x}{r}dX \quad \textbf{(3.120)}$$

$$L=\int_{x-\sqrt{r^2-(h-y)^2}}^{x+\sqrt{r^2-(y_1-y)^2}}\frac{r}{\sqrt{r^2-(X-x)^2}}dX \quad \textbf{(3.121)}$$

$$H_c=Fxh \text{ and } Ns=\frac{H_c}{\frac{c}{\gamma}}$$

Table 3.25. Comparison of results between limit analysis and other limit equilibrium methods.

Slope Angle	Friction Angle	Stability Factor (Ns) for α = 0			
		LIMIT EQUILIBRIUM			Limit Analysis
β	**ϕ**	**Slices**	**ϕcircle**	**Log-spiral**	**Log-spiral**
90	**0**	3.83	3.83	3.83	3.83
	5	4.19	4.19	4.19	4.19
	10	5.02	5.02	-	5.02
	25	6.06	6.06	6.06	6.06
75	**0**	4.57	4.57	4.57	4.56
	5	5.13	5.13	-	5.14
	15	6.49	6.52	-	6.57
	25	8.48	8.54	-	8.58
60	**0**	5.24	5.24	5.24	5.25
	5	6.06	6.18	6.18	6.16
	15	8.33	8.63	8.63	8.63

(Table 3.25) contd.....

	25	12.20	12.65	12.82	12.74
45	**0**	5.88	5.88*	5.88*	5.53*
	5	7.09	7.36	-	7.35
	15	11.77	12.04	-	12.05
	25	20.83	22.73	-	22.90
30	**0**	6.41*	6.41*	6.41*	5.53*
	5	8.77*	9.09*	-	9.13*
	15	20.84	21.74	-	21.69
	25	83.34	111.1	-	119.93
15	**0**	6.90*	6.90*	6.90*	5.53*
	5	13.89*	14.71*	14.71	14.38*
	10	-	43.62	-	45.49

* Critical failure surface passes below toe.

To determine the stability number from either limit analysis or Swedish method, a computer program is required to be developed. Cheng has developed a program SLOPE which will determine the critical solution based on limit analysis as well as Swedish method by pattern search, simulated annealing method and FSQP (Feasible sequential quadratic programming) method. It should be mentioned that pattern search is slow in computation as all combinations of variables are tried. For FSQP, different initial choices must be chosen as this problem has many local minima by nature (random number generation for several trials in the program SLOPE). On the other hand, the simulated annealing method can avoid the limitations of the previous two methods. The program together with simple graphics support and source code (all the methods and computational technique) can be obtained from the author at ceymchen@polyu.edu.hk. This program is suitable for student to learn the application of limit analysis to slope stability problem. The stability numbers determined by program SLOPE using limit analysis are compared with the classical results from limit equilibrium in Table **3.25**. It is observed that except for isolated cases, the results from limit analysis are very close to those by the limit equilibrium method. The results from limit analysis are comparable to those by the limit equilibrium method, yet there is no iteration or convergence problem which will occur in the limit equilibrium method. The disadvantage of limit analysis is that it is very difficult to be applied to general problem with arbitrary geometry and nonhomogeneous medium. For such condition, the finite element limit analysis approach will be more appropriate which will be discussed in chapter 5.

To determinate the lateral earth pressure of a fill on a retaining wall when frictional forces act on the back of the wall, upper-bound technique of limit analysis can be conveniently applied to this problem to obtain the solutions for the general lateral earth pressure problems including the effect of wall friction. Three failure mechanisms are selected for limit analysis to determine the active and passive soil pressures which are:

(a) Two-Wedge Mechanism

(b) Log-Spiral Mechanism

(c) Arc-Sandwich Mechanism

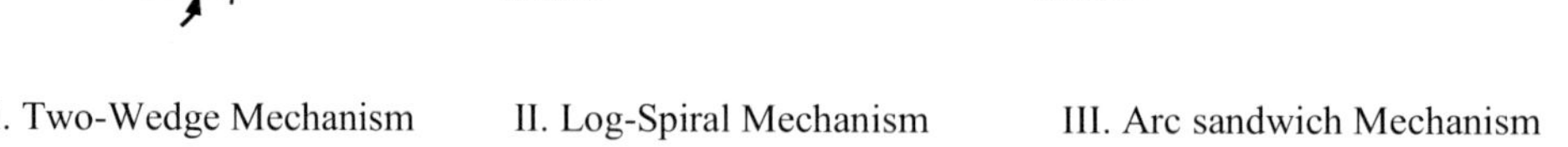

I. Two-Wedge Mechanism II. Log-Spiral Mechanism III. Arc sandwich Mechanism

Fig. (3.29). Failure mechanism used in computer program "LEP".

The formulations as considered in Fig. (**3.29**) are similar to that by Chen and Liu (1990). For a Coulomb soil (c-ϕ) with uniform surcharge at the back of retaining wall, the lateral pressure can be expressed as:

$$P_a = \tfrac{1}{2} \; H^2K_a + qHK_{aq} - 2cHK_{ac}$$

$$P_p = \tfrac{1}{2} \; H^2K_p + qHK_{pq} + 2cHK_{pc} \quad \textbf{(3.122)}$$

Where

$K_{a\gamma}$, $K_{p\gamma}$: for the case of cohesionless soil with no surcharge loading. (c=0, q=0)

K_{ac}, K_{pc}: for the case of cohesive weightless soil with no surcharge loading. (γ=0, q=0)

K_{aq}, K_{pq}: for the case of cohesionless weightless soil with surcharge loading (c=0, γ=0)

In finding the solution of a particular mechanism, two following work equations should be formed first:

i. the external rate of work due to the external applied loads and soil weight.
ii. the internal dissipation of energy in the plastically deformed region.

The solution can then be found by equating the above two work done equations, and the analytical solutions are obtained by the symbolic mathematics option in Mathcad. There are totally six long equations for the active and passive pressure coefficients for the cases in (Fig. **3.29**), and interested readers can obtain the long equations and detailed procedures for derivations from the author (too long and tedious to include here). Since the equations are very long and tedious, to obtain the maximum active pressure and minimum passive pressure, a computer program as shown in the Appendix of this chapter has been developed. The computer will compute the three coefficients corresponding to active and passive pressure coefficients for the three mechanisms in Fig. (**3.29**) for wall with arbitrary inclination and wall friction. A more complete program including simple graphics support can be obtained from the author at ceymchen@polyu.edu.hk. After detailed study, it is found that mechanism III as shown in Fig. (**3.29c**) is always not the controlling mechanism (except for limited cases). In fact, when the wall friction is high, this mechanism will produce a very poor result which is shown in (Table **3.26**). It should also be pointed out that according to upper bound analysis, different failure mechanisms should be considered, and the critical solution from different mechanism will then be a good estimate of the solution. In practice, it is impossible to derive the equations for arbitrary failure mechanism, and the finite element limit analysis as discussed in chapter 5 has to be adopted for such purpose.

Table 3.26. Comparison between three mechanisms on passive pressure coefficients for zero ground slope behind the wall (Passive Case).

Angle of internal Friction φ	Wall Friction angle δ	Passive Pressure Coefficient ($K_{p\gamma}$) α = 70 I	II	III	α = 90 I	II	III	α = 110 I	II	III
10	0	1.03601	1.3601	1.3601	1.4203	1.4203	1.4203	1.7406	1.7405	1.7401
	5	1.4547	1.4547	1.4547	1.5599	1.5599	1.5601	1.9620	1.9611	1.9549
	10	0.5380	1.5379	1.5373	1.6822	1.6815	1.6749	2.1677	2.1632	2.1402
20	0	0.7482	1.7482	1.7482	2.0396	2.0396	2.0396	2.9060	2.9053	2.9308
	10	2.0761	2.0761	2.0761	2.5830	2.5825	2.6076	3.9226	3.9108	3.9416
	20	2.4369	2.4356	2.4662	3.1812	3.1690	3.1860	5.1189	5.0373	4.9446
30	0	2.2743	2.2743	2.2743	3.0000	3.0000	3.0000	5.0876	5.0844	5.3229
	15	3.1597	3.1597	3.1616	4.7133	4.7083	4.9702	9.0294	8.9195	10.2399

(Table 3.26) contd.....

	30	4.4258	4.4117	4.7562	7.2428	7.1021	8.3140	15.4732	14.3971	17.6223
40	**0**	3.0180	3.0180	3.0180	4.5989	4.5989	4.5989	9.7218	9.7045	10.6598
	20	5.3139	5.3137	5.3380	10.1156	10.0698	11.7720	26.6162	25.4523	68.9161
	40	10.0473	9.8821	15.5175	22.7146	20.9146	67.9121	73.5860	56.5992	560.7383

Angle of internal Friction ϕ	**Wall Friction angle δ**	**Passive Pressure Coefficient (K_{pc})**								
		α = 70			**α = 90**			**α = 110**		
		I	**II**	**III**	**I**	**II**	**III**	**I**	**II**	**III**
10	**0**	0.7148	0.7041	0.7041	1.4213	1.4203	1.4203	3.3385	3.3477	3.4152
	5	0.8305	0.8208	0.8208	1.7475	1.7472	1.7575	4.1555	4.0971	4.2084
	10	0.9785	0.9698	0.9698	2.1399	2.1364	2.1663	5.0775	4.9357	5.1074
20	**0**	0.8990	0.8830	0.8830	2.0415	2.0396	2.0396	5.7244	5.6759	6.2804
	10	1.3069	1.2931	1.2931	3.3384	3.3320	3.5129	9.4341	9.1039	10.8707
	20	2.0410	2.0396	2.0396	5.2905	5.1909	6.0054	14.9920	13.5283	17.5555
30	**0**	1.1229	1.0983	1.0983	3.0036	3.0000	3.0000	10.4710	10.3406	13.5474
	15	2.2262	2.2509	2.2509	7.3033	7.2352	8.7517	26.1481	24.0652	48.5650
	30	5.2976	5.2651	5.8777	16.4250	15.0533	30.2558	61.8636	47.0593	147.8946
40	**0**	1.3950	1.3554	1.3554	4.6066	4.5989	4.5989	21.6380	21.2036	32.6936
	20	4.6053	4.5989	4.5989	19.9034	19.2927	32.6936	102.1064	84.4295	1469.7900
	40	17.8485	16.8482	41.5915	78.4460	59.3991	1420.2319	493.5247	258.1019	94912.523

Angle of internal Friction ϕ	**Wall Friction angle δ**	**Passive Pressure Coefficient (K_{pq})**								
		α = 70			**α = 90**			**α = 110**		
		I	**II**	**III**	**I**	**II**	**III**	**I**	**II**	**III**
10	**0**	1.3623	1.3601	1.3601	1.4204	1.4203	1.4203	1.7114	1.7094	1.7159
	5	1.4550	1.4547	1.4547	1.5469	1.5457	1.5507	1.8730	1.8603	1.8737

(Table 3.26) contd.....

	10	1.5283	1.5254	1.5323	1.6353	1.6212	1.6350	1.9984	1.9636	1.9826
20	**0**	1.7544	1.7482	1.7482	2.0401	2.0396	2.0396	2.8058	2.7984	2.8884
	10	2.0762	2.0761	2.0761	2.5219	2.5153	2.5915	3.5164	3.4511	3.6963
	20	2.3870	2.3709	2.4626	2.9487	2.8724	3.1015	4.2156	3.9985	4.4243
30	**0**	2.2878	2.2743	2.2743	3.0012	3.0000	3.0000	4.8007	4.7773	5.3143
	15	3.1568	3.1568	3.1616	4.4705	4.4389	4.9679	7.3738	7.0687	9.6774
	30	4.2063	4.1274	4.7562	6.1881	5.8038	8.1812	10.8034	9.4999	15.9370
40	**0**	3.0463	3.0180	3.0180	4.6019	4.5989	4.5989	8.8706	8.7918	10.6598
	20	5.2830	5.2815	5.3380	9.0834	8.9157	11.7720	18.7419	17.0444	67.6366
	40	8.9657	8.5264	15.5175	16.8084	14.3934	67.7120	39.6483	29.0104	521.7122

Two-Wedge Case = I, Log-Spiral Case = II, Arc-Sandwich Case = III.

Although the true failure mechanism will deviate slightly from a log-spiral mechanism for the effect of self-weight which is illustrated in Fig. (**3.6**), the use of log-spiral mechanism can still provide good solution for normal engineering applications which is illustrated in Table **3.27**. In carrying out limit analysis or even slip line analysis, there is one important point which should be noted. In the evaluation of the individual lateral earth pressure coefficient, it is assumed that other effects are neglected in the computation (direct superposition). For the limit analysis program LEP as shown in the Appendix or the slip line program SLIP/KA/KP by the author, it is also possible to carry out a global analysis in that the effects due to self-weight of soil, cohesive strength and surcharge can be lumped together. By doing so, the concept of earth pressure coefficients will be absent. Cheng (2002) has demonstrated that the direct superposition of the three effects is a good approximation of the global analysis for normal engineering problems. The computer code as shown in the Appendix actually carries out the global analysis besides the evaluation of the individual earth pressure coefficients. Besides slope stability and lateral earth pressure problem, limit analysis has also been used for bearing capacity problem by Chen (1975) which will not be repeated here. The results for the bearing capacity factors by limit analysis are found to be the same as that by limit equilibrium or slip line method for the factors N_c and N_q. For the terms N_γ, the differences between the results from slip line method and limit analysis are also small (Cheng and Au 2005).

Table 3.27. Comparison of $K_{p\gamma}$ between limit analysis and slip line solutions for a vertical wall with no back slope.

$\phi=\delta$ (°)	LA	Slip line
10	1.675	1.660
20	3.170	3.052
30	7.102	6.569
40	20.915	18.216

3.6. OVERALL DISCUSSION

As demonstrated in this chapter, stability of a geotechnical system can be evaluated by means of slip line method, limit equilibrium method and limit analysis. There are no major differences between the results from the three different methods if the problem is properly considered. In fact, the author has also illustrated that the three classical stability problems can be unified under the ultimate condition, and each problem can be viewed as the inverse of the other problem, depending on the angle of view of the engineer. Slip line method and limit analysis are however suitable only for simple problems with regular geometry and homogeneous soil condition. On the other hand, limit equilibrium method is an approximate lower bound method which relies on the assumption of an interslice force function. To evaluate this function, the extremum approach by Cheng *et al.*, (2010) has to be adopted which is a time consuming computation. Luckily, for most of the practical problems, the classical limit equilibrium method can quickly provide a solution which is accurate and good enough for normal engineering applications. This method is also easily understood by most of the engineers and is well adapted to problems with irregular geometry and nonhomogeneous condition. It is not surprising that while no commercial program is available for slip line method or limit analysis, there are many commercial programs targeted towards different applications using the limit equilibrium method.

The most critical difference between the three different methods is the application in three-dimensional condition. Currently, there is no application of the slip line method (exclude axi-symmetric problem) and very few successful applications of the limit analysis for three-dimensional application (Cheng *et al.,* 2015a). The extension of these two methods to three-dimensional applications is very difficult, and Cheng and Zhu (2002) have proposed an improved three-dimensional Bishop's method. On the other hand, the extension of limit equilibrium method to three-dimensional is relatively easy, and the most critical problem of three-dimensional limit equilibrium formulation is the sliding direction which is solved effectively by Cheng and Yip (2007a). The general application of three-dimensional limit equilibrium method is mainly limited by the difficulty in defining a precise three-dimensional ground condition (expensive and time consuming) and the search for critical three-dimensional solution. For the latter limitation, it is limited by the lack of a general equation which can approximate all the possible three-dimensional slip surfaces with limited control variables. Cheng *et al.,* (2005) have proposed the use of NURBS function for such purpose, but even this function will not be able to model all the possible real world three-dimensional failure surfaces. In this respect, the finite element method which is discussed in chapter 5 will be a simpler solution for three-dimensional stability analysis.

ACKNOWLEDGEMENTS

The present work is supported by the GRF project PolyU 5128/13E, Hong

Kong SAR Government.

REFERENCES

Anderheggen E. and Knopfel H. (1972), Finite element limit analysis using linear programming, International Journal of Solids and Structures, 8(12), 1413-1431.

Arai K. and Tagyo K. (1985), Determination of noncircular slip surfaces giving the minimum factor of safety in slope stability analysis, Soils and Foundations, 25, 43–51.

Assadi A. and Sloan S.W. (1990), Undrained stability of shallow square tunnel, Journal of the Geotechnical Engineering Division, ASCE, 117(8), 1152–1173.

Baker R. and Garber M. (1978), Theoretical analysis of the stability of slopes, Geotechnique, 28, 395–411.

Baker R. (1980), Determination of the critical slip surface in slope stability computations, International Journal of Numerical and Analytical Methods in Geomechanics, 4:333-359.

Baker R. (2003), Sufficient conditions for existence of physically significant solutions in limiting equilibrium slope stability analysis, International Journal of Solids and Structures, vol.40 (13-14), 3717-3735.

Berezantzev V. G. (1958), Earth pressure on the cylindrical retaining wall." Proc., Brussels Conf. on Earth Pressure Problems, vol. 2, p.21-27.

Bishop A.W. (1955), The use of the slip circle in the stability analysis of earth slopes, Geotechnique, 5(1), 7–17.

Bolton H.P.J., Heymann G. and Groenwold A. (2003), Global search for critical fail- ure surface in slope stability analysis, Engineering Optimization, 35(1), 51–65.

Booker J.R., Zheng X. (2000), Application of the theory of classical plasticity to the analysis of the stress distribution in wedges of a perfectly frictional material, Modelling in Geomechanics, John Wiley, New York.

Bottero A., Negre R., Pastor J. and Turgeman S. (1980), Finite element method and limit analysis theory for soil mechanics problems, Computer Methods in Applied Mechanics and Engineering, 22, 131-149.

Castilo E. and Luceno A. (1980), Evaluation of variational methods in slope analysis. Proceedings of International Symposium on Landslides, New-Delhi, India, vol. 1, pp. 255–258.

Castilo E. and Luceno A. (1982), A critical analysis of some variational methods in slope stability analysis. International Journal for Numerical and Analytical Methods in Geomechanics, 6, 195–209.

Celestino T.B. and Duncan J.M. (1981), Simplified Search for Non-Circular Slip Surface, Proceedings of 10th International Conference on Soil Mechanics and Foundation Engineering, Stockholm, Sweden, pp. 391–394.

Chen W.F. and Giger M.W. (1971), Limit analysis of stability of slopes. Journal of the Soil Mechanics and Foundations Division, Proceedings of the American Society of Civil Engineers, Vol. 97, pp. 19–26.

Chen W.F. (1975), Limit Analysis and Soil Plasticity. Elsevier, US.

Chen W.F. and Liu X.L. (1990), Limit analysis in soil mechanics, New York, Elsevier.

Chen Z. and Morgenstern N.R. (1983), Extensions to generalized method of slices for stability analysis, Canadian Geotechnical Journal, 20(1), 104–109.

Chen Z. and Shao C. (1983), Evaluation of minimum factor of safety in slope stability analysis, Canadian Geotechnical Journal, 25(4), 735–748.

Chen Z. (1998), On Pan's principles of rock and soil stability analysis, Journal of Tsinghua University (Sci & Tech), 38, 1–4 (in Chinese).

Chen Z.Y., Wang X.G., Haberfield C., Yin J.H. and Wang Y.J. (2001a), A three- dimensional slope stability analysis method using the upper bound theorem. Part I: Theory and methods. International Journal of Rock Mechanics and Mining Sciences, 38(3), 369–378.

Chen Z.Y., Wang J., Wang Y.J., Yin J.H. and Haberfield C. (2001b), A three-dimensional slope stability analysis method using the upper bound theorem. Part II: Numerical approaches, applications and extensions. International Journal of Rock Mechanics and Mining Sciences, 38(3), 379–397.

Cheng Y.M. (2002), Slip Line Solution and Limit Analysis for Lateral Earth Pressure Problem, the Ninth Conference on Computing in Civil and Building Engineering, April 3-5, Taipei, Taiwan, p.311-314.

Cheng Y.M. and Zhu L.J. (2002), An Improved Three Dimensional Bishop's method in Slope Stability Analysis, Chinese Journal of Geotechnical Engineering, vol.24, no.6, 706-709.

Cheng Y.M. (2003a), Locations of Critical Failure Surface and some Further Studies on Slope Stability Analysis, Computers and Geotechnics, vol.30, 255-267.

Cheng Y.M. (2003b), Seismic lateral earth pressure coefficients by slip line method, Computers and Geotechnics, vol. 30 no.8, p.661-670.

Cheng Y.M. and Zhu L.J. (2004), Unified Formulation for Two Dimensional Slope Stability Analysis and Limitations in Factor of Safety Determination, Soils and Foundations, 121-128, vol.44 (6) : 121-128.

Cheng Y.M. and Hu Y.Y. (2005), Active earth pressure for circular shaft lining by Simplified Slip Line Solution with general tangential stress coefficient, Chinese Journal of Geotechnical Engineering, vol.27, no.1, 110-115.

Cheng Y.M. and Au S.K. (2005), Slip line solution of bearing capacity problems with inclined ground, Canadian Geotechnical Journal, vol.42, 1232-1241.

Cheng Y.M., Liu H.T. and Au S.K. (2005), Location of critical three-dimensional non-spherical failure surface with applications to highway slopes, Computers and Geotechnics, no. 32, 387-399.

Cheng Y.M. and Yip C.J. (2007a), Three-dimensional Asymmetrical Slope Stability Analysis -Extension of Bishop's and Janbu's Techniques, Journal of Geotechnical and Geoenvironmental Engineering, ASCE, 133(12):1544-1555.

Cheng Y.M., Hu Y.Y. and Wei W.B. (2007b), Axi-symmetric active earth pressure by Methods of Characteristics – Theory and Numerical Formulation, International Journal on Geomechanics, ASCE, 7(1), 1:15.

Cheng Y.M. and Li L. (2007), Particle swarm optimization algorithm for non-circular failure surface in two dimensional lope stability analysis, Computers and Geotechnics, 34(2):92-103.

Cheng Y.M. (2007), Global optimization analysis of slope stability by simulated annealing with dynamic bounds and Dirac function, Engineering Optimization, 39(1):17-32.

Cheng Y.M., Li L. and Chi S.C. (2007a), Studies on six heuristic global optimization methods in the location of critical slip surface for soil slopes, Computers and Geotechnics, vol34, 462-484.

Cheng Y M, Lansivaara T, Wei W.B. (2007b), Two-dimensional Slope Stability Analysis by Limit Equilibrium and Strength Reduction Methods, Computers and Geotechnics, vol.34, 137-150, 2007.

Cheng Y.M., Zhao Z.H. and Wang J.A. (2008a), Factor of safety in accordance with Pan's extremum principle by optimization method, Chinese Journal of Rock Mechanics and Engineering, 27(4): 782-788.

Cheng Y.M., Hu Y.Y., Au S.K. and Wei W.B. (2008b), Active pressure for circular cut with Berezantzev's and Prater's Theories, Soils and Foundations, 48(5):621-632

Cheng Y.M., Lansivaara T. and Siu J. (2008c), Impact of Convergence on Slope Stability Analysis and Design, Computers and Geotechnics, 35(1), pp.105-115.

Cheng Y.M., Li L. and Chi S.C., Wei W.B. (2008d), Determination of critical slip surface using artificial fish swarms algorithm, Journal of Geotechnical and Geoenvironmental Engineering, ASCE, 134(2):244-251.

Cheng Y.M., Li. L., Lansivaara T., Chi S.C. and Sun Y.J. (2008e), Minimization of factor of safety using different slip surface generation methods and an improved harmony search minimization algorithm, Engineering Optimization, vol.40(2), 95-115.

Cheng Y.M., Zhao Z.H. and Sun Y.J. (2010), Evaluation of interslice force function and discussion on convergence in slope stability analysis by the lower bound method, Journal of Geotechnical and Geoenvironmental Engineering, ASCE, 136(8): 1103-1113.

Cheng Y.M., Li D.Z., Li L., Sun Y.J., Baker R. and Yang Y. (2011), Limit equilibrium method based on approximate lower bound method with variable factor of safety that can consider residual strength, Computers and Geotechnics, 38, 628-637.

Cheng Y.M., Li L., Sun Y.J. and Au S.K. (2012), A coupled particle swarm and harmony search optimization algorithm for difficult geotechnical problems, Structural and Multidisciplinary Optimization, 45:489-501.

Cheng Y.M., Lansivaara T., Baker R. and Li N. (2013a) The use of internal and external variables and extremum principle in limit equilibrium formulations with application to bearing capacity and slope stability problems, Soils and Foundation, 53(1):130-143.

Cheng Y.M., Li D.Z., Li N., Li Y.Y. and Au S.K. (2013b), Solution of some engineering partial differential equations governed by the minimal of a functional by global optimization method, Journal of Mechanics, 29(3), 493-506.

Cheng Y.M. and Lau C.K. (2014) Slope stability analysis and stabilization, 2nd edition, Spon Press

Cheng Y.M., Li N. and Yang X.Q. (2015a), Three Dimensional Slope Stability Problem with a Surcharge Load, Natural Hazards and Earth System Sciences (to publish).

Cheng Y.M., Li L. and Liu L.L. (2015b), Simplified Approach for Locating the Critical Probabilistic Slip Surface in Limit Equilibrium Analysis, Natural Hazards and Earth System Sciences (to publish).

Cheng Y.M. and Li N. (2015), Equivalence between Bearing Capacity, Lateral Earth Pressure and Slope Stability Problems by Slip-line and Extremum Limit Equilibrium Methods, Canadian Geotechnical Journal (under review)

Collins I.F. (1974), A note on the interpretation of Coulomb's analysis of the thrust on a rough retaining wall in terms of the limit theorems of limit plasticity, Geotechnique, 24, 106–108.

De Jong D.J.G. (1957), Graphical method for the determination of slip-line fields in soil mechanics, Ingenior, vol. 69, p.61-65.

De Jong D.J.G. (1980), Application of calculus of variation to the vertical cut-off in cohesive frictionless soil, Geotechnique, 30(1):1–16.

De Jong D.J.G. (1981), Variational fallacy, Geotechnique, 31(4):289–290.

Donald, I.B., and Chen, Z. 1997. Slope stability analysis by the upper bound approach: fundamentals and methods. Canadian Geotechnical Journal, 34: 853–862.

Drescher A. (1983), Limit plasticity approach to piping in bins." ASME Trans. J. Appl. Mech., vol.50, 549–553.

Drescher A. (1986), Kinematics of axisymmetric vertical slopes at collapse, Int. J. Numer. Analyt. Meth. Geomech., vol.10, 431–444.

Drescher A. and Detournay E. (1993), Limit load in translational failure mechanisms for associative and non-associative materials. Geotechnique, 43(3), 443–456.

Drucker D.C. and Prager W. (1952), Soil mechanics and plastic analysis or limit design, Quarterly of Applied Mathematics, vol.10, 157–165.

Drucker, D.C., Greenberg, W. and Prager, W. (1951), The safety factor of an elastic plastic body in plane strain, Transactions of the ASME, Journal of Applied Mechanics, 73, 371.

Duncan J.M. (1996), State of the art: Limit equilibrium and finite element analysis of slopes, Journal of Geotechnical Engineering, ASCE, 122(7), 577–596.

Fredlund D.G. and Krahn J. (1984), Analytical methods for slope analysis, International Symposium on Landslides, pp. 229–250.

Greco V.R. (1996), Efficient Monte Carlo technique for locating critical slip surface, Journal of Geotechnical Engineering, ASCE, 122, 517–525.

Hansen, J.B. (1970), A revised and extended formula for bearing capacity, Danish Geotechnical Institute, Bulletin 28, Copenhagen.

Hill, J. M. and Cox, G. M. (2000), Cylindrical cavities and classical rat-hole theory occurring in bulk materials. Int. J. Numer. Analyt. Meth. Geomech., vol. 24, 971–990.

Houlsby G. T. and Wroth C. P. (1982), Direct solution of plasticity problems in soils by the method of characteristics, Proc., 4th Int. Conf. on Numerical Methods in Geomechancis, Edmonton, Alta., Canada, Vol. 3, 1059–1071.

Izbicki R.J. (1981), Limit plasticity approach to slope stability problems, Journal of the Geotechnical Engineering Division, Proceedings of the American Society of Civil Engineers, 107, 228–233.

Kerisel J. and Absi E. (1990), Active and Passive Earth Pressure tables, A.A. Ba1kema.

Janbu, N. (1957), Earth pressure and bearing capacity by generalized procedure of slices, Proc. 4th International Conference on Soil Mechanics, p207-212.

Janbu N. (1973), Slope stability computations, in: R.C. Hirschfield and S.J. Poulos (eds.), Embankment-Dam Engineering, pp. 47–86, John Wiley.

Jenike, A. J. and Yen, B. C. (1962). Slope stability in axial symmetry, Proc., 5th Symp. on Rock Mechanics, Univ. of Minnesota, Pergamon, New York, 689–711.

Karel K. (1977a), Application of energy method, Journal of the Geotechnical Engineering Division, Proceedings of the American Society of Civil Engineers, 103, 381–397.

Karel K. (1977b), Energy method for soil stability analyses, Journal of the Geotechnical Engineering Division, Proceedings of the American Society of Civil Engineers, 103, 431–445.

Kotter F. (1903), Die Bestimmung des Druckes an gekrmmten Gleitflchen, eine Aufgabe aus der Lehre vom Erddruck, Berlin Akad. Wiss., Berlin, 229–233.

Kumar J. (2001), Seismic passive earth pressure coefficients for sands. Canadian Geotechnical Journal vol.38, p.876–81.

Kumar J. and Chitikela S. (2002), Seismic passive earth pressure coefficients using the method of characteristics. Canadian Geotechnical Journal vol. 39, p.463–71.

Li D. Z. and Cheng Y. M. (2012). Lower Bound Limit Analysis Using Nonlinear Failure Criteria, 2012 International Conference on Structural Computation and Geotechnical Mechanics. Procedia Earth and Planetary Science City: Kun Ming China, pp. 170-174.

Li D.Z., Cheng Y.M., Wang J.A., Yang Y. and Li N. (2013), Application of finite-element-based limit analysis with mesh adaptation in geotechnical engineering, Chinese Journal of geotechnical Engineering, 35(5):922-929.

Lowe J. and Karafiath L. (1960), Stability of Earth Dams Upon Drawdown, Proceedings of the 1st Pan-American Conference on Soil Mechanics and Foundation Engineering, Mexico City, Vol. 2, pp. 537–552.

Lyamin A.V. and Sloan S.W. (1997), A comparison of linear and nonlinear programming formulations for lower bound limit analysis, in: S. Pietruszczak and

G.N. Pande (eds.), Proceedings of the 6th International Symposium on Numerical Models in Geomechanics, pp. 367–373, Balkema, Rotterdam.

Malkawi A.I.H, Hassan W.F. and Sarma S.K. (2001), Global search method for locating general slip surface using Monte Carlo techniques, Journal of Geotechnical and Geoenvironmental Engineering, 127, 688–698.

Martin C.M. (2004), User Guide for ABC – Analysis of Bearing Capacity (v1.0), Department of Engineering Science, University of Oxford.

Meyerhof, G.G. (1963), Some recent research on the bearing capacity of foundations, Canadian Geotechnical Journal, 1: 16-26.

Michalowski R.L. (1989), Three-dimensional analysis of locally loaded slopes, Geotechnique, 39(1), 27–38.

Michalowski R.L. (1995), Slope stability analysis: A kinematical approach, Geotechnique, 45(2), 283–293.

Michalowski R.L. (1997). An estimate of the influence of soil weight on bearing capacity using limit analysis. Soils and Foundations, 37 (4) 57-64.

Mononobe N. and Matsuo H. (1929), On the determination of earth pressure during earthquakes, Proceeding of the World Engineering Congress, Tokyo, Vol.9, 179-187.

Morgenstern N.R. and Price V.E. (1965), The analysis of stability of general slip surface, Geotechnique, 15(1), 79–93.

Morgenstern N.R. (1992), The Evaluation of Slope Stability – A 25-Year Perspective, Stability and Performance of Slopes and Embankments – II, Geotechnical Special Publication No. 31, ASCE.

Morrison E.E. Jr. and Ebeling R.M. (1995), Limit equilibrium computation of dynamic passive earth pressure, Canadian Geotechnical Journal, vol32, 481-487.

Mroz Z. and Drescher A. (1969), Limit plasticity approach to some cases of flow of bulk solids, Journal of Engineering for Industry, Transactions of the American Society of Mechanical Engineers, 51, 357–364.

Nash D. (1987), A comparative review of limit equilibrium methods of stability analysis, in: M.G. Anderson and K.S. Richards (eds.), Slope Stability, pp. 11–75, New York: John Wiley & Sons.

Nguyen V.U. (1985), Determination of critical slope failure surfaces, Journal of Geotechnical Engineering, ASCE, 111(2), p.238–250.

Novotortsev V.I. (1938), Experience with the application of the theory of plasticity to problems of determination of the bearing capacity of foundation beds of structures, Izv. Nauchno-Issled. Inst. Anzh. Geol., 22, p.115-128.

Pan Jiazheng (1980), Analysis of Stability and Landslide for Structures, Hydraulic Press, Beijing, pp. 25–28 (in Chinese).

Prandtl L. (1920), Uber die Harte plasticher Korper, Nachrichten von der Koniglichen Gesellschaften, Gottingen, Math.-phys. Klasse, p.74-85.

Prater, E. G. (1977). An examination of some theories of earth pressure on shaft lining." Can. Geotech. J., vol. 14(1), p.91–106.

Reissner H. (1924), Zum Erddruckproblem, Proc. 1st Int. Congress for Applied Mechanics, eds. C.B. Biezeno and J.M. Burgers, Delft, p.295-311.

Revilla J. and Castillo E. (1977), The calculus of variations applied to stability of slopes, Geotechnique, 27, 1–11.

Sloan S.W. (1988), Lower bound limit analysis using finite elements and linear programming, International Journal for Numerical and Analytical Methods in Geomechanics, 12(1), 61-77.

Sloan S.W. (1989), Upper bound limit analysis using finite-elements and linear-programming, International Journal for Numerical and Analytical Methods in Geomechanics, 13(3), 263-282.

Sokolovskii, V.V. 1965, Statics of Granular Media, Pergamon Press.

Spencer E. (1967), A method of analysis of the stability of embankments assuming parallel inter-slice forces. Geotechnique, vol. 17, p11–26.

Soubra A. H. (2000), Static and seismic passive earth pressure coefficients on rigid retaining structures, Canadian Geotechnical Journal, vol.37, 463-478.

Vesic A.S. (1973), Analysis of ultimate loads for shallow foundations, Journal of Soil Mechanics and Foundations Division, ACSE, 99(SM1): 45-73.

Wei W.B., Cheng Y.M. and Li L. (2009), Three-dimensional slope failure by strength reduction and limit equilibrium methods, Computers and Geotechnics, 36:70-80.

Wei W.B. and Cheng Y.M. (2009), Soil nailed slope by strength reduction and limit equilibrium methods, Computers and Geotechnics, 37:602-618.

Wei W.B. and Cheng Y.M. (2009), Strength reduction analysis for slope reinforced with one row of piles, Computers and Geotechnics, 36:1176-1185.

Wei W.B. and Cheng Y.M. (2010), Stability analysis of slope with water flow by strength reduction method, Soil and Foundations, 50(1), 83-92.

Yu H.S., Salgado R., Sloan S.W. and Kim J.M. (1998), Limit analysis versus limit equilibrium for slope stability, Journal of Geotechnical and Geoenvironmental Engineering, ASCE, 124(1), 1–11.

Zhang X. (1999), Slope stability analysis based on the rigid finite element method, Geotechnique, 49(5), 585–593.

Zheng H., Zhou C.B. and Liu D.F. (2009). A robust solution procedure for the rigorous methods of slices. Soils and Foundations 49(4): 537 – 544.

Zhu D.Y., Lee C.F., Qian Q.H., Zou Z.S. and Sun F. (2001), A new procedure for computing the factor of safety using the Morgenstern–Price's method, Canadian Geotechnical Journal, 38(4), 882–888.

Zhu D.Y., Lee C.F. and Jiang H.D. (2003), Generalised framework of limit equilibrium methods for slope stability analysis, Geotechnique, 53(4), 377-395.

Zolfaghari A.R., Heath A.C. and McCombie P.F. (2005), Simple genetic algorithm search for critical non-circular failure surface in slope stability analysis, Computers and Geotechnics, 32(3), 139–152.

CHAPTER 4

Numerical Methods – Finite Element and Distinct Element Methods

Abstract: In this chapter, the basic theory about the two most important numerical methods in stability analysis are introduced. After that, these two methods are applied to different cases, and some laboratory tests are also used for comparison. In general, it is found that both methods are useful, and each method has its own merits and limitations.

Keywords: Distinct element, Failure mechanism, Finite element, Numerical method.

Most of the problems in geotechnical engineering are difficult to be solved by nature. In particular, the irregular geometric domain, nohomogeneous ground conditions, the presence of water, external and internal loadings and structural elements and the complex mechanical response of soil and soil-structure interaction have created extreme difficulty in the analysis of real engineering problems. Towards this, the use of numerical method will be more appropriate. Currently, the use of finite element, finite difference, boundary element and distinct element methods are the most popular methods which are adopted by many engineering geotechnical programs. Boundary element method is most suitable for linear problem with elastic constitutive behavior, and it is not commonly adopted in geotechnical problems. The capabilities of finite difference and finite element methods are comparable, but there are only few general finite difference programs in the world due to various technical difficulties. Finite element method is the most popular numerical tool to solve many types of differential equations in various disciplines. Currently, there are hundreds of finite

Y. M. Cheng, H. Wong, C. J. Leo & C. K. Lau

element programs available to the engineers for various applications, but there are also many limitations of FEM to geotechnical problems which include: (1) difficult to be applied for very large displacement problems, (2) generation and loss of contacts are difficult to be modelled, (3) fracture is difficult to be handled (but not impossible), (4) flow of the materials. Towards these limitations, the finite difference based distinct element method may be more suitable, though there are also many fundamental limitations in DEM which are still difficult to be resolved. There are also coupled methods based on DEM and FEM, and DEM is applied in special highly stressed/fractured region while FEM is applied to the general medium. In general, FEM and DEM are the two most important numerical methods in geotechnical engineering, and these methods will be introduced in this chapter. After the basic introduction to these two numerical methods, applications to some stability problems will be discussed.

4.1. PLANE STRAIN AND PLANE STRESS FOR CONTINUOUS PROBLEMS

For simplicity, two-dimensional problem will be discussed here. The extension of two-dimensional analysis to three-dimensional analysis is possible and is covered in many books which will not be repeated here. A problem is two-dimensional if the field quantities such as stress and displacement depend on only the two coordinates (x, y). In this sense, there are strictly no two-dimensional problems because every structure or loading system is three-dimensional in the real world. Many engineering problems can however be simplified approximately to a plane problem. Such approximation will greatly reduce the computational effort and can yield results satisfying practical requirements at the same time. Generally speaking, there are two kinds of elastic plane problems (Barber 2010): plane strain and plane stress, which will be introduced in the following subsections.

4.1.1. Plane Strain

To illustrate the problem, Fig. (**4.1**) shows a section of a structure with its cross-section independent of its length, except for the body force. On the surface of the structure, there may be surface force or constraint. It should be noted that both the internal and external forces are parallel to the cross-section and the values of them are invariable with the length normal to the section.

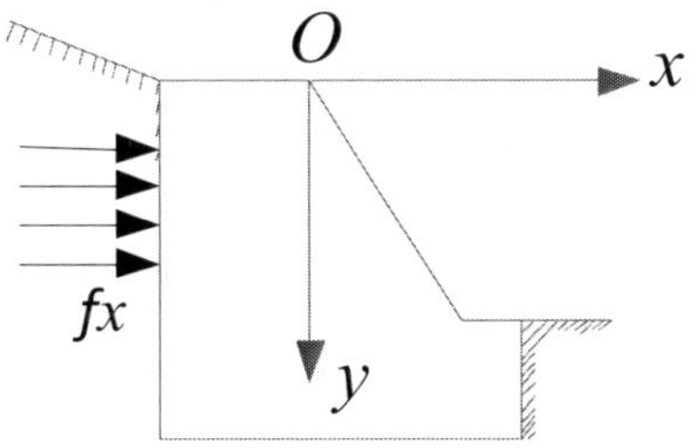

Fig. (4.1). Cross-section of an infinite long structure with external force.

Assume the length of the structure to be infinite, and consider any one of the cross-sections as the *xy*-plane, to which *z*-axis is perpendicular. All the field quantities (*e.g.*, stress, strain, and displacement, *etc.*) depend only on the two coordinates (*x, y*). Additionally, any cross-section is a symmetrical surface so that any point in this surface will only have *x*-displacement and *y*-displacement (*i.e.*, *z*-displacement is 0). For this condition, it follows that

$$\gamma_{zx} = \gamma_{zy} = \varepsilon_z = 0 \quad \textbf{(4.1)}$$

In view of the stress-strain relations, we can deduce that

$$\tau_{zx} = \tau_{xz} = 0, \tau_{zy} = \tau_{yz} = 0, \gamma_{xz} = \gamma_{yz} = 0 \quad \textbf{(4.2)}$$

Therefore, all the shear strain components equal to 0 except those related to *xy*-plane (*i.e.*, $\varepsilon_x, \varepsilon_y$ and γ_{xy}). This two-dimensional state is known as the plane strain condition.

It should be noted that while the normal strain is 0, the normal stress in the *z*-direction is not 0, which can be deduced according to the stress-strain relations as follows

$$0 = \varepsilon_z = \frac{1}{E}\left[\sigma_z - \nu(\sigma_x + \sigma_y)\right]$$

$$\sigma_z = \nu(\sigma_x + \sigma_y) \tag{4.3}$$

Similarly, by substituting σ_{zz} to the stress-strain relations we can deduce

$$\varepsilon_x = \frac{1-\nu^2}{E}\left(\sigma_x - \frac{\nu}{1-\nu}\sigma_y\right) \tag{4.4}$$

$$\varepsilon_y = \frac{1-\nu^2}{E}\left(\sigma_y - \frac{\nu}{1-\nu}\sigma_x\right) \tag{4.5}$$

$$\gamma_{xy} = \frac{2(1+\nu)}{E}\tau_{xy} \tag{4.6}$$

For simplicity, the stress-strain relations for 2-D plane strain problem can be expressed in a matrix form as:

$$\{\boldsymbol{\sigma}\} = [\boldsymbol{D}]\{\boldsymbol{\varepsilon}\} \tag{4.7}$$

$$[\boldsymbol{D}] = \frac{E_0}{1-\nu_0^2}\begin{bmatrix} 1 & \nu_0 & 0 \\ \nu_0 & 1 & 0 \\ 0 & 0 & \frac{1-\nu_0}{2} \end{bmatrix} \tag{4.8}$$

where $[\boldsymbol{D}]$ is called the elastic matrix with $E_0 = \frac{E}{1-\nu^2}$ and $\nu_0 = \frac{\nu}{1-\nu}$.

4.1.2. Plane Stress

Plane stress is an approximate solution, in contrast to plane strain condition, which is practically exact for many problems. In other words, plane strain condition is a special solution of the complete three-dimensional equations of elasticity, whereas plane stress is only approached in the limit as the thickness of the loaded body tends to zero.

Plane stress problem is usually related with an identical thickness plate which has external force acting on the edge only, and the external force is assumed to be parallel to the plate surface and independent of the thickness of the plate as well as body force. As shown in Fig. (**4.2**) which is a deep beam plane stress problem. Suppose the thickness of the plate is h, and take the intermediate plane of the plate as xy-plane with any line perpendicular to it as z-axis.

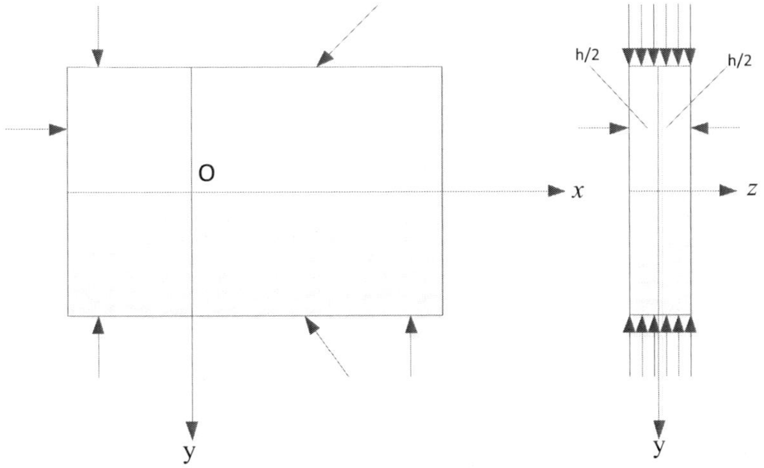

Fig. (4.2). Deep beam with identical thickness.

It is argued that if the two bounding z-planes of a thin plate are sufficiently close in comparison with the other dimensions, and if they are also free of tractions, the stresses on all parallel z-planes will be sufficiently small to be negligible, and we can write

$$\tau_{zx} = \tau_{zy} = \sigma_z = 0 \tag{4.9}$$

Obviously, we can deduce

$$\tau_{xz} = \tau_{yz} = 0 \tag{4.10}$$

Thus, there are only three stress components ($\sigma_x, \sigma_y\ and\ \tau_{xy}$) remain in plane stress condition, and all of them depend only on the two-dimensional coordinates (*x, y*). The stress-strain relations can be expressed as (Barber 2010)

$$\varepsilon_x = \frac{1}{E}\left(\sigma_x - \nu\sigma_y\right) \tag{4.11}$$

$$\varepsilon_y = \frac{1}{E}\left(\sigma_y - \nu\sigma_x\right) \tag{4.12}$$

$$\gamma_{xy} = \frac{2(1+\nu)}{E}\tau_{xy} \tag{4.13}$$

For simplicity, the stress-strain relations for 2-D plane stress problem can be expressed in matrix form as follows:

$$\{\boldsymbol{\sigma}\} = [\boldsymbol{D}]\{\boldsymbol{\varepsilon}\} \tag{4.14}$$

$$[\boldsymbol{D}] = \frac{E_0}{1-\nu_0^2}\begin{bmatrix} 1 & \nu_0 & 0 \\ \nu_0 & 1 & 0 \\ 0 & 0 & \frac{1-\nu_0}{2} \end{bmatrix} \tag{4.15}$$

where $[\boldsymbol{D}]$ is called elastic matrix with $E_0 = E$ and $\nu_0 = \nu$.

4.1.3. Relationship Between Plane Strain and Plane Stress

The solution of a problem under either the plane strain or plane stress assumptions involves finding a two-dimensional stress field, defined in terms of $\sigma_x, \sigma_y, \tau_{xy}$, which satisfies the equilibrium equations:

$$\frac{\partial\sigma_x}{\partial x} + \frac{\partial\tau_{xy}}{\partial y} + f_x = 0 \tag{4.16}$$

$$\frac{\partial\tau_{yx}}{\partial x} + \frac{\partial\sigma_y}{\partial y} + f_y = 0 \tag{4.17}$$

or in index notation

$$\frac{\partial\sigma_{ij}}{\partial x_i} + f_i = 0 \tag{4.18}$$

and for which the corresponding strains $\varepsilon_x, \varepsilon_y, \gamma_{xy}$, satisfy the only non-trivial compatibility equation

$$\nabla^2(\sigma_x + \sigma_y) = 0 \qquad \textbf{(4.19)}$$

The equilibrium and compatibility equations are the same in both formulations, and the only difference between these two cases is the relation between the stress and strain components, which for normal stresses are given by eqs. (4.4) and (4.5) for plane strain condition and (4.11) and (4.12) for plane stress condition. The relation between the shear stress τ_{xy} and the shear strain γ_{xy} is the same for both formulations and is given by eqs. (4.6) or (4.13). Thus, from a mathematical perspective, the plane strain solution looks like the plane stress solution for a material with different elastic constants and vice versa. Actually, it can be easily verified that equation (4.8) can be obtained from equation (4.15) by making substitutions

$$E_0 = \frac{E}{1-\nu^2}, \nu_0 = \frac{\nu}{1-\nu} \qquad \textbf{(4.20)}$$

This substitution also leaves the shear stress-shear strain relation unchanged as required.

4.2. FUNDAMENTALS OF FEM

Except for some simple problems with regular geometry and loading, it is very difficult to solve most of the boundary value problems in engineering directly with the yield of analytical solutions. Towards this, the use of numerical method seems indispensable, and the finite element is one of the most popular methods used by the engineers (Rao 2011, Zienkiewicz *et al.*, 2011). There are two fundamentals approaches to FEM which are the weighted residual method (WRM) and variational principle, but there are also other less popular principles which may be more effective under certain special cases. In finite element analysis of an elastic problem, solution is obtained from the weak form of the equivalent integration for the differential equations by WRM as an approximation. Alternatively, different approximate approaches (*e.g.*, collocation method, least square method and Galerkin method) for solving differential equations can be obtained by choosing different weights based on the WRM, and Galerkin method appears to be the most popular approach in general.

Specifically, in elasticity for instance, the principle of virtual work (including both principle of virtual displacement and virtual stress) is considered to be the weak form of the equivalent integration for the governing equilibrium equations. Furthermore, the aforementioned weak form of equivalent integration on the basis of Galerkin method can also be evolved to a variation of a functional if the differential equations have some specific properties such as linearity and self-adjointness. Principle of minimum potential energy and complementary energy are two variational approaches equivalent to the fundamental equations of elasticity.

Since displacement is usually the basic unknown quantity in FEM, only the principle of virtual displacement and minimum potential energy will be introduced in the following parts. In this case, the FEM introduced herein is also called displacement finite element method (DFEM). There are other ways to form the basis of FEM with advantages in some cases, but these approaches are less general and will not be discussed here.

4.2.1. Principle of Virtual Displacement

Principle of virtual displacement is the weak form of the equivalent integration for equilibrium equations and force boundary conditions. Given the equilibrium equations and force boundary conditions in index notation

$$\sigma_{ij,j} + f_i = 0, \quad (\text{in domain } V) \tag{4.21}$$

$$\sigma_{ij}n_j - T_i = 0, \quad (on\ domain\ boundary\ S_\sigma) \tag{4.22}$$

In WRM, without loss of generality, the variation of true displacement δu_i and its boundary value (*i.e.*, $-\delta u_i$) can be selected as the weight functions in the equivalent integration

$$\int_V \delta u_i(\sigma_{ij,j} + f_i)\,\mathrm{d}V - \int_{S_\sigma} \delta u_i\left(\sigma_{ij}n_j - T_i\right)\mathrm{d}S = 0 \tag{4.23}$$

The weak form of equation (4.23) is given as:

$$\int_V (-\delta\varepsilon_{ij}\sigma_{ij} + \delta u_i f_i)\,\mathrm{d}V + \int_{S_\sigma} \delta u_i T_i\,\mathrm{d}S = 0 \tag{4.24}$$

It can be seen clearly from eq. (4.24) that the first item in the volume integral indicates the work done by the stresses under the virtual strain (*i.e.*, internal virtual work), while the remaining items indicate the work done by the body force and surface force under the virtual displacement (*i.e.*, external virtual work). In other words, the summation of the internal and external virtual work is equal to 0, which is called the principle of virtual displacement. Under this case, we can conclude that a force system will satisfies equilibrium equations if the summation of the work done by it under any virtual displacement and strain is equal to 0.

4.2.2. Principle of Minimum Potential Energy (PMPE)

Based on equation (4.24), by substituting equations (4.14) for σ_{ij}, we can deduce that

$$\int_V (\delta\varepsilon_{ij}D_{ijkl}\varepsilon_{kl} - \delta u_i f_i)\,\mathrm{d}V + \int_{S_\sigma} \delta u_i T_i\,\mathrm{d}S = 0 \tag{4.25}$$

Due to the symmetry of the constitutive matrix D_{ijkl}, we can further obtain

$$(\delta\varepsilon_{ij})D_{ijkl}\varepsilon_{kl} = \delta\left(\tfrac{1}{2}D_{ijkl}\varepsilon_{ij}\varepsilon_{kl}\right) = \delta U(\varepsilon_{mn}) \tag{4.26}$$

where $U(\varepsilon_{mn})$ is the unit volume strain energy. Given the assumptions in linear elasticity

$$-\delta\phi(u_i) = f_i\delta u_i, -\delta\psi(u_i) = T_i\delta u_i \tag{4.27}$$

equation (4.25) is further simplified to

$$\delta\Pi_P = 0 \tag{4.28}$$

Π_P is the total potential energy of the system, which is equal to the summation of the potential energy of deformation and external force and can be expressed as

$$\Pi_P = \Pi_P(u_i) = \int_V \left(\tfrac{1}{2}D_{ijkl}\varepsilon_{ij}\varepsilon_{kl} - f_i u_i\right) \mathrm{dV} - \int_{S_\sigma} T_i u_i \,\mathrm{d}S \quad \textbf{(4.29)}$$

Equation (4.28) shows that, among all the potential displacements, the total potential energy of system will take stationary value at the real displacement, and it can be further verified that this stationary value is exactly the minimum value which is the principle of minimum potential energy.

4.3. GENERAL EXPRESSIONS AND IMPLEMENTATION PROCEDURE OF FEM

The solution of a general continuum problem by FEM always follows an orderly step-by-step process which is easy to be programmed and used by the engineers. For illustration, a three-node triangular element for plane problems is taken as an example to illustrate the general expressions and implementation procedures of FEM.

4.3.1. Discretization of Domain

The first step in the finite element method is to divide the structure or solution region into subdivisions or elements. Hence, the structure is to be modeled with suitable finite elements. In general, the number, type, size, and arrangement of the elements are critical towards good performance of the numerical analysis. A typical discretization with three-node triangular element is shown schematically in Fig. (**4.3**).

Mesh generation can be a difficult process for a general irregular domain. If only triangular element is to be generated, this is a relatively simple work, and many commercial programs can perform well in this respect. There are also some public domain codes (Easymesh or Triangle) which are sufficient for normal purposes. For quadrilateral or higher elements, mesh generation is not that simple, and it is preferable to rely on the use of commercial programs for such purposes.

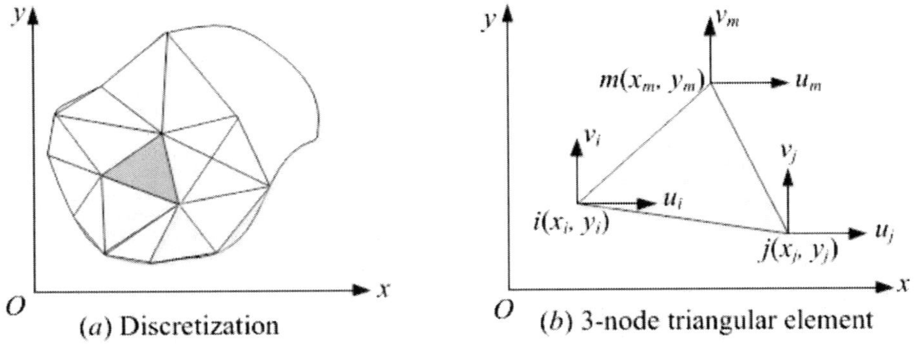

Fig. (4.3). Discretization of a 2-D domain with 3-node triangular element.

4.3.2. Interpolation or Displacement Model

As can be seen from Fig. (**4.3*b***), the nodal number of a typical 3-node triangular element is coded in anticlockwise order (*i.e.*, in the order of i, j and m), and each node has two degrees of freedom (DOFs) or two displacement components which is stored in a column vector in index notation as follows:

$$\boldsymbol{a}_i = \begin{bmatrix} u_i \\ v_i \end{bmatrix} \qquad (i, j, m) \tag{4.30}$$

Totally, each element has 6 nodal displacements, *i.e.*, 6 DOFs. Putting all the displacements in a column vector, we can obtain the element nodal displacement column matrix as

$$\boldsymbol{a}^e = \begin{bmatrix} \boldsymbol{a}_i \\ \boldsymbol{a}_j \\ \boldsymbol{a}_m \end{bmatrix} = [u_i \quad v_i \quad u_j \quad v_j \quad u_m \quad v_m]^{\mathrm{T}} \tag{4.31}$$

In FEM, nodal displacement is chosen as the basic unknowns, so interpolation at any arbitrary point is based on the three nodal displacements of each element, which is called a displacement mode. For a 3-node triangular element, linear polynomial is utilized, and the element displacement in both *x*-direction and *y*-direction are

$$u = \beta_1 + \beta_2 x + \beta_3 y \tag{4.32}$$

$$v = \beta_4 + \beta_5 x + \beta_6 y \tag{4.33}$$

Obviously, displacements of all the three nodes should satisfy equations (4.32) and (4.33). By substituting the six nodal displacement components into these equations, it is easy to obtain another form of displacement mode as

$$u = N_i u_i + N_j u_j + N_m u_m \tag{4.34}$$

$$v = N_i v_i + N_j v_j + N_m v_m \tag{4.35}$$

where

$$N_i = \frac{1}{2A}(a_i + b_i x + c_i y) \qquad (i, j, m). \tag{4.36}$$

In equation (4.34), N_i, N_j and N_m denote the interpolation function or shape function for the three nodes, respectively. A is the area of the element, and $a_i, b_i, c_i \cdots, c_m$ are constants related to the coordinates of the three nodes. Similarly, equations (4.34) and (4.35) can also be expressed in the form of matrix as

$$\boldsymbol{u} = \begin{bmatrix} u \\ v \end{bmatrix} = \begin{bmatrix} N_i & 0 & N_j & 0 & N_m & 0 \\ 0 & N_i & 0 & N_j & 0 & N_m \end{bmatrix} \begin{bmatrix} u_i \\ v_i \\ u_j \\ v_j \\ u_m \\ v_m \end{bmatrix} = \boldsymbol{N a}^e \tag{4.37}$$

where $\boldsymbol{N}$ is shape function matrix and $\boldsymbol{a}^e$ is the element nodal displacement vector. For the geometric equations, element strains are

$$\boldsymbol{\varepsilon} = \begin{bmatrix} \varepsilon_x \\ \varepsilon_y \\ \gamma_{xy} \end{bmatrix} = \mathbf{L}\boldsymbol{u} = \mathbf{L}\boldsymbol{N}\boldsymbol{a}^e = \mathbf{L}[\boldsymbol{N}_i \quad \boldsymbol{N}_j \quad \boldsymbol{N}_m]\boldsymbol{a}^e$$
$$= [\boldsymbol{B}_i \quad \boldsymbol{B}_j \quad \boldsymbol{B}_m]\boldsymbol{a}^e = \boldsymbol{B}\boldsymbol{a}^e \tag{4.38}$$

where $\mathbf{L}$ is the differential operator and $\boldsymbol{B}$ is the element strain displacement matrix which can be given as

$$\boldsymbol{B}_i = \mathbf{L}\boldsymbol{N}_i = \begin{bmatrix} \frac{\partial}{\partial x} & 0 \\ 0 & \frac{\partial}{\partial y} \\ \frac{\partial}{\partial y} & \frac{\partial}{\partial x} \end{bmatrix} \begin{bmatrix} N_i & 0 \\ 0 & N_i \end{bmatrix} = \begin{bmatrix} \frac{\partial N_i}{\partial x} & 0 \\ 0 & \frac{\partial N_i}{\partial y} \\ \frac{\partial N_i}{\partial y} & \frac{\partial N_i}{\partial x} \end{bmatrix} \quad (i,j,m) \tag{4.39}$$

Substitute equation (4.38) in equation (4.7),

$$\boldsymbol{\sigma} = \begin{bmatrix} \sigma_x \\ \sigma_y \\ \tau_{xy} \end{bmatrix} = \boldsymbol{D}\boldsymbol{\varepsilon} = \boldsymbol{D}\boldsymbol{B}\boldsymbol{a}^e = \boldsymbol{S}\boldsymbol{a}^e \tag{4.40}$$

where

$$\boldsymbol{S} = \boldsymbol{D}\boldsymbol{B} = \boldsymbol{D}[\boldsymbol{B}_i \quad \boldsymbol{B}_j \quad \boldsymbol{B}_m] = [\boldsymbol{S}_i \quad \boldsymbol{S}_j \quad \boldsymbol{S}_m] \tag{4.41}$$

S is called the element stress matrix. It should be noted that both the strain and stress matrices are constant for each element, because in 3-nodes triangular element the displacement mode is a first order function, and differentiate this function will gives a constant function.

4.3.3. Stiffness Equilibrium Equation (SEE) of FEM Derived from PMPE

For elastic plane problems, the total potential energy Π_P in equation (4.29) can be expressed in matrix formulation as follows:

$$\Pi_P = \int_\Omega \frac{1}{2}\boldsymbol{\varepsilon}^{\mathrm{T}}\boldsymbol{D}\boldsymbol{\varepsilon}t \,\mathrm{d}x\mathrm{d}y - \int_\Omega \boldsymbol{u}^{\mathrm{T}}\boldsymbol{f}t \,\mathrm{d}x\mathrm{d}y - \int_{S_\sigma} \boldsymbol{u}^{\mathrm{T}}\boldsymbol{T}t \,\mathrm{d}S \tag{4.42}$$

where ***t***, ***f***, and ***T*** denote the thickness, body force and surface force, respectively. For a FEM problem, the total potential energy is the summation of that from all the elements. Therefore, substituting equations (4.37) and (4.38) to equation (4.42) gives

$$\Pi_P = \sum_e \Pi_P^e = \sum_e \left(\boldsymbol{a}^{e\mathrm{T}} \int_{\Omega_e} \frac{1}{2} \boldsymbol{B}^{\mathrm{T}} \boldsymbol{D} \boldsymbol{B} t \, \mathrm{d}x\mathrm{d}y \, \boldsymbol{a}^e \right) -$$

$$\sum_e \left(\boldsymbol{a}^{e\mathrm{T}} \int_{\Omega_e} \boldsymbol{N}^{\mathrm{T}} \boldsymbol{f} t \, \mathrm{d}x\mathrm{d}y \right) - \sum_e \left(\boldsymbol{a}^{e\mathrm{T}} \int_{S_\sigma{}^e} \boldsymbol{N}^{\mathrm{T}} \boldsymbol{T} t \, \mathrm{d}x\mathrm{d}y \right) \tag{4.43}$$

Equation (4.43) can be viewed as

$$\boldsymbol{K}^e = \int_{\Omega_e} \boldsymbol{B}^{\mathrm{T}} \boldsymbol{D} \boldsymbol{B} t \, \mathrm{d}x\mathrm{d}y \, , \boldsymbol{P}_f^e = \int_{\Omega_e} \boldsymbol{N}^{\mathrm{T}} \boldsymbol{f} t \, \mathrm{d}x\mathrm{d}y$$

$$\boldsymbol{P}_S^e = \int_{S_\sigma{}^e} \boldsymbol{N}^{\mathrm{T}} \boldsymbol{T} t \, \mathrm{d}x\mathrm{d}y \, , \boldsymbol{P}^e = \boldsymbol{P}_f^e + \boldsymbol{P}_S^e \tag{4.44}$$

where $\boldsymbol{K}^e$ and $\boldsymbol{P}^e$ are named as the element stiffness matrix and equivalent element nodal load matrix, respectively. Substitute equations (4.44)-(4.47) to equation (4.43), the total potential energy of the structure can be simplified as

$$\Pi_P = \boldsymbol{a}^T \frac{1}{2} \sum_e (\boldsymbol{K}^e) \, \boldsymbol{a} - \boldsymbol{a}^T \sum_e (\boldsymbol{P}^e) \tag{4.45}$$

Given

$$\boldsymbol{K} = \sum_e \boldsymbol{K}^e \, , \boldsymbol{P} = \sum_e \boldsymbol{P}^e \tag{4.46}$$

equation (4.45) is further simplified as

$$\Pi_P = \frac{1}{2} \boldsymbol{a}^T \boldsymbol{K} \, \boldsymbol{a} - \boldsymbol{a}^T \boldsymbol{P} \, \boldsymbol{a} \tag{4.47}$$

where $\boldsymbol{K}$ and $\boldsymbol{P}$ are global stiffness matrix and global nodal load matrix, respectively.

For PMPE, the variation of Π_P *is* equal to 0 and the unknown variable is $\boldsymbol{a}$, thus equation (4.28) gives

$$\frac{\partial \Pi_P}{\partial \boldsymbol{a}} = 0 \tag{4.48}$$

which finally comes to the SEE of FEM as

$$\boldsymbol{K}\,\boldsymbol{a} = \boldsymbol{P} \tag{4.49}$$

From equation (4.46), we know that the global stiffness matrix and the global load matrix are the assemblage of the element stiffness matrices and equivalent element nodal load matrices respectively. Specifically, in order to solve equation (4.49), element stiffness matrix, element equivalent nodal load vector, global stiffness matrices and global nodal load vector are all determined together with some given displacement boundary conditions. Without the provision of adequate boundary condition, the system is singular as rigid body motion will produce no stress in the system and such mode will be present in the SEE.

4.3.4. Derivation of Element Stiffness Matrices (ESM)

For a 3-nodes triangular element, the element strain matrix $\boldsymbol{B}$ is constant, thus equation (4.44) gives

$$\boldsymbol{K}^e = \boldsymbol{B}^{\mathrm{T}}\boldsymbol{D}\boldsymbol{B}tA = \begin{bmatrix} \boldsymbol{K}_{ii} & \boldsymbol{K}_{ij} & \boldsymbol{K}_{im} \\ \boldsymbol{K}_{ji} & \boldsymbol{K}_{jj} & \boldsymbol{K}_{jm} \\ \boldsymbol{K}_{mi} & \boldsymbol{K}_{mj} & \boldsymbol{K}_{mm} \end{bmatrix} \tag{4.50}$$

of which the sub-matrix

$$\boldsymbol{K}_{ij} = \begin{bmatrix} k_{ij}{}^{xx} & k_{ij}{}^{xy} \\ k_{ij}{}^{yx} & k_{ij}{}^{yy} \end{bmatrix} \tag{4.51}$$

$\boldsymbol{K}_{ij}$ indicates the ith nodal force along x and y-directions in the Cartesian coordinate system when the displacement of the jth node is unit along x and y-directions, which can be easily obtained. Moreover, the element stiffness matrix is symmetric, and the computational memory required in a FEM program can be reduced by using this property.

It should be noted that for a higher order triangular element (*e.g.*, 6 nodes triangular element) or quadrilateral element for which higher order terms are involved, the strain matrix B is not constant any more so that the element stiffness matrix need to be evaluated by numerical integration (direct integration is seldom adopted). Towards this, numerical integration methods such as the Gaussian integration or the Newton-Cotes integration can be utilized.

4.3.5. Assembling of ESMs and ENLMs

For a FEM process, we need to solve equation (4.49) which is the global equilibrium equation. Most of the elements in the matrix $\boldsymbol{K}$ are 0 simply because each node is only shared by a few surrounding elements. In view of that, a rectangular matrix can represent the global stiffness matrix (which is a square matrix), and the half band width D can be defined as

$$D = (1 + NDIF) \times NDOF \quad \textbf{(4.52)}$$

where *NDIF* denotes the largest absolute difference between the element node numbers among all the element in the finite element mesh.

In conclusion, the properties of the global stiffness matrix can be summarized as: symmetric, banded distribution, singularity and sparsity. Among all the properties, singularity will vanish by introducing appropriate boundary conditions to equation (4.49) to eliminate the rigid body motion. Also, other properties like banded distribution should be fully taken into consideration to reduce the computational memory and enhance the computation efficiency.

4.3.6. Isoparametric Element and Numerical Integration

Most of the engineering structure is not regular in shape, and some of them even have very complicated boundary shapes. Although the use of triangular element can always fit a complicated boundary, the accuracy of this element is low in general. To cope with the irregular boundary shape with a higher accuracy in analysis, one of the most common approaches is the use higher order element, and the isoparametric formulation is the most commonly used at present. Consider an arbitrary 4-node quadrilateral element as an example which is schematically shown in (Fig. **4.4**). If we can find the transformation from Fig. (**4.4a**) to Fig. (**4.4b**), then it will become easier to carry numerical integration with complicated shapes for an arbitrary element. Fig. (**4.4a**) is defined in the Cartesian coordinate system, while Fig. (**4.4b**) is defined in the local coordinate system (or natural coordinate system) within a specific domain (*i.e.*, $\xi, \eta \in (-1,1)$). The relation between these two kinds of coordinate system can be described as

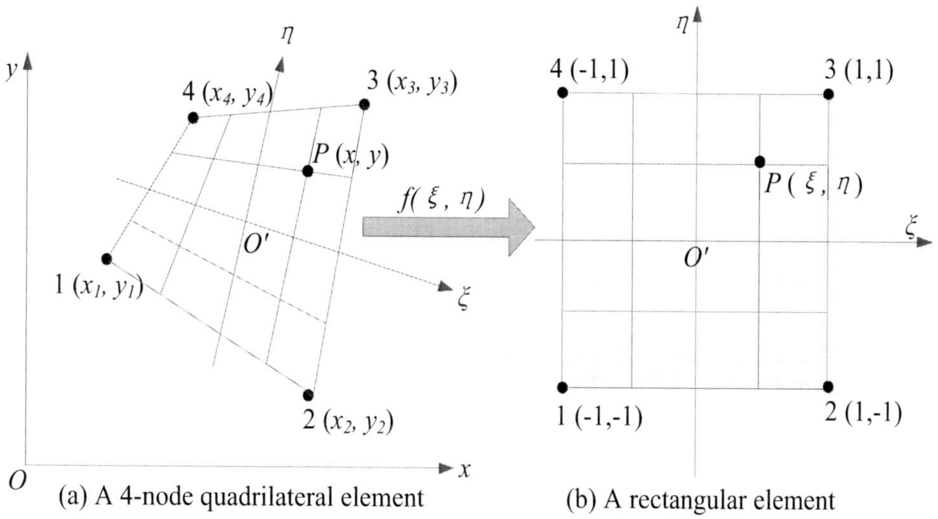

Fig. (4.4). Isoparametric transition.

$$\begin{Bmatrix} x \\ y \end{Bmatrix} = f \begin{Bmatrix} \xi \\ \eta \end{Bmatrix} \tag{4.53}$$

which can be further modified by the interpolation function at nodes in the local coordinate system as follows:

$$x = \sum_{i=1}^{m} N_i' x_i \qquad y = \sum_{i=1}^{m} N_i' y_i \tag{4.54}$$

where (x_i, y_i)are coordinates in Cartesian coordinate system corresponding to the ith node in local coordinate system, $\boldsymbol{N'}_i$ is interpolation function of the ith node in local coordinate system, and m is the number of nodes chosen to transform the coordinates. Therefore, the regular element in the natural coordinate system can be transformed to the irregular element in Cartesian coordinate system. The former element is called the parent element while the latter is called the subelement. Specifically, equation (4.55) can be further expanded as

$$\begin{Bmatrix} x \\ y \end{Bmatrix} = \begin{bmatrix} N_1 & 0 & N_2 & 0 & N_3 & 0 & N_4 & 0 \\ 0 & N_1 & 0 & N_2 & 0 & N_3 & 0 & N_4 \end{bmatrix} \begin{Bmatrix} x_1 \\ y_1 \\ x_2 \\ y_2 \\ x_3 \\ y_3 \\ x_4 \\ y_4 \end{Bmatrix} \tag{4.55}$$

Using the same interpolation functions, the element displacement model can be written as

$$\begin{Bmatrix} u \\ v \end{Bmatrix} = \begin{bmatrix} N_1 & 0 & N_2 & 0 & N_3 & 0 & N_4 & 0 \\ 0 & N_1 & 0 & N_2 & 0 & N_3 & 0 & N_4 \end{bmatrix} \begin{Bmatrix} u_1 \\ v_1 \\ u_2 \\ v_2 \\ u_3 \\ v_3 \\ u_4 \\ v_4 \end{Bmatrix} \quad \textbf{(4.56)}$$

where $\boldsymbol{J}$ denotes Jacobi matrix while the interpolation functions are given by

$$N_1 = \tfrac{1}{4}(1-\xi)(1-\eta), \quad N_2 = \tfrac{1}{4}(1+\xi)(1-\eta)$$

$$N_3 = \tfrac{1}{4}(1+\xi)(1+\eta), \quad N_2 = \tfrac{1}{4}(1-\xi)(1+\eta) \quad \textbf{(4.57)}$$

As mentioned before, during the derivation of the element stiffness matrix and the equivalent load vector, the derivative of the shape function and the integration in element surface or volume in Cartesian coordinate system are required. Since the shape functions adopted herein are expressed in natural coordinates, therefore, derivative and integration transformation relationships are essential when isoparametric element is used.

4.3.7. Derivative and Integral Transformation

According to the law of partial differential,

$$\frac{\partial N_i}{\partial \xi} = \frac{\partial N_i}{\partial x}\frac{\partial x}{\partial \xi} + \frac{\partial N_i}{\partial y}\frac{\partial y}{\partial \xi},$$

$$\frac{\partial N_i}{\partial \eta} = \frac{\partial N_i}{\partial x}\frac{\partial x}{\partial \eta} + \frac{\partial N_i}{\partial y}\frac{\partial y}{\partial \eta}, \quad \textbf{(4.58)}$$

or in matrix form

$$\begin{Bmatrix} \dfrac{\partial N_i}{\partial \xi} \\ \dfrac{\partial N_i}{\partial \eta} \end{Bmatrix} = \begin{bmatrix} \dfrac{\partial x}{\partial \xi} & \dfrac{\partial y}{\partial \xi} \\ \dfrac{\partial x}{\partial \eta} & \dfrac{\partial y}{\partial \eta} \end{bmatrix} \begin{Bmatrix} \dfrac{\partial N_i}{\partial x} \\ \dfrac{\partial N_i}{\partial y} \end{Bmatrix} = J \begin{Bmatrix} \dfrac{\partial N_i}{\partial x} \\ \dfrac{\partial N_i}{\partial y} \end{Bmatrix} \tag{4.59}$$

Inverse of equation (4.59) gives

$$\begin{Bmatrix} \dfrac{\partial N_i}{\partial x} \\ \dfrac{\partial N_i}{\partial y} \end{Bmatrix} = J^{-1} \begin{Bmatrix} \dfrac{\partial N_i}{\partial \xi} \\ \dfrac{\partial N_i}{\partial \eta} \end{Bmatrix} \tag{4.60}$$

where

$$J = \begin{bmatrix} \dfrac{\partial x}{\partial \xi} & \dfrac{\partial y}{\partial \xi} \\ \dfrac{\partial x}{\partial \eta} & \dfrac{\partial y}{\partial \eta} \end{bmatrix} = \begin{bmatrix} \sum_{i=1}^{4} \dfrac{\partial N_i}{\partial \xi} x_i & \sum_{i=1}^{4} \dfrac{\partial N_i}{\partial \xi} y_i \\ \sum_{i=1}^{4} \dfrac{\partial N_i}{\partial \eta} x_i & \sum_{i=1}^{4} \dfrac{\partial N_i}{\partial \eta} y_i \end{bmatrix}$$

$$= \begin{bmatrix} \dfrac{\partial N_1}{\partial \xi} & \dfrac{\partial N_2}{\partial \xi} & \dfrac{\partial N_3}{\partial \xi} & \dfrac{\partial N_4}{\partial \xi} \\ \dfrac{\partial N_1}{\partial \eta} & \dfrac{\partial N_2}{\partial \eta} & \dfrac{\partial N_3}{\partial \eta} & \dfrac{\partial N_4}{\partial \eta} \end{bmatrix} \begin{bmatrix} x_1 & y_1 \\ x_2 & y_2 \\ x_3 & y_3 \\ x_4 & y_4 \end{bmatrix} \tag{4.61}$$

For an infinitely small element, the area under the Cartesian coordinate system and the natural coordinate system are related by

$$\mathrm{d}s = \mathrm{d}x\mathrm{d}y = |\boldsymbol{J}|\mathrm{d}\xi\mathrm{d}\eta, \tag{4.62}$$

where $|\boldsymbol{J}|$ is determinant of the Jacobian matrix $\boldsymbol{J}$. Therefore, element stiffness matrix and equivalent nodal load matrix in equation (4.44) can be transformed to

$$K^e = \int_{\Omega_e} B^{\mathrm{T}} D B t\, |J| \mathrm{d}\xi \mathrm{d}\eta\, , P_f^e = \int_{\Omega_e} N^{\mathrm{T}} f |J| \mathrm{d}\xi \mathrm{d}\eta$$

$$P_S^e = \int_{S_\sigma^e} N^{\mathrm{T}} T |J| \mathrm{d}\xi \mathrm{d}\eta \tag{4.63}$$

For solving the integral equation, usually Gaussian integration method is employed. In practice, both two and three integration points along each direction of integration are commonly used. Since the discretized system is usually over-stiff, it is commonly observed that the use of two integration points along each direction of integration will slightly reduce the stiffness of the matrix and give better results as compared with the use of three integration points. The use of exact integration is possible for some elements, but such approaches are usually tedious and are seldom adopted. The advantage in using the exact integration is that the integration is not affected by the shape of the element while the transformation as shown in eq. (4.63) may be affected if the poor shape of the element is poor. The author has developed many finite element programs for teaching and research purposes which can be obtained at ceymchen@polyu.edu.hk. The programs available include plane stress/strain problem, thin/thick plate bending problem, consolidation in 1D and 2D (Biot), seepage problem, slope stability problem, pile foundation problems and others.

4.4. DISTINCT ELEMENT METHOD

In practical applications, limit equilibrium method based on the method of slices or method of columns and strength reduction method based on the finite element method or finite difference method are used for many types of stability problems. These two major analysis methods take the advantage that the *in situ* stress field which is usually not known with good accuracy is not required in the analysis. The uncertainties associated with the stress-strain relation can also be avoided by a simple concept of factor of safety or the determination of the ultimate limit state. In general, this approach is sufficient for engineering analysis and design. If the condition of the system after failure has initiated is required to be assessed, these two methods will not be applicable. Even if the *in situ* stress field and the stress-strain relation can be defined, the post-failure collapse is difficult to be assessed using the conventional continuum based numerical method, as sliding, rotation and collapse of the slope involve very large displacement or even separation without the requirement of continuity.

The most commonly used numerical methods for continuous systems are the FDM, the FEM and the boundary element method (BEM). The basic assumption adopted in these numerical methods is that the materials concerned are continuous throughout the physical processes. This assumption of continuity requires that at all points in a problem domain, the material cannot be torn open or broken into pieces. All material points originally in the neighborhood of a certain point in the problem domain remain in the same neighborhood throughout the whole physical process. Some special algorithms have been developed to deal with material fractures in continuum mechanics based methods, such as the special joint elements by Goodman (1976) and the displacement discontinuity technique in BEM by Crouch and Starfield (1983). However, these methods can only be applied with limitations (Jing and Stephansson, 1993):

1. large-scale slip and opening of fracture elements are prevented in order to maintain the macroscopic material continuity;
2. the amount of fracture elements must be kept to relatively small so that the global stiffness matrix can be maintained well-posed, without causing severe numerical instabilities; and
3. complete detachment and rotation of elements or groups of elements as a consequence of deformation are either not allowed or treated with special algorithms.

Before a slope starts to collapse, the factor of safety serves as an important index in both the LEM and SRM to assess the stability of the slope. The movement and growth after failure has launched which is also important in many cases cannot be simulated on the continuum model, and this should be analyzed by the distinct element method (DEM).

In continuum description of soil material, the well-established macro constitutive equations whose parameters can be measured experimentally are used. On the other hand, a discrete element approach will consider that the material is composed of distinct grains or particles that interact with each other. The commonly used distinct element method is an explicit method based on the finite difference principles which is originated in the early 1970s by a landmark work on the progressive movements of rock masses as 2D rigid block assemblages (Cundall, 1971). Later, the works by Cundall are developed to the early versions of the UDEC and 3DEC codes (Cundall, 1980; Cundall and Hart, 1985, 1992). The method has also been developed for simulating the mechanical behavior of granular materials (Cundall and Strack, 1979), with a typical early code BALL (Cundall, 1978) which later evolved into the codes of the PFC group for 2D and 3D problems of particle systems (Itasca, 1995). Through continuous developments and extensive applications over the last three decades, there has accumulated a great body of knowledge and a rich field of literature about the distinct element method. The main trend in the development and application of the method in rock engineering is represented by the history and results of the code groups UDEC/3DEC. Currently, there are many open source (Oval, LIGGGHTS, ESyS, Yade, ppohDEM, Lammps) as well as commercial DEM programs, but it general, this method is still limited to basic research instead of practical application as there are many limitations which include: (1) difficult to define and determine the micro-parameters; (2) there are still many drawbacks in the use of matching with the macro response to determine the micro-parameters; (3) not easy to set up a computer model; (4) not easy to include structural element or water pressure; (5) extremely time-consuming to perform an analysis; (6) post-processing is not easy or trivial. It should also be noted that DEM can be formulated by an energy based implicit integration scheme which is the Discontinuous Deformation Analysis (DDA) method. This method is similar in many respect to the force based explicit integration scheme as mentioned previously.

In DEM, the packing of granular material can be defined from statistical distributions of grain size and porosity, and the particles are assigned normal and shear stiffness and friction coefficients in the contact relation. Two types of bonds can be represented either individually or simultaneously; these bonds are referred to the contact and parallel bonds respectively (Itasca, 1995). Although the individual particles are solid, these particles are only partially connected at the contact points which will change at different time step. Under low normal stresses, the strength of the tangential bonds of most granular materials will be weak and the material may flow like a fluid under very small shear stresses. Therefore, the behaviour of granular material in motion can be studied as a fluid-mechanical phenomenon of particle flow where individual particles may be treated as 'molecules' of the flowing granular material. In many particle models for geological materials in practice, the number of particles contained in a typical domain of interest will be very large, similar to the large numbers of molecules.

One of the primary objectives of the particle model is the establishment of the relations between microscopic and macroscopic variables/parameters of the particle systems, mainly through micromechanical constitutive relations at the contacts. Compared with a continuum, particles have an additional degree of freedom of rotation which enables them to transmit couple stresses, besides forces through their translational degrees of freedom. At certain moment, the positions and velocities of the particles can be obtained by translational and rotational movement equations and any special physical phenomenon can be traced back from every single particle interactions. Therefore, it is possible for DEM to analyze large deformation problems and flow process which will occur after slope failure has initiated. The main limitation of DEM is that there is great difficulty in relating the microscopic and macroscopic variables/ parameters, hence DEM is mainly tailored towards qualitative instead of quantitative analysis.

DEM runs according to a time-difference scheme in which calculation includes the repeated application of the law of motion to each particle, a force-displacement law to each contact, and a contact updating scheme. Generally, there are two types of contact in the program which are the ball-wall contact and the ball-ball contact. In each cycle, the set of contacts is updated from the known particles and known wall positions. Force-displacement law is firstly applied on each contact, and new contact force is then calculated according to the relative motion and constitutive relation. Law of motion is then applied to each particle to update the velocity, the direction of travel based on the resultant force, and the moment and contact acting on the particles. Although every particle is assumed as rigid material, the behavior of the contacts is characterized using soft contact approach in which finite normal stiffness is taken to represent the stiffness which exists at the contact. The soft contact approach allows small overlap between the particles which can be easily observed. Stress on particles is then determined from this overlapping through the particle interface.

4.4.1. General Formulation of DEM

The PFC runs according to a time-difference scheme in which calculation include the repeated application of the law of motion to each particle, a force-displacement law to each contact, and a contact updating a wall position. Generally, there are two types of contact exist in the program which are ball to wall contact and ball to ball contact. In each cycle, the set of contacts is updated from known particle and known wall position. Force-displacement law is first applied on each contact. New contact force is calculated and replaces the old contact force. The force calculations are based on pre-set parameters like normal stiffness, density and friction. Next, law of motion is applied to each particle to update its velocity, direction of travel based on the resultant force, moment and contact acting on particle. Force-displacement law is then applied to continue the circulation.

4.4.2. The Force-Displacement Law

The force-displacement law is described for both the ball-ball and ball-wall contacts. The contact arises from contact occurring at a point. For the ball-ball contact, the normal vector is directed along the line between the ball centers. For the ball-wall contact, the normal vector is directed along the line defining the shortest distance between the ball center and the wall. The contact force vector Fi is composed of normal and shear component in single plane surface

$$F_i = F_{ij}^n(t) + F_{ij}^s(t + \Delta t) \quad \textbf{(4.64)}$$

The force acting on particle i in contact with particle j at time t is given by

$$F_{ij}^n(t) = k_n(r_i + r_j - l_{ij}(t)) \quad \textbf{(4.65)}$$

where r_j and r_i stand for particle i and particle j radii, $l_{ij}(t)$ is the vector joining both centres of the particles and k_n represent the normal stiffness at the contact. The shear force acting on particle i during a contact with particle j is determined by:

$$F_{ij}^s(t+\Delta t) = \pm\min(F_{ij}^s(t) + k_s\Delta s_{ij}, f\,|\,F_{ij}^n(t+\Delta t)\,|) \quad \textbf{(4.66)}$$

where f is the particle friction coefficient, ks represent the tangent shear stiffness at the contact. The new shear contact force is found by summing the old shear force (min $F_{ij}(t)$) with the shear elastic force. Δs_{ij} stands for the shear contact displacement-increment occurring over a time step Δt.

$$\Delta s_{ij} = V_{ij}^s \Delta t \quad \textbf{(4.67)}$$

where V_{ij}^s is the shear component of the relative velocity at contact between particles i and j over the time step Δt

4.4.3. Law of Motion

The motion of the particle is determined by the resultant force and moment acting on it. The motion induced by resultant force is called translational motion. The motion induced by resulting moment is rotational motion. The equations of motion are written in vector form as follow:

- (Translational motion)

$$\sum_j F_{ij} + m_i g + F_i^d = m_i x_i'' \quad \textbf{(4.68)}$$

- (Rotational motion)

$$\sum_j r_i F_{ij} + M_i^d = I_r \theta_i^{''} \quad \textbf{(4.69)}$$

where xi" and θi" stand for the translational acceleration and rotational acceleration of particles i. *Ir* stands for moment of inertia. Fi^d and Mi^d stand for the damping force and damping moment. Unlike finite element formulation, there are now three degree of freedom for 2D problem and six degree of freedom for 3D problems. In Cundall and Strack's explicit integration distinct element approach, solution of the global system of equations is avoided by considering the dynamic equilibrium of the individual particles rather than solving the entire system simultaneously. That means, Newton's law of motion is applied directly. This approach also avoids the generation and storage of the large global stiffness matrix that will appear in finite element analysis. On the other hand, the implicit DDA approach will generate a global stiffness matrix which is even larger than that in finite element analysis, as the rotation is involved directly in the stiffness matrix.

In a typical DEM simulation, if there is no yield by contact separation or frictional sliding, the particles will vibrate constantly and the equilibrium is difficult to be achieved. To avoid this phenomenon which is physical incorrect, numerical or artificial damping is usually adopted in many DEM codes, and two most common approaches to damping are the mass damping and non-viscous damping. For mass damping, the amount of damping that each particle "feels" is proportional to its mass, and the proportionality constant depends on the eigenvalues of the stiffness matrix. This damping is usually applied equally to all the nodes. As this form of damping introduces body forces which may not be appropriate in flowing regions, it may influence the mode of failure. Alternatively, Cundall (1987) proposed an alternative method where the damping force at each node is proportional to the magnitude of the out-of-balance-force, with a sign to ensure that the vibrational modes are damped rather than the steady motion. This form of damping has the advantage that only accelerating motion is damped and no erroneous damping forces will arise from steady-state motion. The damping constant is also non-dimensional and the damping is frequency independent. As suggested by Itasca (2004), an advantage of this approach is that it is similar to the hysteretic damping, as the energy loss per cycle is independent of the rate at which the cycle is executed. While damping is one way to overcome the non-physical nature of the contact constitutive models in DEM simulations, it is quite difficult to select an appropriate and physically meaningful value for the damping. For many DEM simulations, particles are moving around each other and the dominant form of energy dissipation is for frictional sliding and contact breakages. The choice of damping may affect the results of computations. Currently, most of the DEM codes allow the use of automatic damping or manually prescribed the damping if necessary.

To capture the inherent non-linearity behavior of the problem (with generation and removal of contacts, non-linear contact response and stress-strain behavior and others) the displacement and contact forces in a given time-step must be small enough so that

in a single time step, the disturbances cannot propagate from a particle further than its nearest neighbours. For most of the DEM programs, this can be achieved automatically, and the default setting is usually good enough for normal cases. It is however sometimes necessary to manually adjust the time-step in some special cases when the input parameters are unreasonably high or low. Most of the DEM codes use the central difference time integration algorithm which is a second order scheme in timestep.

4.4.4. Measuring Logic

If the local results in DEM are analyzed, it is found that there will be large fluctuations with respect to both locations and time. Such results are not surprising, as the results are highly sensitive to the interaction between particles and hence the time step under which the results are monitored. It can be viewed that such local results can be meaningless unless the results are monitored over a long time span or region. A number of quantities in a DEM model are defined with respect to a specified measurement circle. These quantities include coordinate number, porosity, sliding fraction, stress and strain rate. The coordination number and stress are defined as the average number of contacts per particle. Only particles with centroids that are contained within the measurement circle are considered in computation. In order to account for the additional area of particles that is being neglected, a corrector factor based on the porosity is applied to the computed value of stress.

Since measurement circle is used, stress in particle is described as the two in-plane force acting on each particle per volume of particle. Average stress is defined as the total stress in particle divided by the volume of measurement circle. Thus, shape of particle is regardless of the average stress measurement because the reported stress is easily scaled by volume unity. The reported stress is interpreted as the stress per volume of measurement circle.

4.4.5. Contact Constitutive Models

The constitutive model acting at a particular contact consists of three parts: a stiffness model, a slip model and a bonding model. The stiffness model provides an elastic relation between the contact force and relative displacement. The slip model enforces a relation between the shear and normal contact forces such that the two contacting balls may slip relative to one another. The bonding model serves to limit the total normal and shear forces that the contact can carry by enforcing the bond-strength limits. For ball to ball contact, the ideal force displacement relation is given by Hertz (1881) which relates the normal stiffness to one-third order of the normal force and two-third order of the shear stiffness of the grain material, while the shear stiffness is a function of the normal stiffness which is controlled by the Poisson ratio. Such relation is not easy to be used in computation, and normal stiffness is commonly assumed to be constant for many problems for simplicity.

To determine the contact forces, the contact detection and contact resolution must be evaluated which are actually computationally intensive. An elegant contact detection algorithms is required to keep track of the neighbor list which can keep on changing with time. Contact resolution requires the accurate calculation of the contact geometry and kinematics, overlap depth/separation, relative tangential motion, overlap area or volume. A contact constitutive model is then used to calculate the contact forces from these information. In most of the DEM codes, the contact is simplified to be a single point. The strains at the contacting particles and the nonuniform stress distributions within the particles are usually not explicitly considered in the DEM simulation. The overlap between the particles is considered to represent completely the effect of the deformation.

The basic DEM formulation provides no resistance to rotation at the contact points while the non-convex and rough surfaces for real soil particles contacts provide resistance to rotation at the contact points. While contact constitutive models have been proposed to account for the rolling resistance at the contact points, the energy dissipation during spin or resistance to spin motion is rarely considered in the current DEM models. The parallel bond model which is available in some commercial programs considers these components of motion by a cement between the particles. There are also various rotation resistance models which have been proposed by various researchers (Iwashita and Oda 1998, Jiang *et al.,* 2005) with success.

4.4.6. Model generation

To generate a DEM model which is later subjected to loading, there are four major groups of methods which are shown in Fig. (**4.5**). In the section expand method, balls were generated and expanded within small box without the top wall, and the process repeats until the total area is filled. In rain method, the balls are dropped into the section until falling rain has filled the section. Simple expand and rain method appears to be more promising in that it provide a balance between performance and model generation time. It should be noted that for generation of a large model, it can be extremely time-consuming in the computation. The generation of a suitable model is however vital to the study of granular materials. For example, the void ratio, particle size distribution, combinations of several particles to form irregular or nonconvex particles are extremely important towards the understanding of the more realistic behavior of granular materials.

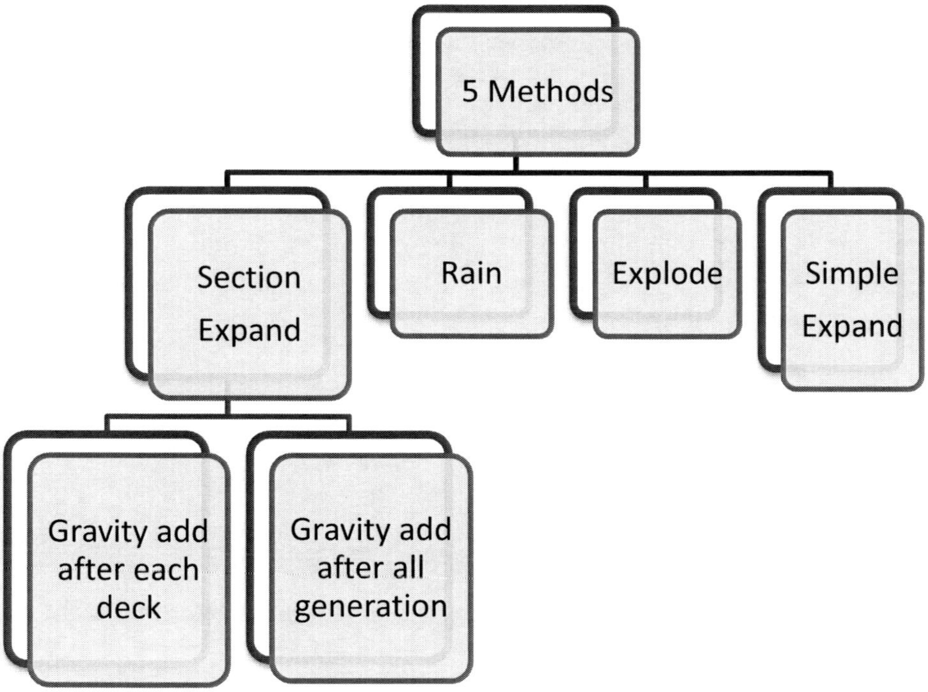

Fig. (4.5). Generation of a DEM model.

4.5. FINITE ELEMENT ANALYSIS OF SLOPE STABILITY PROBLEMS

At present, there are two major applications of the finite element analysis in slope stability problems. The first approach (which is available in limited commercial programs) is to perform an elastic (or elasto-plastic) stress analysis by applying the body force from the weight of soil to the slope. Once the equilibrium stresses are determined, the local factors of safety at any specific point can be determined easily from the stresses and the Mohr-Coulomb criterion. The global factor of safety can also be defined in a similar way by determining the ultimate shear force and the actual driving force along the failure surface, and an optimization algorithm is required to search for the critical slip surface with the minimum factor of safety. It is interesting to note that both the factor of safety and the location of the critical failure surface from such analysis are usually close to that by the limit equilibrium method or the second finite element approach. If elasto-plastic instead of elastic stress analysis is carried out for finite element slope stability analysis, one precaution should be noted. If the deformation is too large so that the finite element mesh is greatly distorted, the geometric non-linear effect may induce a major effect on the results (Cheng and Lau 2014). The second finite element slope stability approach is the strength reduction method (SRM) which is an option available in many commercial programs. In the SRM, the gravity load vector for a material with unit weight γ_s is determined from eq. (4.70) as:

$$\{f\} = \gamma_s \int [N]^T dv \tag{4.70}$$

where {f} is the equivalent body force vector and [N] is the shape factor matrix. The material parameters c' and ϕ' are reduced according to

$$c_f = c'/F \ ; \qquad \phi_f = \tan^{-1}\{\tan(\phi'/F)\} \tag{4.71}$$

The factor of safety F keeps on changing (some program uses a bracket approach) until the ultimate state of the system is attained, and the corresponding factor of safety will be the factor of safety of the slope. The ultimate state/termination criterion is usually based on one of the following criteria:

1. the nonlinear equation solver cannot achieved convergence after a pre-set maximum number of iteration;
2. there is a sudden increase in the rate of change of displacement in the system;
3. a failure mechanism has developed.

The location of the critical failure surface is then determined from the contour of the maximum shear strain or the maximum shear strain rate. It should be noted that a sudden increase in the rate of change may not be obtained in some cases, particularly when soil nails are present. Under such cases, the system appears to be able to tolerate a much large displacement without major change in the factor of safety.

The main advantages of the SRM as compared with the classical limit equilibrium as discussed in chapter 3 are that: (i) the critical failure surface is found automatically from the localized shear strain distribution, and it is usually possible to visualize the slip surface from the shear strain contour plot; (ii) it requires the assumption on flow rule but not the assumption on the interslice shear force distribution; (iii) it is applicable to many complex conditions, and is usually convenient to be used for three-dimensional problem as compared with LEM (but the reverse for two-dimensional problem).

4.5.1. Three-Dimensional Slope Stability Analysis by Finite Element

Although all the slope failures are three-dimensional (3D) in nature, two-dimensional (2D) analysis is still adopted in most of the slope stability analysis because of various practical reasons and difficulties. Most of 3D LEM methods including those by Hovland (1977), Chen and Chameau (1982), Zhang (1988), Ugai (1988), Lam and Fredlund (1993), Chang

(2002) adopt the assumption of symmetrical slip surface with a certain sliding direction. Such methods are basically the extension of 2D methods, where the interslice force relationship is extended to intercolumn force assumptions and the corresponding equilibrium equations are considered under 3D framework. Among these works, Lam and Fredlund (1993) in their development of a general LEM method have found that dominating intercolumn force functions are X/E and V/P with regard to the normal and vertical shear forces on the xy- and yz- plane (xz- and yz- plane in this study) respectively. Huang *et al.,* (2002) in their general method for 3D slope stability analysis have also involved the intercolumn force, but their formulation suffers from several limitations which are discussed by Cheng and Yip (2007). Cheng and Yip (2007), on the other hand, developed an asymmetric model prescribing only one sliding direction for the whole failure mass. The convergence problem under transverse load in the Huang and Tsai formulation (2000) has been overcome under this new formulation, and Cheng and Yip (2007) have demonstrated that this approach is equivalent to rotation of sliding axis until the minimum factor of safety is determined. The advantage of this formulation is that there is no need to carry out the axis rotation explicitly, which will save tremendous computations with no loss of accuracy. 3D NURBS surface and the simulated annealing method proposed by Cheng *et al.,* (2005) are incorporated to locate the 3D critical slip surface. Up to present, there is no commercial 3D LEM program which can be used conveniently and accurately. Most of the commercial programs adopt an elliptical surface to describe the critical slip surface, as there is no simple way to use few parameters to describe a general 3D slip surface which is valid for all cases.

The limitations of 2d lem can be overcome easily by 3d srm, as there is not a need to generate a 3d slip surface in 3d srm. There is also not a need to perform optimization search for the critical 3d slip surface which is a very difficult task. 3d srm is robust in application, provided that sufficient computer time is allowed in the modelling. Many researchers have considered 3d stability under different complex conditions of slope. Ugai and leshchinsky (1995) have included a pseudo-static seismic force component in their 3d srm analysis for vertical cuts. Zheng *et al.,* (2005) have used program ansys to conduct an extensive srm analysis in slope, tunnel and ultimate bearing capacity of foundations. Griffiths and marquez (2007) have considered both vertical and inclined boundaries to investigate the constraint effect of slopes with finite length for several 3d slope examples by srm. Deng *et al.,* (2007) have also conducted 3d srm to analyze the stability of a pre-existing landslide with multiple sliding directions.

For many slopes, due to route selection, geology, neighbor constructions and other necessary considerations, the slopes are curved in the geometric layout. For the previous 3d analysis methods, there are only limited works on the effect of curvature on the stability of slope. Rassam and williams (1999) have conducted a survey on the curvature effect on fill slope stability with concave and convex faces by configuring the axi-symmetric option in flac2d. To investigate the three-dimensional geometry effect for both convexity and concavity conditions, 3d srm analyses will be conducted in this section. Intercolumn force function will also be investigated to supplement the intercolumn force function which is an outstanding work up to the present.

4.5.2. Failure Mechanism of Curvilinear Slope

Intuitively, curvature is the amount by which a geometric object deviates from being flat or straight. In general mathematics, curvature is in the inverse of the radius of curvature of 2D circle or 3D surface. Curvature can be classified as concavity and convexity. The curvature of plane surface can be regarded as zero with infinite radius of curvature.

By changing the orientation relative to the axis of symmetry, both convex and concave slopes can be obtained by extension of 2D slope using the basic section as shown in Fig. (**4.6**). The friction angle of soil is prescribed as 20°, slope angle is 45°, height of slope is 6m, soil unit weight of 19kN/m^3 and cohesive strength of 20kPa or 4kPa are assumed. For pure convex and concave slopes with no transverse load, the problem domain is axi-symmetric. Similar to 2D plane-strain analysis, the tangential force or stress vector is zero in theory and only stress and force on the radial plane are considered as significant. Therefore, the radial intercolumn force function corresponds to f(x) and tangential intercolumn force function corresponds to f(y). For simplicity and clarity in this study, a slope is defined as locally convex when the curvature is positive or locally saddle when the curvature is negative according to the theorem of Gauss curvature. The radius of rotation about the axis of symmetry (R) is accordingly positive for convex and is negative for concave, and cylindrical coordinate system O(r, θ, z) will be adopted instead of the Cartesian coordinate system O(x,y,z) for the investigation of intercolumn force function, and f(r) and f(t) represent the radial and tangential intercolumn force function. By such definition, plane slope is the extreme situation of curvilinear slope with $R=\infty$.

In general, the slip surfaces for 2d and 3d analyses are very similar for the present situation. The curvature seems to have not a noticeable influence on the 3d slope failure mechanism when the critical slip surface and the overall displacement vector are observed as shown in Fig.(**4.6a**), that is the critical slip surface is similar and the displacement is pointing to the slope toe in the radial direction in both convex and concave slope. However when the displacement vector at middle portion where influence by boundary condition can be eliminated are investigated as shown in Fig. (**4.7b**), the displacement vectors in convex slope is mainly downstream and only slight sliding out could be found at the slope toe while for concave slope the soil mass sliding out is obvious. Combined with the results of shear stress on middle section shown in Fig. (**4.7c**), the failure mechanism behind might be explained as: (1) failure soil mass in the convex slope evenly slides down from the concentrated ground at crest and through the diverging geometry until reaching the toe where the shear stress is mainly mobilized; noticeable shear stress mobilized could be found near the crest because possibly the concentrated ground at crest restricts the trigger of failure; (2) the converging geometry of concave slope forms arching effect; such effect increases shear stress significantly in almost whole failure mass and results in accumulation of soil hence larger upheaval displacement at toe as compared with that in convex slope (shown in Fig. **4.8**); the converging geometry towards downstream restricts the develop of failure. These restrictions from curvilinear geometry may be regarded as the reason why curvilinear slope obtains higher/slightly higher safety than plane slope (see Table **4.1**). Furthermore, it should be noted that the diverging geometry in convex slope, compared to converging geometry in concave slope, imposes no positive influence on the restriction of failure development especially at middle height of slope as we have already spotted in the stress analysis (see Fig. **4.7c**). This result accounts for the fact that the factor of safety of convex slope (1.86) is about 10% lower than that of concave slope (2.05).

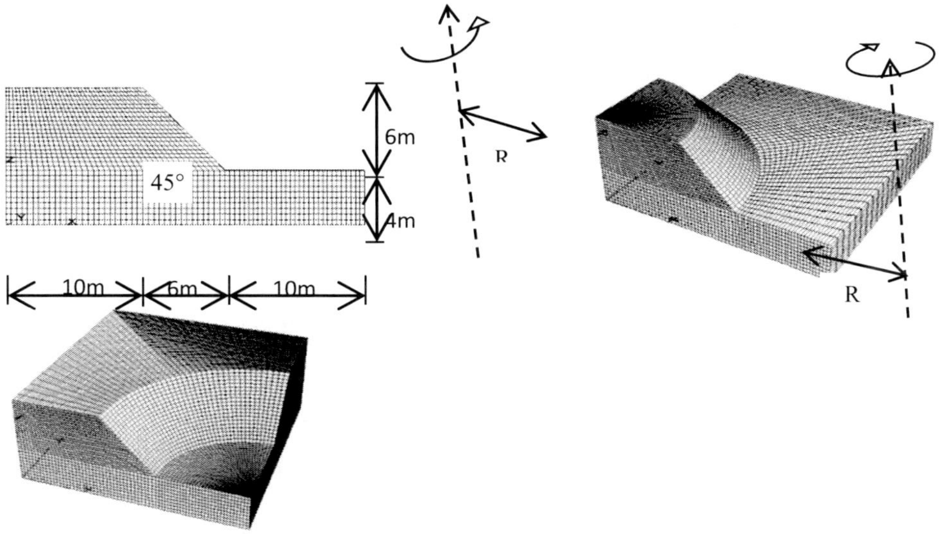

(**a**) Plane slope (basic section) (**b**) convex slope (**c**) concave slope

Fig. (4.6). Geometry of plane, convex and concave slopes.

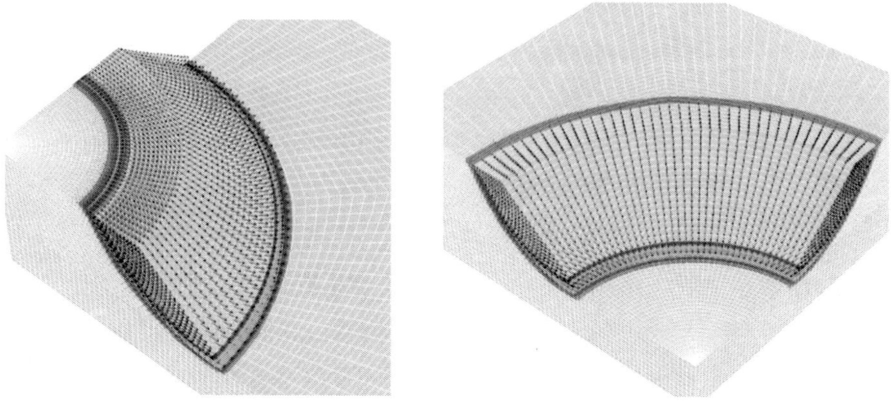

convex(FOS=1.86) concave(FOS=2.05)

(a) critical slip surface and overall displacement

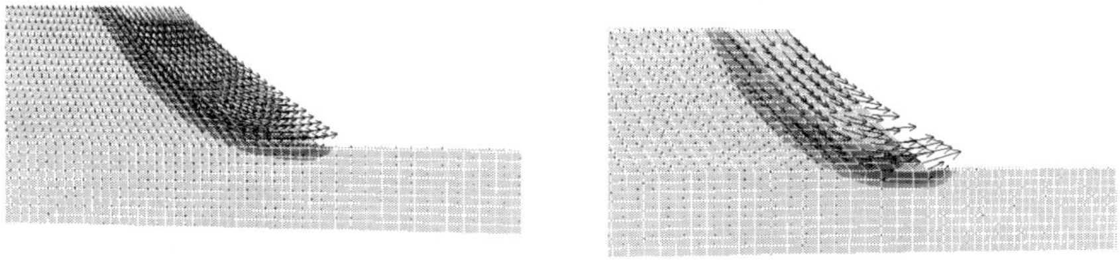

convex(FOS=1.86) concave(FOS=2.05)

(b) section view of displacement at middle portion

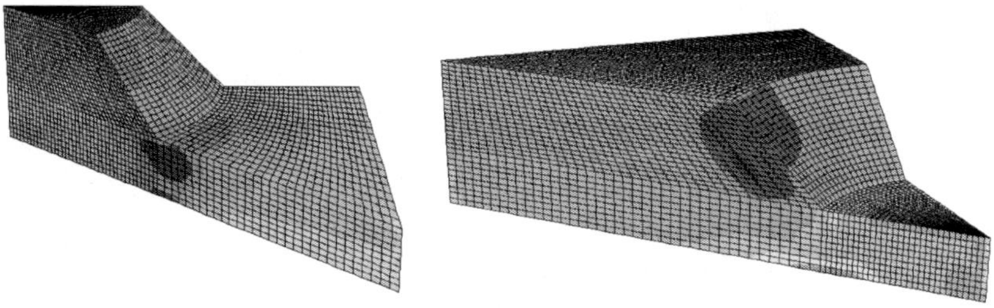

(c) shear stress distribution near the middle portion

Fig. (4.7). Failure mode of simple curvilinear slope (R=10m, c=20kPa).

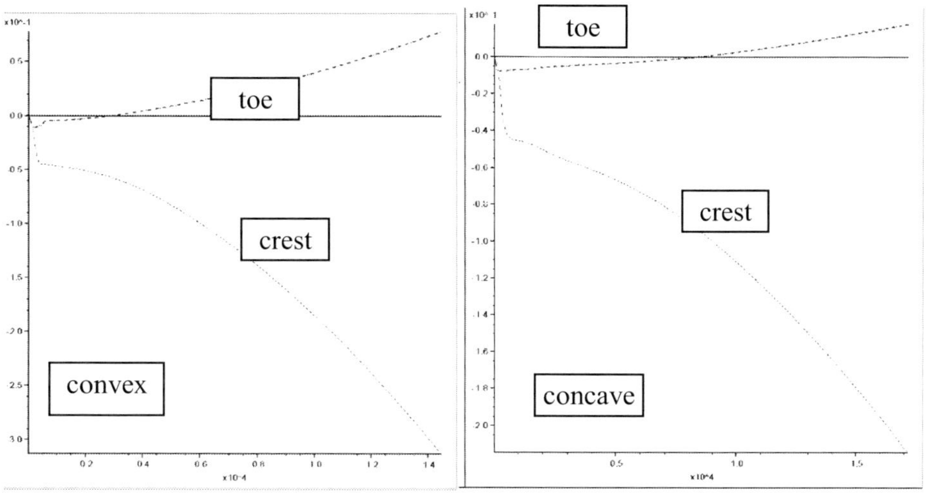

Fig. (4.8). Displacement history of simple curvilinear slope (R=10m).

Table 4.1. Summary of factor of safety and λ of simple slope with no loading.

Geometry	Plane		Convex				Concave			
R	∞	∞	R=+10		R=+20		R=-10		R=-20	
c	c=20kPa	c=4kPa	c=20kPa	c=4kPa	c=20kPa	c=4kPa	c=20kPa	c=4kPa	c=20kPa	c=4kPa
FOS	1.82	0.49	1.86	0.49	1.82	0.49	2.05	0.52	1.94	0.51
λrmax	0.7	0.7	1.7	1.0	1.7	0.9	0.36	0.4	0.36	0.4

In addition, different curvatures are also studied. Factors of safety of slope with different radii of curvature and different cohesive strength are investigated and the results are shown in Fig. (**4.9**) and Table **4.1** with reference to the slope with plane facing as well. Different curvatures (0.2, 0.1, 0.667, 0.05) correspond to different radius of rotation about the symmetric axis (see Fig. **4.6**), *i.e.* R=5m, r=10m, r=15m, and r=20m, respectively. As shown in Fig. (**4.9**), the effect of curvature is beneficial on factor of safety but is noticeable and important only when the curvature is significant. That means, unless the radius of curvature is small, the effect of curvature on factor of safety is not critical for most of the highway slopes if the soil is homogeneous. This result is however not necessarily true if the soil is nonhomogeneous, but then the effect of inhomogeneity will be more important than the effect of curvature.

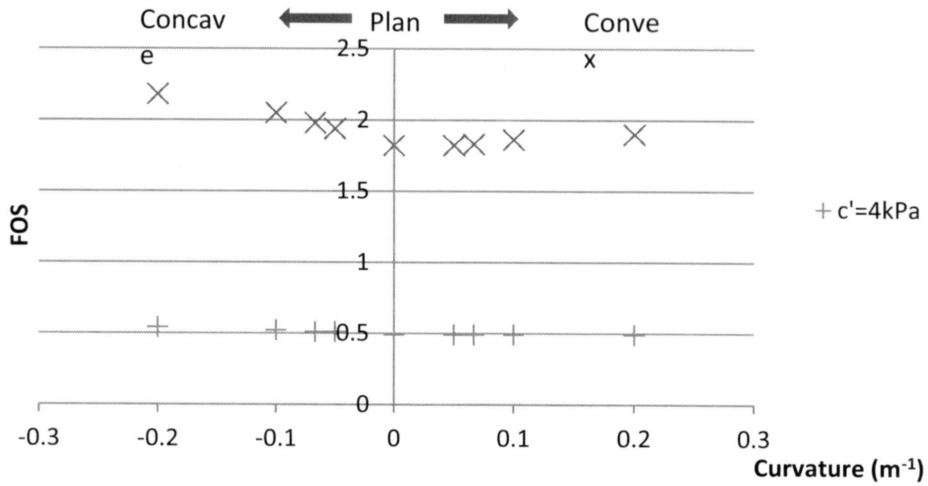

Fig. (4.9). Factor of safety of slopes with different curvature.

If the geometry of slope is complex, the slope profile is difficult to be defined by a single radius of curvature and thus the failure is complicated. Fig. (**4.10**) shows two types of complex slope with similar basic section and properties as previous parameters. Type 1 complex slope combines directly convex and concave portions while type 2 complex slope adopts a plane portion to connect both curvilinear portions. We expect the stability of complex slope would be enhanced by the existence of the curvilinear portions and weakened by existence of plane portion. As shown in Fig. (**4.10**), the factor of safety for type 1 complex slope is 2.0 and 1.89 for type 2 complex slope. Compared with the results of the simple slope with c=20kpa (fos=2.05 for concave slope, fos=1.86 for convex slope and fos=1.82 for plane slope), factor of safety for complex slope ranges among the maximum and minimum factors of safety of constituent portions. These results agree well with the anticipation and can be further explained by studying the critical slip surface. Due to the local stabilization effect from the curvilinear portion, the plastic zone initiated at the portions with low stability only partly develops in the portion with high stability and is not continuous, so there is only local failure in the complex slope.

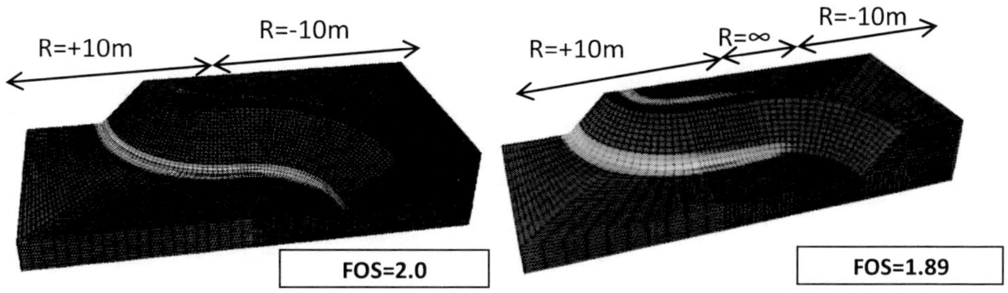

(a) Type 1 complex slope (b) Type 2 complex slope

Fig. (4.10). Critical slip surface of complex slope (c=20kPa).

From the above investigations on simple slope, for homogeneous slope with 3d curvature, the failure is symmetric about the rotation axis with an axi-symmetrical shear failure. For the same radius of curvature, the concave geometry shows higher factor of safety than the convex geometry which is an indication of the arch action introduced by the obvious confining action of the geometry. When different radii are concerned, the concave geometry has more positive effect to the stability when the radius of curvature is small. Both curvilinear geometries are beneficial to the slope stability, but such merit is relatively small and is only significant when the curvature is significant.

4.5.3. 3D Intercolumn Force Function on Plane Slope by Stress Analysis

Compared with 2D plane strain analysis, 3D analysis on slope with plane slope is expected to give similar factor of safety and stress distribution if the length of the slope is large enough. The procedure to determine the intercolumn force function is similar to the interslice force function for 2D problems. The intercolumn force function is studied along both x-direction and y-direction as shown in Fig. (**4.11**) and Fig. (**4.12**), where x=0 indicates the exit end of the slip surface. For f(x), it is basically a simple extension of the 2D interslice force function. On the other hand, λ_y and f(y) is practically 0 which is as expected, as the failure is just a symmetric failure without zero sliding direction. In fact, λ_y should be equal to 0 exactly for a plane slope. For this case, f(y) is actually a meaningless item as λ_y is zero. The peak value of f(x) is situated and maintained around the exit end of soil mass near the toe and decreases rapidly at the middle of the soil mass as shown in Fig. (**4.13**). The abnormality is located at the later proportion of the soil mass where tension crack may exist. For this 3D failure which is actually a simple extension of 2D failure with same factor of safety and internal stresses, a 2D analysis is completely sufficiently good, and f(x) is very similar to that by 2D SRM.

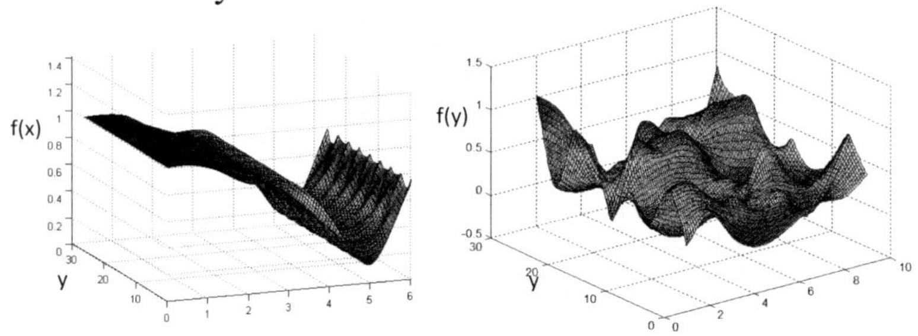

(a) f(x) (λ_{max}=λ_{min}=0.74) (b) f(y) (λ_{max}=1.9e-4), virtually 0

Fig. (4.11). f(x) and f(y) of plane slope (c=20kPa).

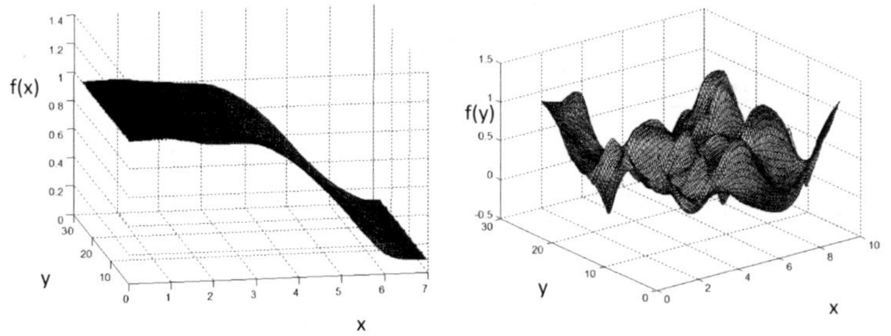

(a) f(x) (λ_{max}=λ_{min}=0.70) (b) f(y) (λ_{max}=4.8e-4), virtually 0

Fig. (4.12). f(x) and f(y) of plane slope (c=4kPa).

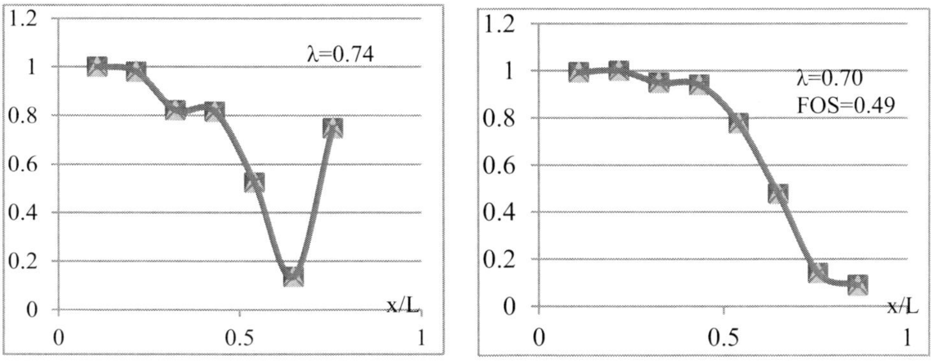

(a) c=20kPa (b) c=4kPa

Fig. (4.13). f(x) of plane slope on xz-plane.

4.5.4. Curvature Effect on the Internal Force Distribution

Fig. (**4.14**) illustrates f(r) and f(t) for a curvilinear slope. It is obvious that there is no solid trend for f(t) while f(r) is still basically similar to that for 2D situation. The f(t) is investigated along different θz-planes as shown in Fig. (**6.10**) where for both convex and concave slope, the distribution of f(t) is very random. This might result from the symmetric slope profile - actually most of the vertical intercolumn shear force is small as seen from the λt shown in Fig. (**4.14**), therefore, f(t) as shown in Fig. (**4.15**) can be neglected in this respect.

In the investigation of f(r) as shown in Fig. (**4.16**), both f(r) of convex and concave slope with R=±10m and c=20kPa is similar on different rz-planes which again due to the property of symmetry. However, it should be noted that f(r) seems to take different form when compared with 2D interslice force function. Further study is carried out among slopes with same properties except for different geometry, or saying different radii of rotation. The results on f(r) is given in Fig. (**4.17**). Compared with the obvious platform of f(r) in the plane slope case, concave slope have peak f(r) at the lower exit end of soil mass and f(r) continuously decreases towards the inner section of soil mass. For the convex slope, f(r) maintains high value for a shorter or longer distance from the toe of slope and then drops rapidly at about x/L=0.6. The main difference of f(r) among different geometries can thus be regarded as the starting locations where the function decreases. It should be noted that unlike plane and convex slope, the concave slope gives slightly higher factor of safety but the location of the maximum intercolumn force function has decreased to about 0.1-0.2 in length. This phenomenon indicates that the internal strength is well mobilized within concave slope. Such merits can also be found when factor of safety is relatively high as shown in Fig. (**4.9**). Table **4.1** summarizes attained factors of safety and maximum λr as well for slopes with different geometries. The relation of maximum λr and slope curvature is plotted in the same form as (Fig. **4.9**). It is of interest that as shown in Fig. (**4.18**) the maximum λr, *i.e.* maximum ratio of shear stress to normal stress on different θz-directional intercolumn surface increases as the curvature.

The thrust line in 2D slope stability analysis has been demonstrated to be located generally at 1/3 or slightly higher of the interslice height from the base of slices and been proved as an alternative to interslice force function. In 3D analysis, the same result can be found in plane slope which are shown in Fig. (**4.19**). The locations of the radial and tangential thrust line of plane slope agree well with the assumption of 1/3. For the thrust line of curvilinear slope, however, the assumption is not always true. The radial thrust line in convex slope is found to be close to the 1/3 line as shown in Fig. (**4.19**). The tangential thrust line, on the other hand, deviates from the assumption evidently at different sections (Fig. **4.20**). As for thrust line location of concave slope shown in Fig. (**4.21**), the radial and tangential thrust line are in the good accordance with each other while the difference between thrust line location and 1/3 line is noticeable at different sections. At the section near to the boundary where the displacement is confined, the thrust line is slightly higher than the 1/3 line but the deviation becomes significant at the middle portion as Fig. (**4.21b**) indicates. That is to say, while location of thrust line is relatively stable in 2D analysis and 3D plane slope, it varies from case to case in 3D slopes influenced by curvature effect and the assumption of 1/3 line should not be regarded anymore as a stable and trustful assumption.

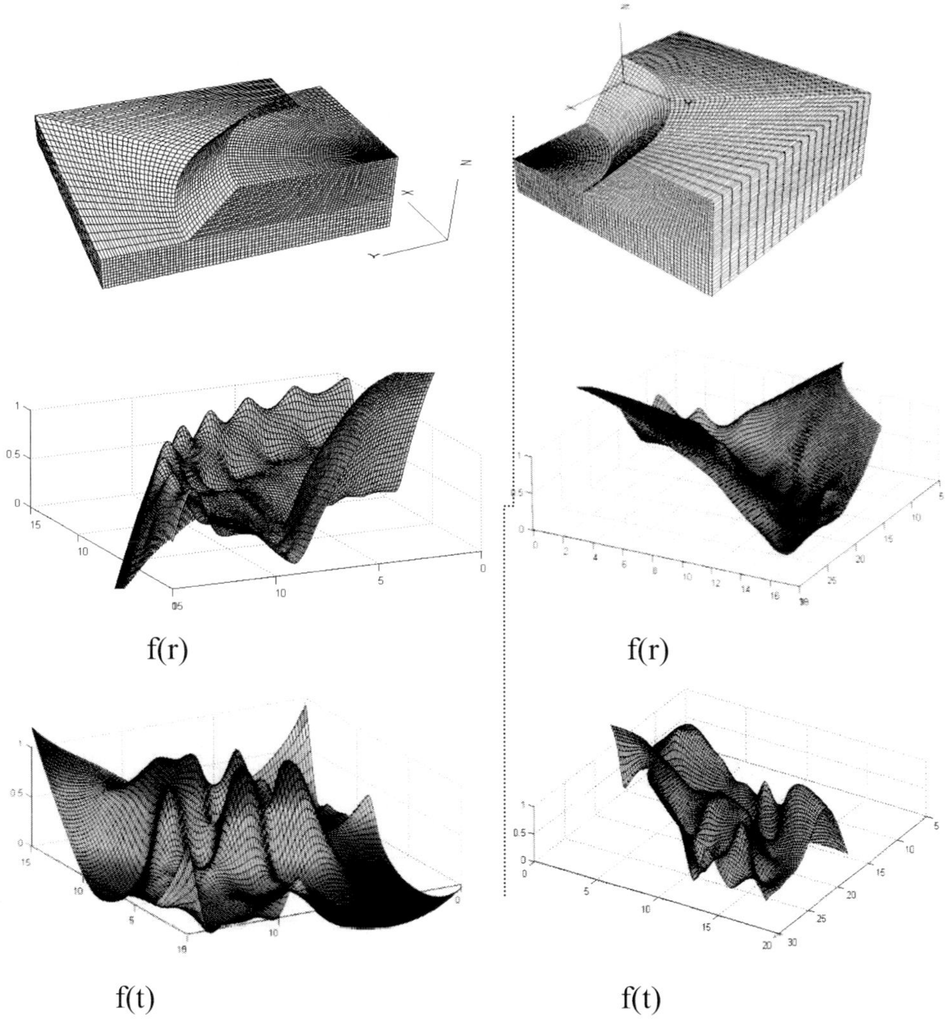

Fig. (4.14). Intercolumn force function f(r) and f(t) for curvilinear slope (R=±10m, c=20kPa).

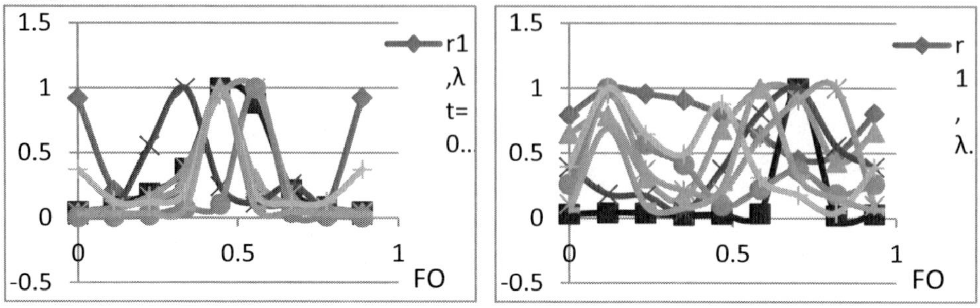

(a) convex slope (b) concave slope

Fig. (4.15). f(t) of curvilinear slope (R=±10m, c=20kPa).

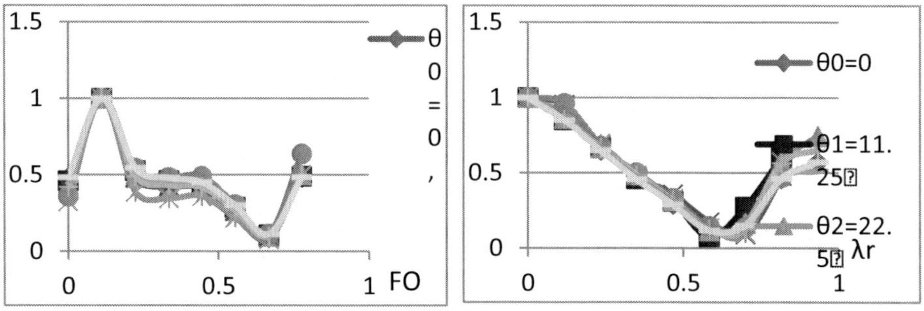

(a) convex slope (b) concave slope

Fig. (4.16). f(r) of curvilinear slope (R=±10m, c=20kPa).

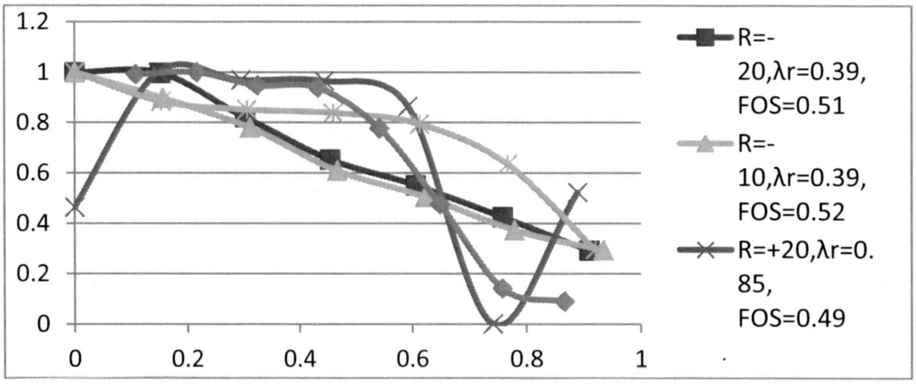

Fig. (4.17). summarized f(r) graph for slope with different R (c=4kPa).

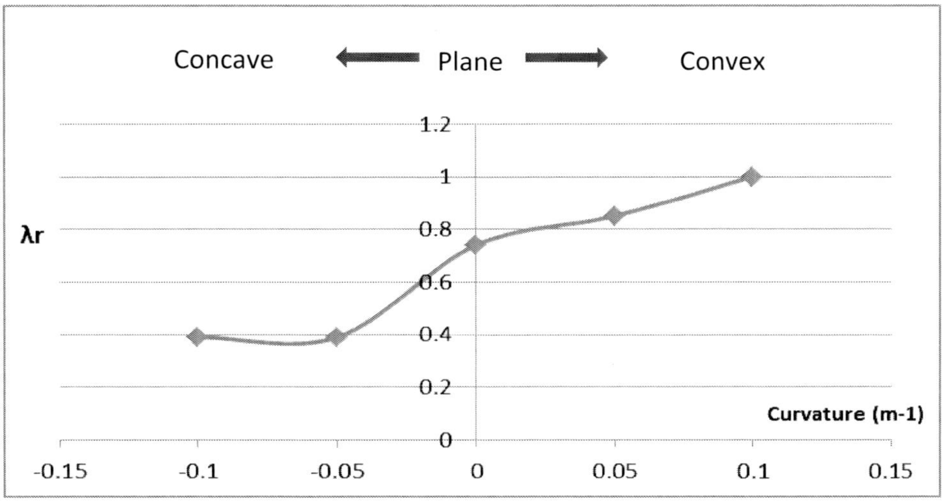

Fig. (4.18). maximum λr of slopes with different curvature.

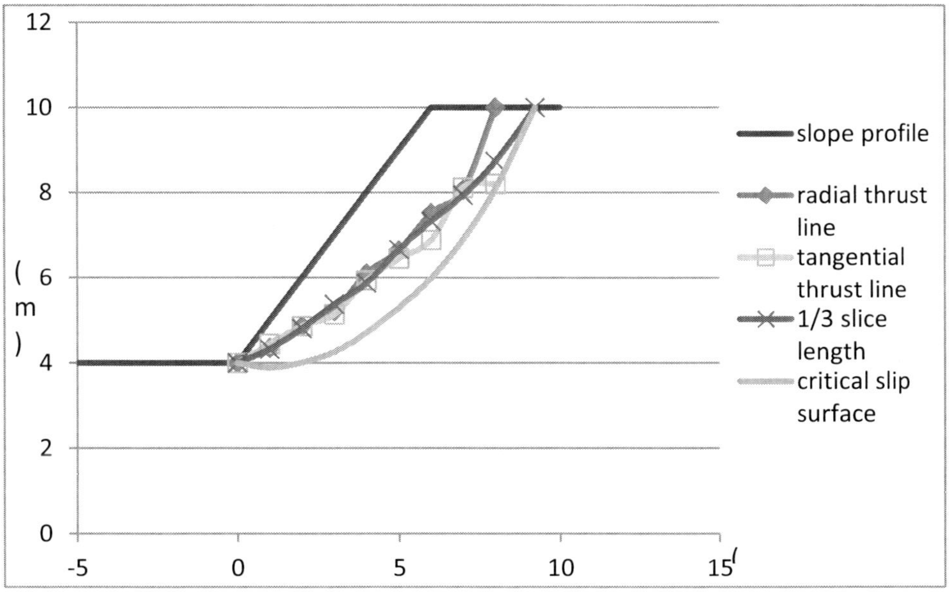

Fig. (4.19). Thrust line of plane slope (R=∞, c=20kPa).

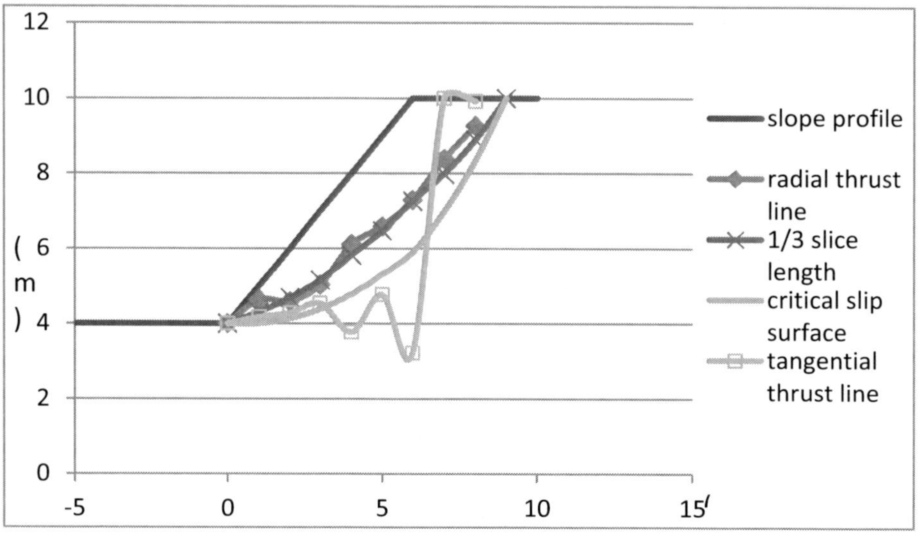

(a) at section near boundary

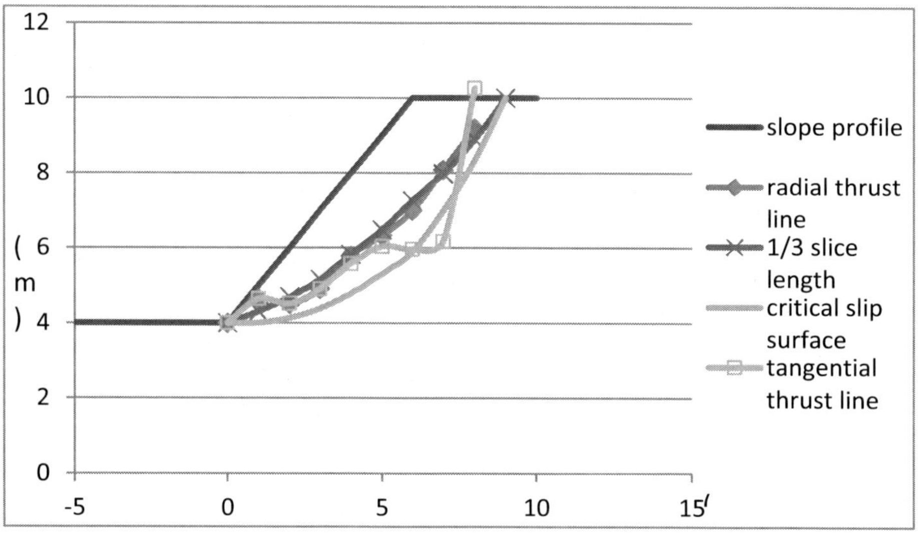

(b) at section of middle portion

Fig. (4.20). Thrust line of convex slope (R=+10m, c=20kPa).

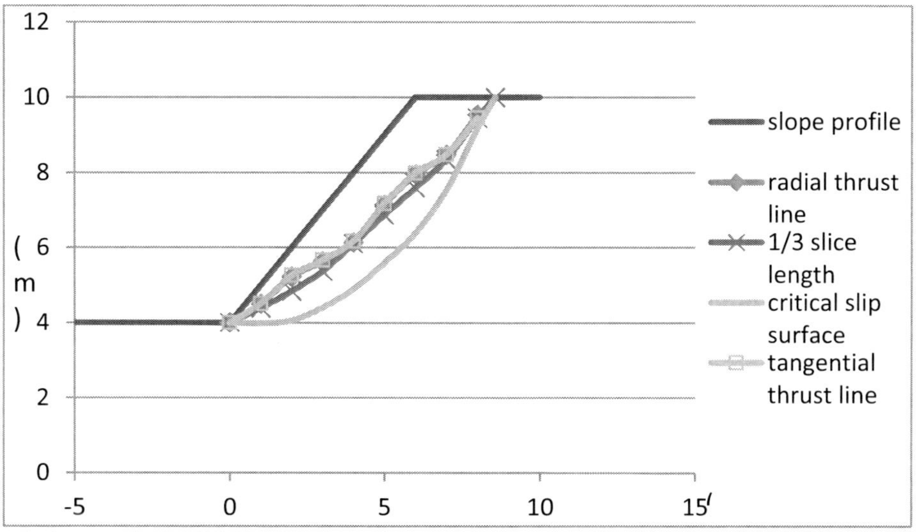

(a) at section near boundary

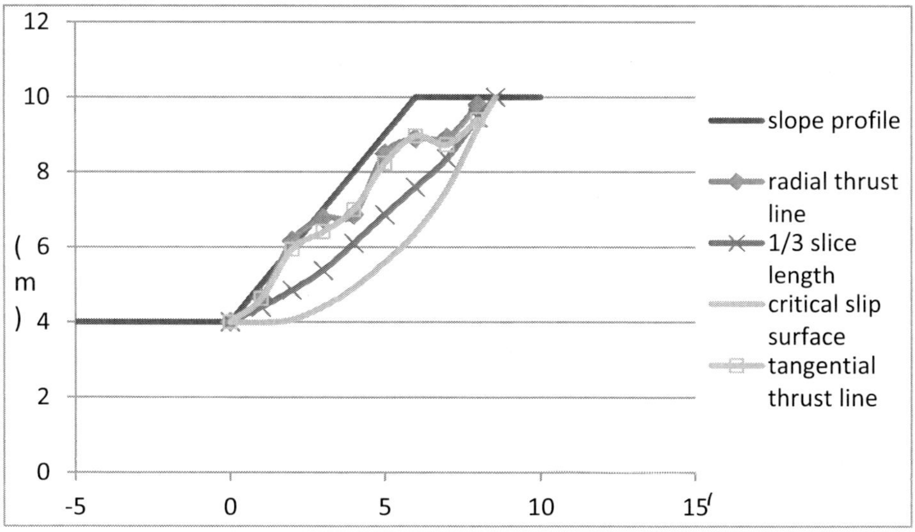

(b) at section of middle portion

Fig. (4.21). Thrust line of concave slope (R=-10m, c=20kPa).

4.5.5. Stability of Locally Loaded Slope with Curvature

If local patch load is superimposed on a curved slope, the difference between concave and convex geometry becomes more apparent. Fig. (**4.22**) illustrates the failure mode when there is a 1m×1m square patch load of 200kpa at the top and 1m away from the rim. It can be seen from Fig. (**4.22(a)**) that the failure becomes local failure and clustered around the loaded district as expected. For the failure at toe for the case of concave slope, there are two failure zones at the toe which are not connected. The factor of safety under such local loading decreases by about 5.9% in convex slope (from 1.86 dropping to 1.75 as listed in Table **4.2**) and 2.4% in concave slope (from 2.05 dropping to 2.0 as listed in Table **4.2**), which indicates that the concave slope is more unstable under local patch load.

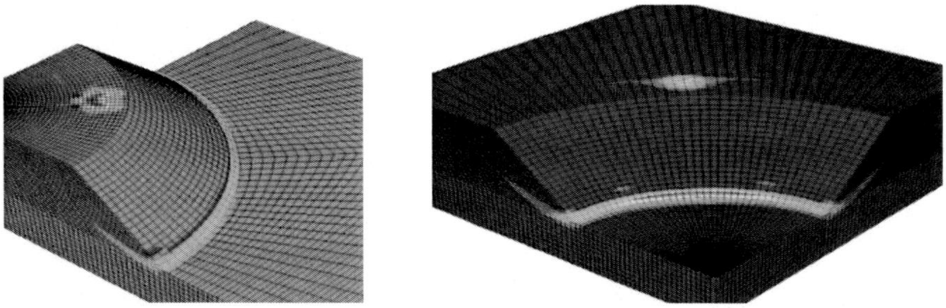

convex(FOS=1.75) concave(FOS=2.00)

(a) overall view

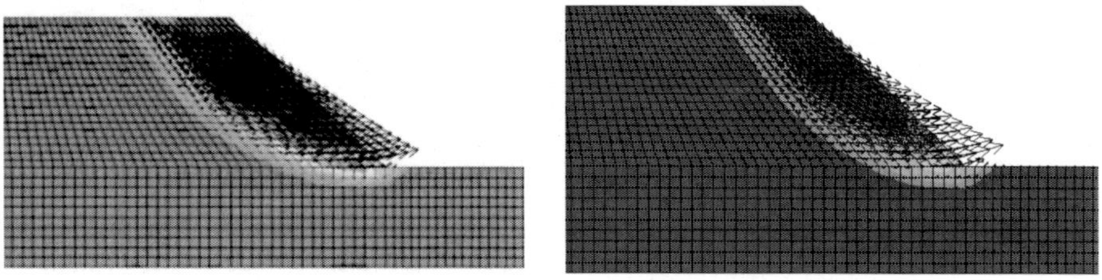

convex(FOS=1.75) concave(FOS=2.00)

(b) section view

Fig. (4.22). Failure mode of locally loaded curvilinear slope (200kPa, R=±10m).

Table 4.2. Summary of factor of safety of different loading and nailing mode (R=±10m).

Geometry	Local Loading	Nailing Mode		
		No Nailing	Radial	Parallel
Convex	0	1.86	2.00	**2.10**
	200kPa	1.75	1.82	**1.87**
Concave	0	2.05	**2.19**	2.16
	200kPa	2.00	**2.16**	2.15

4.6. DISTINCT ELEMENT ANALYSIS OF 3D SLOPE

For finite element/difference based slope stability analysis, the assessment of the factor of safety and critical slip surface can be determined with ease. After the initiation of the failure, the displacement will be very large which is difficult to be assessed by the use finite element method. More critically, there will be separation and loss of continuity which is difficult to be assessed by the conventional numerical method, yet the assessment of the spread of failure is important in many practical applications. Towards this issue, the use of distinct element method will be more suitable. There are however very few applications of DEM in slope stability analysis, and this section is devoted towards this application.

4.6.1. DEM Analysis of 3D Slope with Curvature

Rassam and Williams (1999) conduct a survey on the curvature effect on fill slope stability with concave and convex faces are considered by using SRM in FLAC2D. At present, there are only limited works devoted to the effect of curvature on the factor of safety using LEM and SRM. Practically, there is no previous studies about the post-failure conditions and progressive failure mode of 3D slope with curvature with DEM analysis.

An axisymmetric model is generated for the analysis of a homogeneous concave slopes as shown in (Fig. **4.23**). The radius of curvature of the concave slope is given by r_0/h_0 equals 0.857, where r_0 and h_0 are the radius of curvature and height of slope respectively. The slope angle in the numerical model is 45° with c'=0.5kpa and ϕ'=26.5°. Wall to ball friction in the five faces of boundary in pfc3d model is set to 0.5, which coincides with the soil friction as ball to ball friction in particle flow code. The model is simplified into cube in this numerical analysis so as to reduce the tremendous computer time required for analysis, however, in practical condition, the concave slope section is surrounded by continuous soil mass. That is to say, the boundary wall in numerical modeling is used as other soil particles around in reality. So the boundary condition is the crucial factor in a realistic analysis of the problem. Then the in-situ state of stress in the ground before any excavation or construction is generated by setting the initial conditions which is controlled by gravity in a particle flow model. The in-situ stress state is also an important factor because it will influence the subsequent behavior of the model. Initial stresses, however, are not prescribed directly in pfc3d. The initial stress state must be derived from the initial conditions specified for the compaction of the ball assembly, and it is generated from the gravitational loading in particle flow analysis. Based on the procedures as mentioned above, the initial state of a cut concave slope is generated in Fig. (**4.23**).

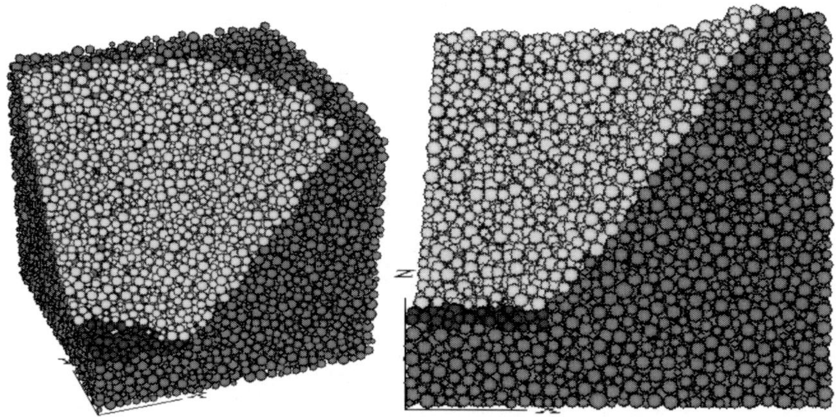

Fig. (4.23). Initial state of a cut concave slope.

The failure process of a concave slope with small cohesion under gravity is analyzed in Fig. (**4.24**). In the beginning, the soils at the two corner side start to collapse while the soil at the middle of the slope crest remains basically stationary. With increasing settlement at the two corners, the soil in the middle part is dragged down gradually as shown in Fig. (**4.24b**). The failure started from partial sliding to global sliding, and in the end the whole slope crest crumbles as illustrated in Fig. (**4.24d**). From the above investigations on simple slope, the 3d curvature effect is basically symmetric about rotation axis with an axi-symmetrical shear failure throughout the corner side of the slope, and a distinct 3d failure surface can be noticed from the displacement graph in Fig. (**4.30d**).

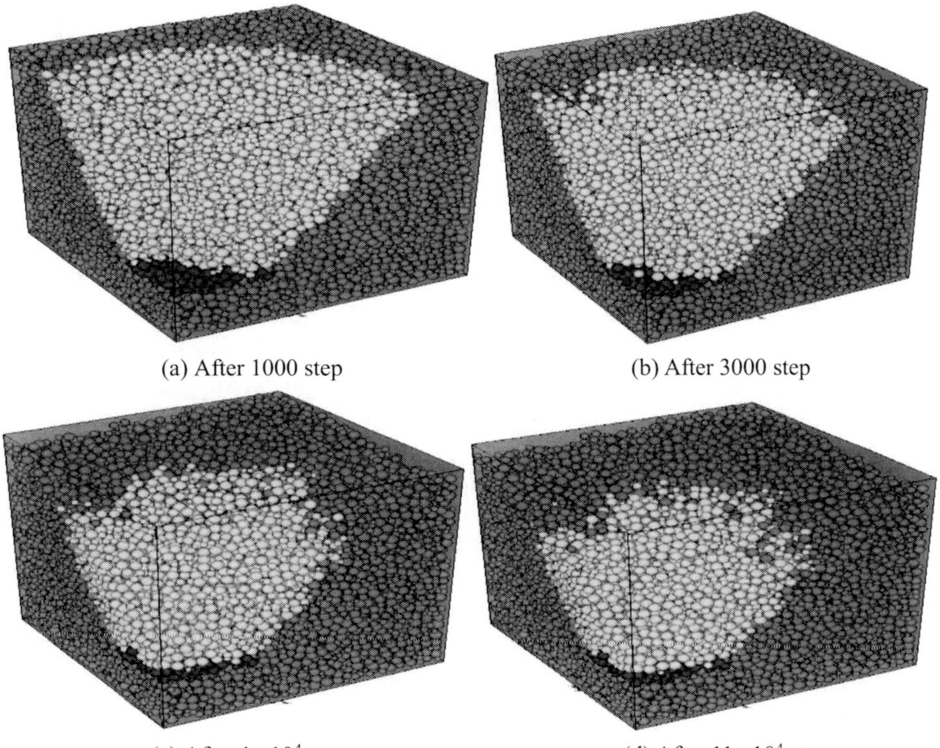

(a) After 1000 step (b) After 3000 step

(c) After 1 x10^4 step (d) After 11 x10^4 step

Fig. (4.24). Failure process of concave slope with lower cohesion under gravity.

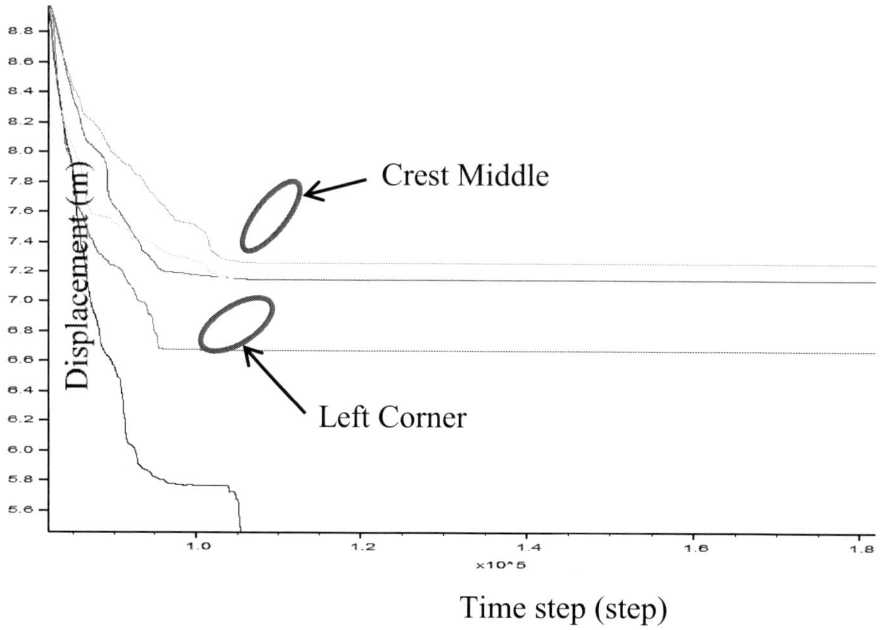

Fig. (4.25). Settlements for balls in the crest left corner *vs* crest middle of concave slope under gravity.

In this slope failure with small cohesion, the soil particles at left corner moves faster so that the curves for the balls at the left corner has steeper gradient than other curves for the particle at crest middle, which can be noticed in Fig. (**4.25**). The soil mass at the two corners experience a deeper falling, due to the influence of the arching effect that brings thrust forces to restrain cracking on middle part of slope crest. Also the impact of side effect is dragging the corner balls away as well as downward.

If a local surcharge is applied on the top of a curvilinear slope, the symmetry of failure will be disturbed and local failure instead of a global failure will take place. In Fig. (**4.26**), a concave slope with a local loading from a raft footing is illustrated. The slope angle is also 45° with c'=3kpa and ϕ'=26.5°, and it has been tested to remain stable under gravity loading with the given soil parameters, so the failure is mainly induced by the application of the local load. A force of 1.0×10^5n is applied on the 2m×2m square footing at the top and 1m away from the crest of the slope. The red balls are connected by parallel bonds of high strength and high stiffness behave as a rigid footing. An external loading force is imposed to each ball comprising the raft in the negative z-direction vertically. The raft is constrained from movement in the x- and y-directions horizontally by initializing velocities to zero and fixing the x- and y-displacement.

The failure mode of this concave slope is shown in Fig. (**4.27**) the footing presses on the soil mass and a cave is formed at the loaded area at the affecting area after 1000 time-step of loading. In dem modeling, force is applied through contact propagation, and this is different from the physical modeling in the laboratory which is a displacement controlled test. Under force controlled analysis, massive displacement happens at the influenced area after a short time of load application, which can be seen as instantaneous disturbance in Fig. (**4.27b**), while other parts remain unchanged due to slow propagation of external loading through contact mode between discrete elements in particle flow code. Because of this hysteresis phenomenon, the failure process between physical model tests and dem numerical model test is different in some ways.

After an obvious hollow zone is generated partially in affecting area, soils at the two side corners remain stable due to high cohesive strength of slope, which is shown in Fig. (**4.27c**). The particles bond tightly, so that no obvious displacement occurs at the corner areas. Disturbance to the soil around the central axis in middle is evoked extremely, which can be noticed in Fig. (**4.28**) that the average velocity curve in red representing the middle part of the crest has tremendous fluctuation compared with slight movement at the corner side. Finally, it can be seen from Fig. (**4.27d**) that the failure turns to be very local around the loaded area as anticipated. Since only local failure has occurred, there is no obvious shear failure through the toe.

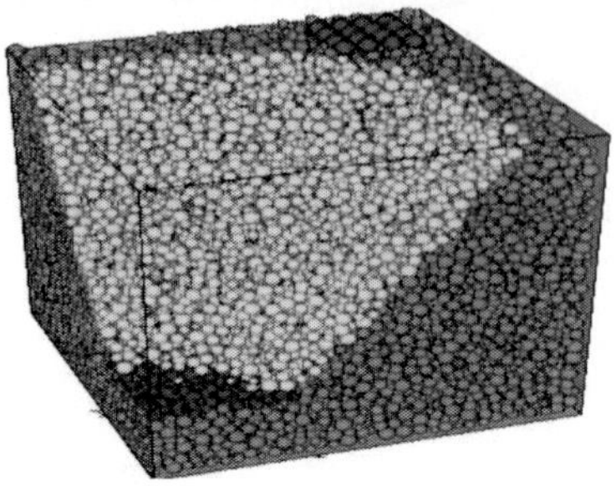

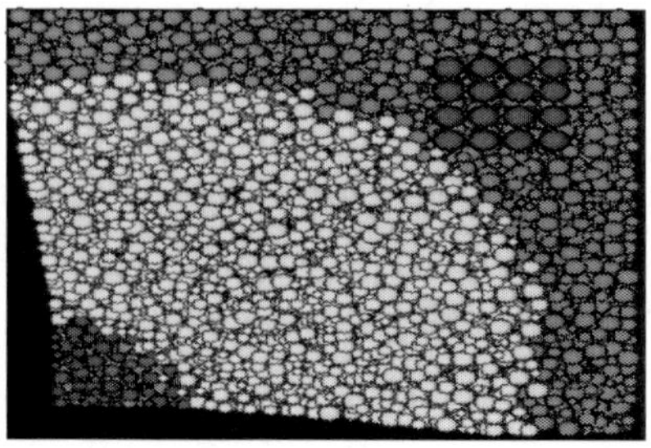

(1) Basic view of slope model in initial state (2) Top view of slope model in X-Y direction

Fig. (4.26). Model of concave slope with local loading impacted by Raft Footing.

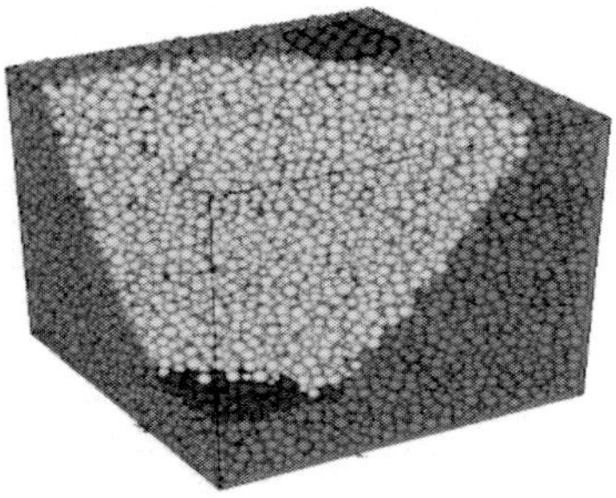

(a) After 600 step

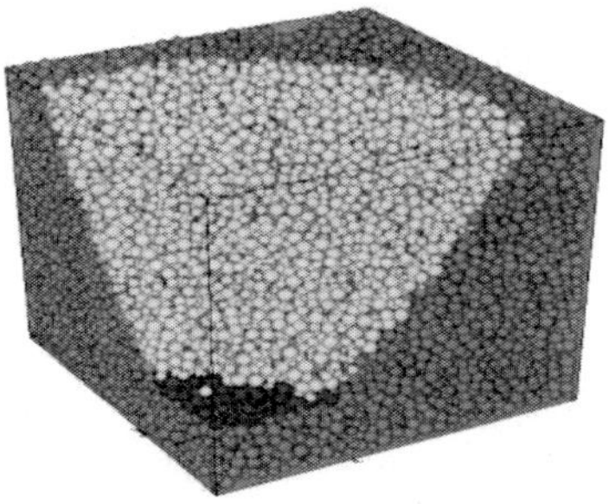

(b) After 1000 step

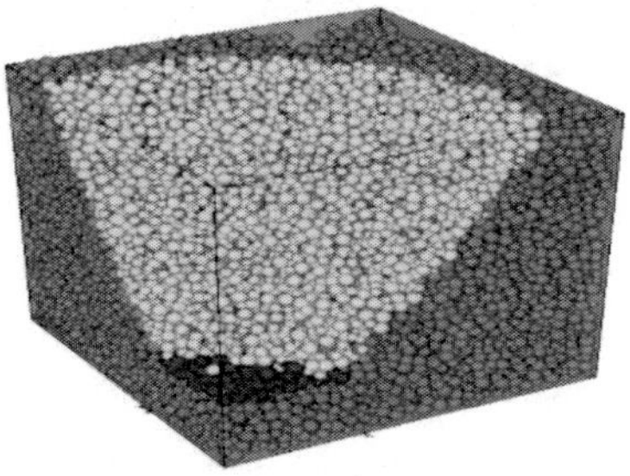

(c) After 6000 step

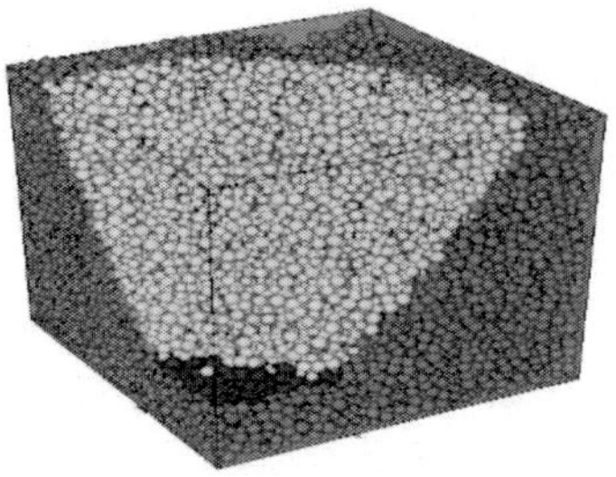

(d) After 11 x10^4 step

Fig. (4.27). Failure process of concave slope with local loading.

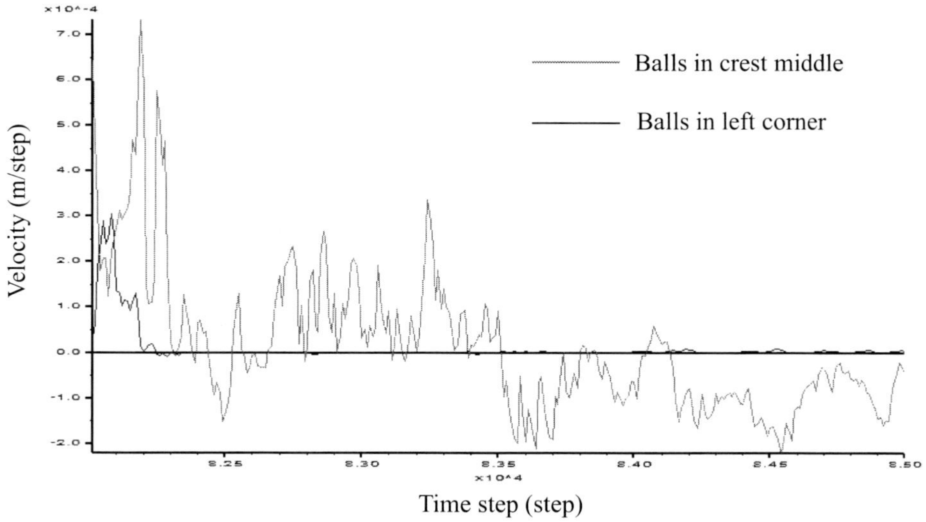

Fig. (4.28). Average velocity history for balls at the left corner and middle part of crest.

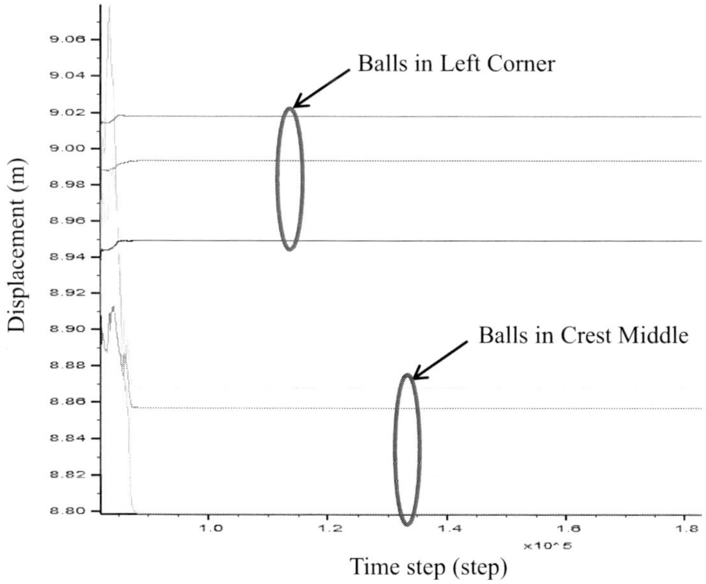

Fig. (4.29). Displacement for balls in the crest left corner *vs* crest middle of concave slope under local loading.

With the external loading applied, the soil particles at the crest middle experience a vertical displacement of about 0.2m in 6000 time-step where this instant model is illustrated in Fig. (**4.27c**). Soil mass at corner side of the slope experience slight upheaval at the same time when the top of the slope is compressed in middle, which can be seen as unchangeable. So distinct difference of two curves is noticed in Fig. (**4.29**). After that, due to the large cohesion strength, soil mass is bonded tightly together as a whole, and slope reaches to equilibrium for the following time.

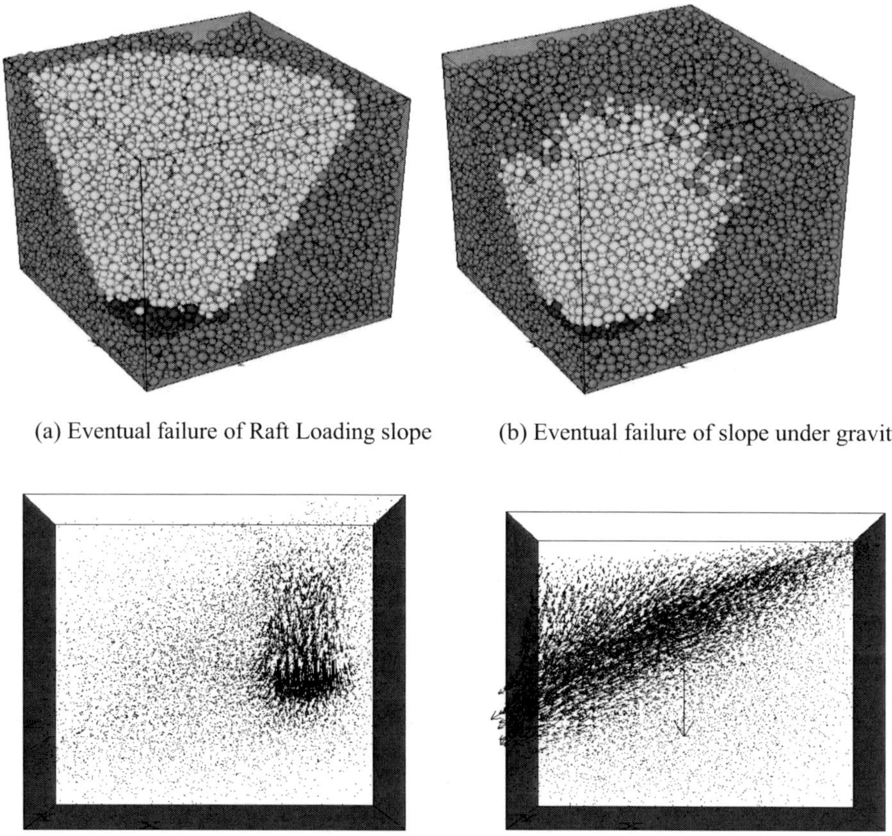

(a) Eventual failure of Raft Loading slope (b) Eventual failure of slope under gravity

(c) Displacement vector of Raft loading slope (d) Displacement vector of slope under gravity

Fig. (4.30). Comparison of concave slope with high and low cohesion strength.

Two basic cases of concave slope is analyzed respectively above, which are slope under gravity with low cohesion strength and slope under raft loading with high bond strength. Comparison of these two cases is made while the eventual failure states and the displacement vectors relations are illustrated in Fig. (**4.30**). From the discussion as mentioned in previous section, the side effect induces the soil particles at the corners of the slope crest move forward faster than the particles at the middle of the crest, while the soil from the corners accumulate at the bottom of slope toe with larger displacement than the soil at the middle as shown in Fig. (**4.30b**) which is the final failure state of the slope under gravity with low cohesion strength, and this result also proves the influence of the arching effect due to the curvature effect. An apparent shear failure surface can be figured out from the displacement vectors results in Fig. (**4.30d**). Meanwhile, an obvious disturbance is imposed by the loading plate for the second case, which is followed by a local collapse for the raft loading slope. There is however no sign of global failure which is noticed in Fig. (**4.30c**) because of the high cohesion strength for the raft loading slope.

For concave slope, due to the influence of the arching effect, only partial/localized failure happens in concave slope under local loading as noticed from (Figs. **4.27** and **4.29**) above. Thrust forces are generated by the soil particles at the two sides, which is the resistance to the slope failure at the middle part. However, different phenomenon occurs in normal plane slope under local loading (with no arching effect) which experiences larger collapse at the middle compared with that at the two sides. An obvious hollow zone appears at the loaded area, and the soil mass near to the disturbance area where the raft footing goes by as well as that in the inclined surface are all driven down and move forward in plane slope, and global collapse is initiated eventually. In the case of plane slope under local loading, soil mass compressed by the loading plate experiences a large settlement of 2m at the middle part of the slope, compared with 1.6m of particle settlement at the left corner as shown in Fig. (**4.31**). This results of settlement are highly different from those in Fig. (**4.29**), but the settlements at crest middle are larger than those at the edges for plane and concave slope. Thus, more attention should be paid to reinforce the area near to the footing in both plane and concave cutting slope.

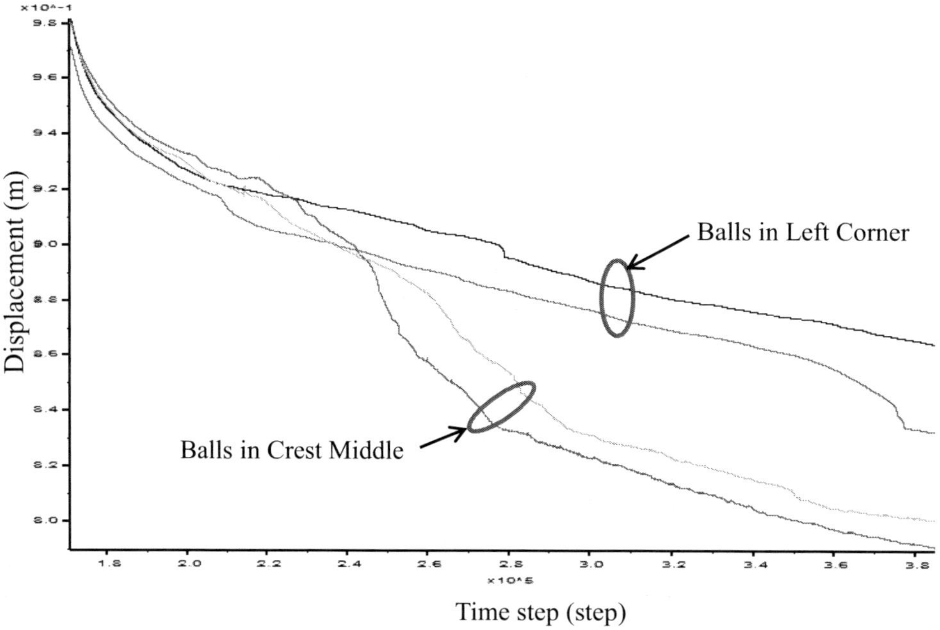

Fig. (4.31). Settlement curves of soil in flat slope between left corner and middle crest.

4.6.2. Laboratory and 3d Dem Analysis of Failure Mechanism of Slope Under External Surcharge

Physical test was carried out in laboratory by the author to investigate the failure process of slope under external local loading. A physical soil tank was built with the layout as shown in Fig. (**4.32**), and the tank was about 1.5m depth x 1.84m wide x 1.2m high. The height of the model slope was 0.7m with a slope angle of 45°. The sectional view of the soil slope model was illustrated in Fig. (**4.33**). Sandy soil with 5% moisture content and particle size distribution as shown in Fig. (**4.34**) was compacted to form a slope model in 7 layers using an electric vibrator. The average bulk density of the soil model was determined as 1672 kg/m^3. The parameters of the river sand used for the physical model were given in (Table **4.3**).

Table 4.3. Shear strength parameters of the river sand.

Cohesion Value (kPa)	Friction Angle (Degree)	Moisture Content (%)	Average Dry Density (kg/m^3)
0.6255	58.61	5.2	1877

Fig. (4.32). Locations of transducers for slope in laboratory test.

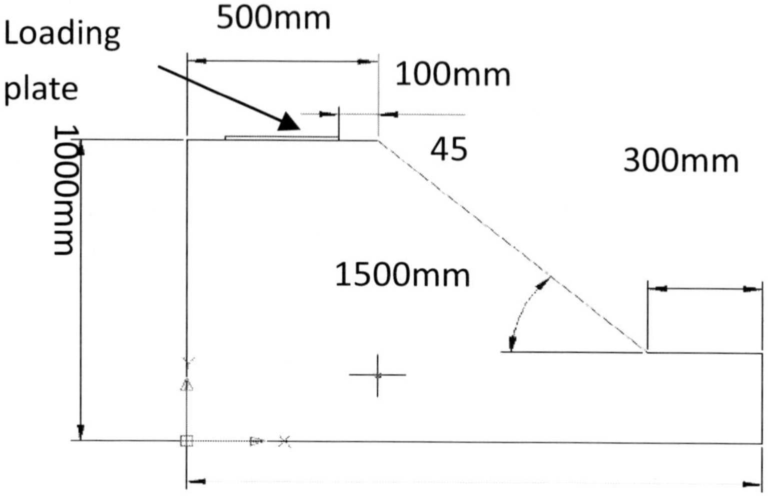

Fig. (4.33). Section view of soil slope model.

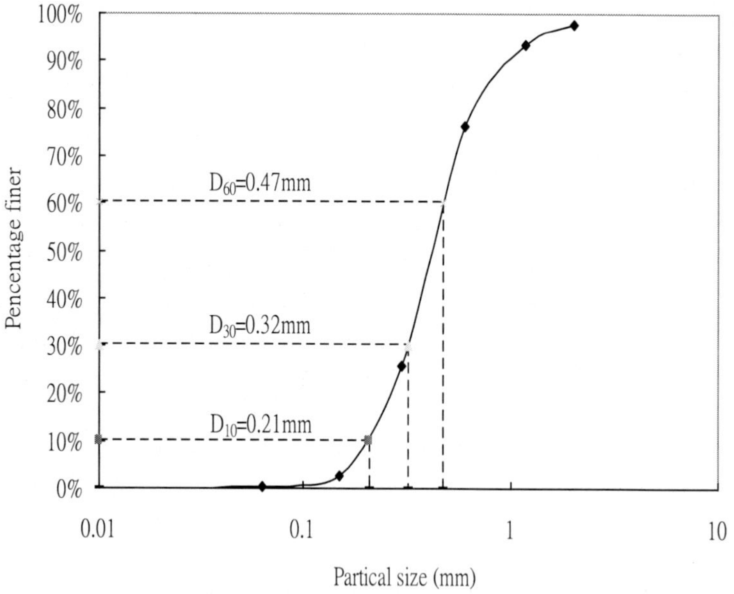

Fig. (4.34). Particle size distribution of the sand in laboratory test.

In this test, the surcharge loading was applied from the hydraulic jack, and the loading was transferred to the steel plate to simulate a local distributed load as shown in Fig. (**4.32**). The loading plate was placed at the center of the slope with 100mm offset away from the slope crest and the back of the frame to reduce the end effect. The loading was applied slowly and the test lasted for about 4 hours with a maximum applied load of 35 kn. Five linear variable differential transducers (lvdt), shown in Figs. (**4.32** and **4.35**), were set up to measure the displacement of soil at different locations which included the upper right (rhs), upper left (lhs), lower right (rhs) and lower left (lhs) on the slope face and at the loading plate on top of the slope. The displacements at different vertical loads are monitored up to failure as shown in Figs. (**4.36** and **4.37**). The 2 pairs of transducers on the slope surface are placed symmetrically with a horizontal spacing of 300 mm. The first and second pairs of transducers are placed at vertical distances of 150 mm and 450 mm from the top of the slope respectively.

Fig. (4.35). LVDT at top and sloping face of the model test.

(a) intermediate process state

(b) sideview of intermediate process state

(c) Final failure state after failure mass

(d) global failure surface after failure mass is removed

(e) Cracks developed at slope surface and crest originated from the loading plate

Fig. (4.36). Failure process of soil slope under increasing loading.

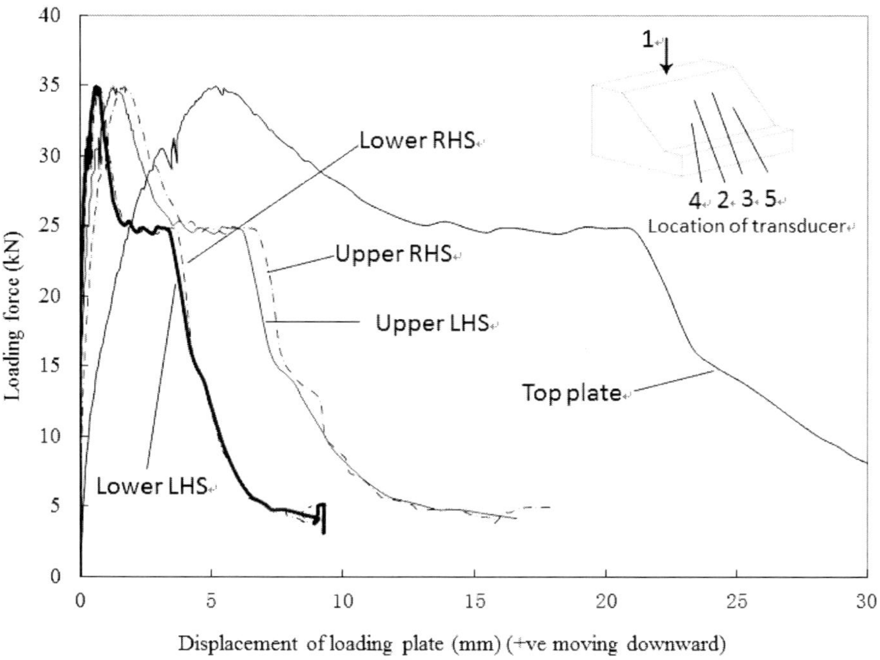

Fig. (4.37). Loading force against the displacement of slope surface.

The intermediate and final failure states of the model slope were illustrated in Fig. (**4.36**), and the results from the five transducers were given in Fig. (**4.37**). From Fig. (**4.36a** and **4.36e**), we could see that cracks developed firstly around the steel loading plate and extended towards each corner side of the slope crest with an angle 45° which were basically in accordance with the classical theory. During the test, flags with different colors were used to locate the time of appearance and the location of cracks during loading which could also be seen in Fig. (**4.36**). From Fig. (**4.36d**), it was noticed that the failure surface was approximately a triangular zone at top plus a log-spiral zone similar to the prandtl mechanism for a bearing capacity problem with an inclined surface below the top triangular zone (cheng and au 2005). The inclination of the failure surface to the horizontal direction in Fig. (**4.36d**) was also close to 74.3° (45°+Φ/2) which was a good illustration of the general shear failure for the present test. The soil mass actually failed gradually instead of a sudden failure in the loading process. As the steel plate kept going down, the inclined slope face was covered with vertical cracks and diagonal cracks which were directed towards the steel plate (see Fig. **4.36e**). The soil mass at the slope surface was drawn down layer by layer to the slope toe by the action of the loading steel plate, and this was the typical face failure phenomenon for cohesionless soil. After four-hours of loading, the sand at the middle top of the slope was highly compressed and a cavity was formed underneath the steel plate which could be noticed after the removal of the steel plate, while the largest cracks were generated at the critical failure surface within the slope body for the global failure (Fig. **4.36d**). The global soil mass was pushed down to the toe of the slope gradually, and the complete physical slope model collapsed eventually which was shown in Fig. (**4.36**).

An ultimate load of 35 kn had been attained at a displacement of about 6 mm as shown in Fig. (**4.37**). For the slope surface, the corresponding displacement at the maximum pressure was about 2 mm and 1 mm at top and bottom of the slope respectively as shown in Fig. (**4.37**), which were much lower than the corresponding displacement at the loading plate. Beyond the peak load, the applied load decreased with increasing jack displacement. From the displacements for the left and right LDTV at upper and lower level, it was clear that the displacements of the slope were basically symmetrical. In Fig. (**4.37**), after the maximum load has been achieved, the loading force decreased with increasing displacement until a displacement of about 13mm for which the load maintained constant for about 9mm. At the beginning of this constant load stage, the local triangular failure zone was fully developed while the failure zones at the two ends of the plate were still not clearly formed. At the end of this constant load stage, the failure zones at the two ends became visually apparent. When the displacements were further increased, the applied load decreased further and the failure zone propagated towards the slope surface until the failure surface as shown in Fig. (**4.36c** and **4.36d**) is obtained, which is the typical global failure as considered in classical stability analysis. At last, the residual load of the test was around 5 kn.

For this test, there were several interesting phenomena worth discussing. The failure profile and cracks first initiated beneath the footing as shown in (Fig. **4.36e**), which was a typical bearing capacity failure with a triangular failure zone. This could also be observed from the upper part of the failure profile as shown in (Fig. **4.36c** and **4.36d**). As the load increased, the failure zone extended and propagated towards the toe of the slope and the final failure surface was shown in (Fig. **4.36d**). It was observed that the failure mechanism of the physical model test is a local triangular failure beneath the bearing plate, and the failure surface propagates towards the slope surface with a curved surface similar to a logspiral curve until a failure mechanism is formed. This type of problem could be considered as a bearing capacity problem as well as a slope stability problem, which was demonstrated to be equivalent by cheng *et al.,* (2013).

To further assess the laboratory test results, the author has adopted the dem in the numerical analysis because the development of cracks, face failure and the final collapse are difficult to be assessed by finite element analysis. In dem, there are several methods of model generation. For the present problem which is relatively simple in geometry and layout, the desired porosity is obtained by the radius expansion method. By using numerical biaxial tests, the micro-mechanical properties of the assembled material in the numerical models are calibrated in order to match with the macroscopic response of the real material in the physical test. Numerical simulations to reproduce the stress-strain and the normal/shear stress relations similar to that by cheng *et al.,* (2009) and cheng *et al.,* (2010) are carried out under the same conditions as the physical experiments such as porosity, boundary conditions and loading. The micro-properties of the river sand as shown in Table **4.4** are determined by varying the micro-properties until the macro-properties obtained numerically match with the experimental results (angle of repose and stress-strain relation). The diameters of the particles in the dem model is kept to be the same as those as given in (Fig. **4.34**). The frictional coefficient of sands is set to 1.638 (corresponding to a friction angle of 58.61° in Table **4.3**). The bond strength is fixed at a value of 6n as referred to cheng *et al.,* (2003). The particle density of the sandy soil is 2650 kg/m^3, while the bulk density for the sand soil is 1650 kg/m^3.

Table 4.4. Microscopic parameters of the sands for particle flow analysis.

Sand	Diameter (mm)	Density of Sphere (kg/m^3)	Friction Coefficient	Normal & Shear Stiffness (N/m^2)	Normal & Shear Bond Strength (N)	Friction Coefficient of the Wall
Tested sand	0.2~0.5	2650	1.638	1e6	6	0

For the three-dimensional dem numerical simulation as shown in (Fig. **4.38**), the dimensions of the numerical model in the particle flow simulation are exactly the same as the physical model. A full scale distinct element analysis is seldom carried out in literature due to the great computer resources requirement, and significant amount of computer time has been used for the analysis in the present work. In this study, two different loading patterns are modeled in order to assess the results of analysis using pfc3d: (1) applying the force on the raft footing, (2) adding velocity on the loading wall, which are illustrated in Fig. (**4.38**). In general, the results from the two methods of simulations are similar, so only the results for simulation 1 are given here.

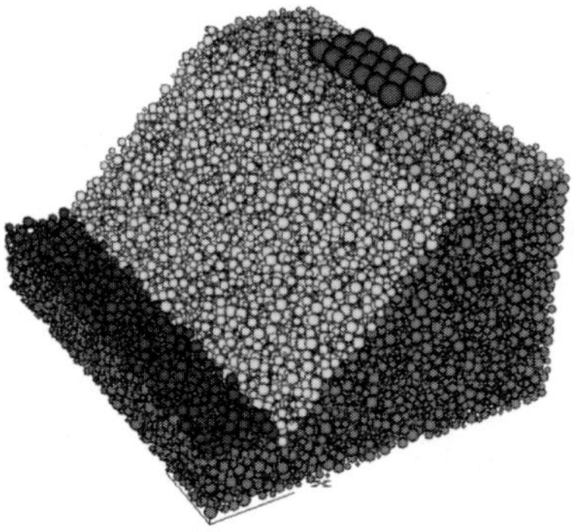

(A)

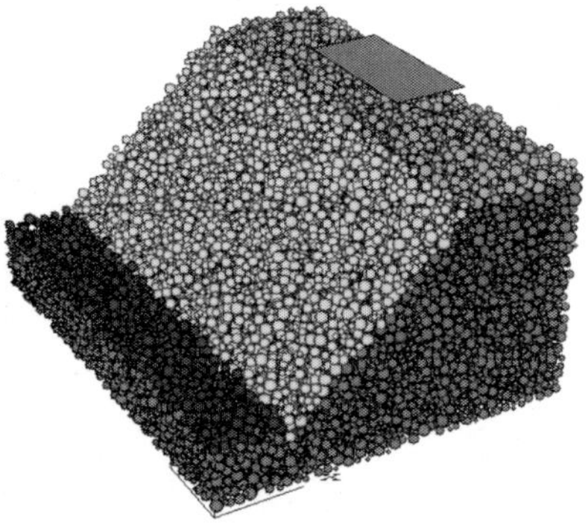

(B)

Fig. (4.38). Two loading patterns of simulation models: (**a**) applying the force on the raft footing; (**b**) adding velocity on the loading wall.

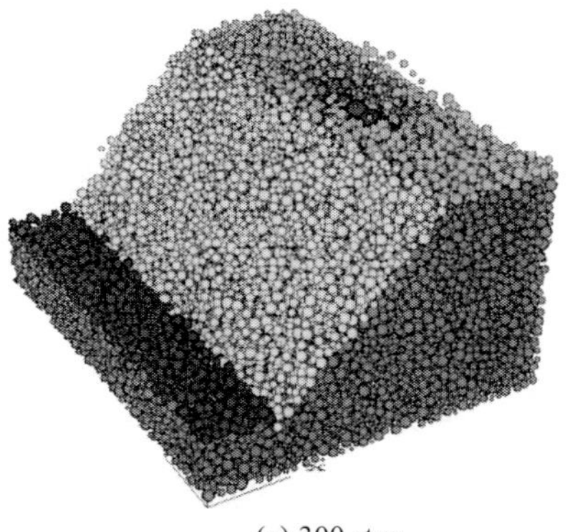

(a) 300 step

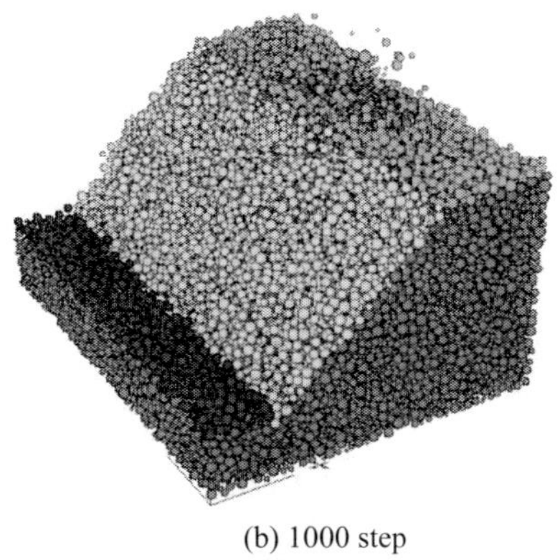

(b) 1000 step

(c) 1.0×10^4 step

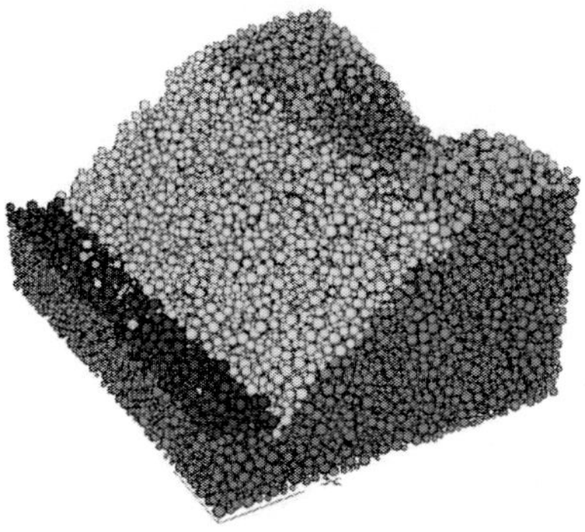

(d) 12×10^4 step

Fig. (4.39). Failure process of numerical slope model under loading raft.

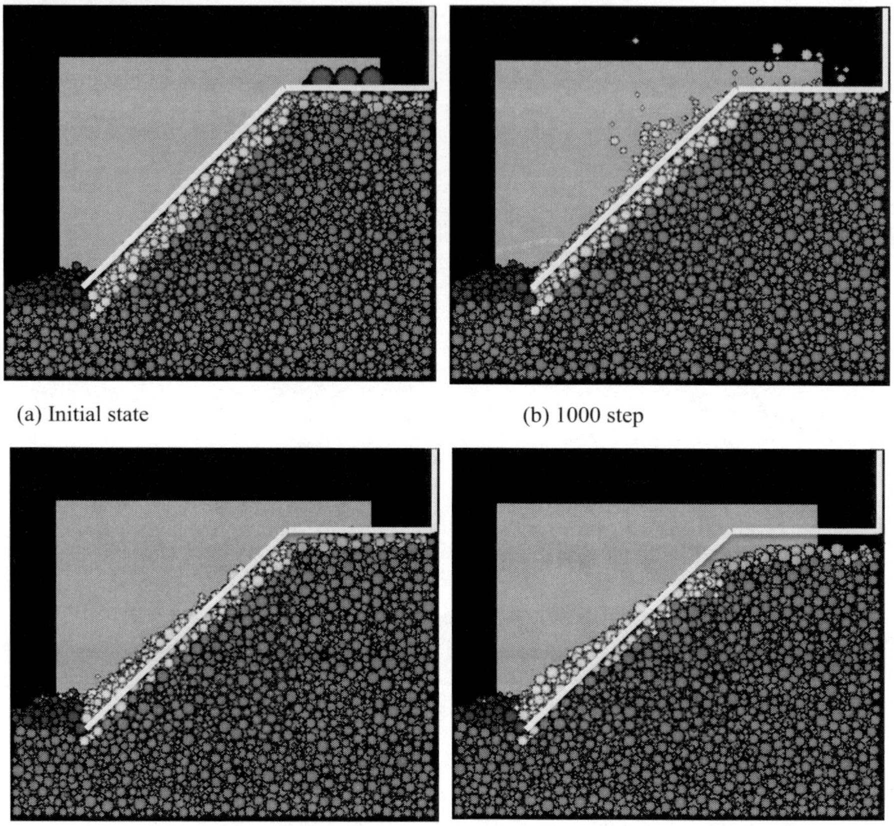

(a) Initial state (b) 1000 step

(c) 1 x 10^4 step (d) 12 x 10^4 step

Fig. (4.40). Failure process of numerical model under loading raft in xy direction.

In the raft footing loading pattern, the footing raft of balls is created by bonding the particle with larger contact bonds, fixing velocity constraints on the balls and modifying their stiffness and friction properties. The raft footing is added to the top of the model, and the footing load is applied in the vertical direction (x-direction) with a magnitude of 3.5 x 10^4n as shown in Fig. (**4.38a**). The parameters of the sands are the same as that in Table **4.4**, where both normal and shear contact bond strength (n_bond, s_bond) of the particles are 6n. The numerical failure process development under loading is illustrated in Fig. (**4.39**) and 4.40. It is noticed that on top of the slope crest, the region below the loaded surface has deformed to form a depression zone from the dem modeling which is also observed from the test as shown in Fig. (**4.39e**). Sand particles are triggered to move down the slope, dragging more and more sands downward. The depression zone develops larger and deeper, accompanying with considerable settlement at the top of the slope, and the inclined slope face moves forward with an upheaval at the slope toe as shown from the xy direction view in Fig. (**4.40(4)**).

Slope with soil particles under higher bond strength is also simulated, where the normal and shear contact bond strength of particles value are increased to 60n, which is ten times larger than that in the former case. The results of the comparison are shown in Fig. (**4.41**). It is observed that noticeable collapse has taken place in case 1 with forward movement of slope body and extruding slope toe. On the contrary, the slope remains stable if the bond strength is 60n under the loading of 500n as shown in Fig. (**4.41b**). For applied load equal to 2000n, both cases of slope experience failure, as the bond strength is completely destroyed by the applied load. It is demonstrated that larger bond strength between soils followed by slower and smaller failure can withstand relatively larger load, so that the soil is more stable under such external load.

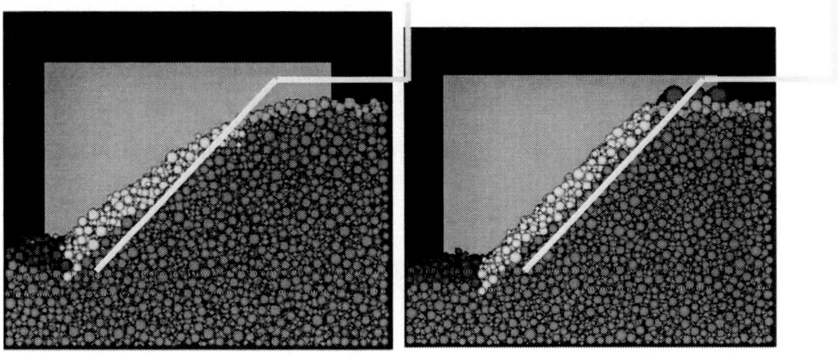

(a) Case 1: bond strength = 6N (b) Case 2: bond strength = 60N

Fig. (4.41). Eventual failure of two modeling cases under loading raft in xy direction (with external load 500n).

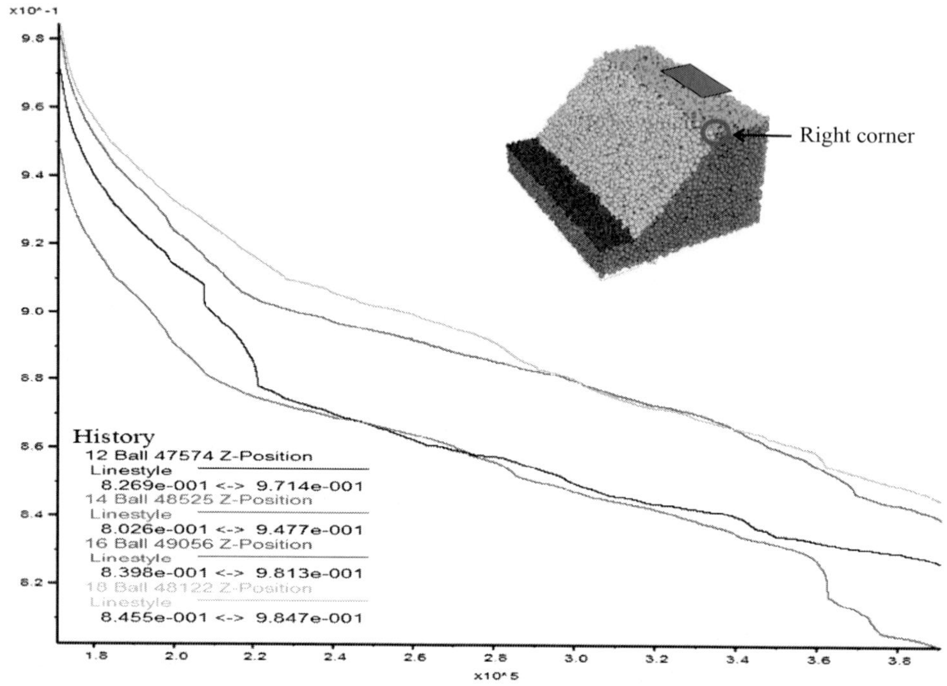

Fig. (4.42). Vertical position history for balls at the right corner of crest.

Soil particles at the left corner, right corner and middle part of the slope crest are selected to monitor the position and velocity change in z-direction (vertical direction) respectively, which are illustrated in Figs. (**4.42-43**). For the right corner of the slope crest as shown in Fig. (**4.42**), the sand at the crest moved down with a high velocity at the beginning, so the displacement increased at a rapid rate within $2x10^5$ time steps. Such a high initial velocity is normal even for laboratory test as the plate is not fully in contact with the soil at the beginning of the test. A small amount of plate movement is required so that the plate is in full contact with all the soil particles at the top of the slope. The increase of the vertical displacement became smooth afterwards with several fluctuations points (typical phenomenon from dem calculation). For the left corner of the slope crest, the sand experience the same trend as that for the right hand side. In theory, the problem is symmetric as the loading plate is located at the center. In microscopic view, there will be minor deviation from symmetry as it is virtually impossible to keep all the soil particles to be truly symmetrical in the generation of the model or for the actual physical model. Such minor deviation from symmetry will however become negligible as the movement of the plate increases. For the slope crest corner as shown in Fig. (**4.42**), it can be noticed that the particle moved down firstly at the corner of the crest where crack is generated at the very beginning of the slope failure process. With further increase in the displacement of the loading plate and when a critical failure surface is generated, an instant collapse is observed from the great fluctuation of these history records.

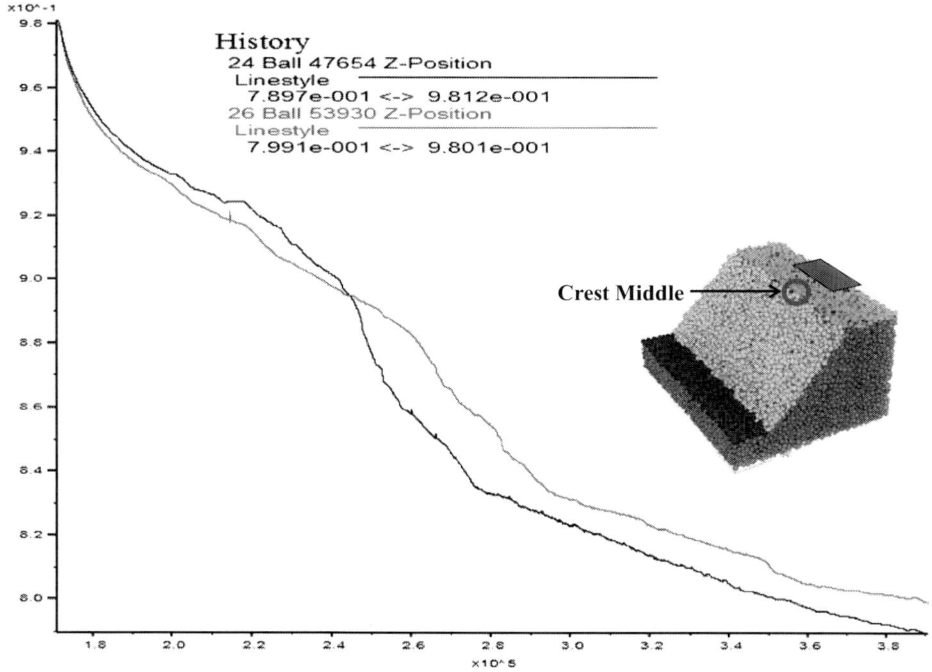

Fig. (4.43). Vertical position history for the balls at the crest middle.

(a) Laboratory test - slope without soil-nail (b) DEM Case1 - bond strength = 6N (c) DEM displacement

Fig. (4.44). Front view of final failure from laboratory test and DEM modeling.

(a) Laboratory test of slope (b) DEM displacement vector with failure line

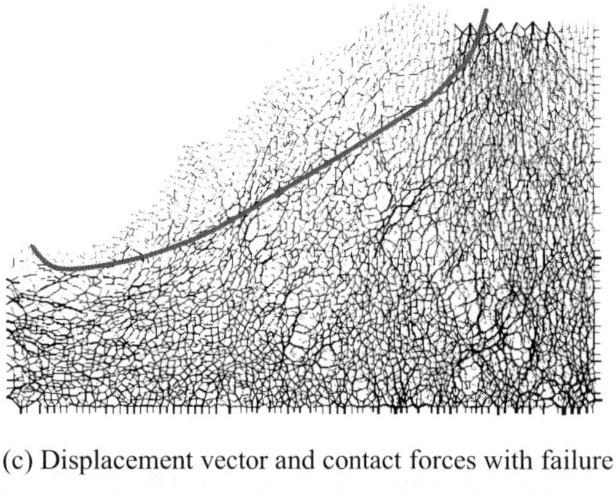

(c) Displacement vector and contact forces with failure line

Fig. (4.45). Side view for the final failure from laboratory test and DEM modelling.

At the middle of the crest of the slope as shown in Fig. (**4.43**), the situation is different in that a more rapid and less moderate trend can be noticed. The displacement reaches to more than 0.2m at the middle of the crest compared with a displacement of almost 0.17m at the crest corner. However, the particles at the corners move faster than those in the middle in the beginning, which are demonstrated by the sharp increase of the settlement in Fig. (**4.42**). Fig. (**4.43**) shows that the loading at the middle of the slope top has pushed the sands at the middle of the slope crest downward and accumulated at the toe of the slope. The front and side view for the final failure of the laboratory model test and dem modeling are shown in Figs. (**4.44** and **4.45**). A steeply inclined failure surface (triangular failure zone underneath the loading plate) which is followed by a curved failure surface can be observed from Fig. (**4.45b** and **4.45c**). The failure mechanisms are marked by the equivalent slip surface similar to that by lem and srm. The slip surface is determined from the velocity/displacement for all the particles. It is noticed that there is a narrow band of particles where the movement is significantly more than the adjoining particles, and the locus of this band is drawn as the slip surface.

The failure profile of the physical test matched reasonably well with that as predicted from the present simulation as shown in Fig. (**4.44** and **4.45a**). At the beginning of the laboratory test, cracks firstly appear along each crest corner in the laboratory test which is the effect due to the side plate of the model. In the particle flow simulation, the loading plate is applied with a small velocity of $5x10^{-5}$ m/step. Besides the triangular failure zone underneath the loading plate, there is an obvious failure zone at the middle of the slope top and the soil moved downward and accumulated at the bottom of the slope which was also observed from the physical test as well as the dem modeling. Based on the observation and comparison above, it can be concluded that dem can give a very good qualitative description about the complete failure process which was not possible with strength reduction method (srm) or limit equilibrium method (lem).

The relations between the applied load and the displacement curves of the slope at different measuring points from laboratory and numerical results are illustrated in Fig. (**4.46**). For the right hand side, three groups of curves of dem results were generated from the measuring points of lower position, upper position of slope surface and middle crest respectively. For the soils at the lower position of the slope surface, there is a rapid increase in the loading applied before the displacement has reached 10mm, which can be noticed in green and light blue curves in the figure. The applied load reaches the peak value of 37kn at about 10mm settlement, and then the loading firstly drops to around 25kn, followed by the second drop down to around 10kn till eventual failure. For the soil at the upper position of the slope surface, the trend is similar to that of lower surface points, but the settlement largely develops to approximately 14mm under maximum loading. For the soil at the middle of the slope crest, the maximum load is achieved at larger displacement of 26mm as compared with the slope surface measuring points, and soil particles experience longer-time settlement with significant displacement value in the whole failure process.

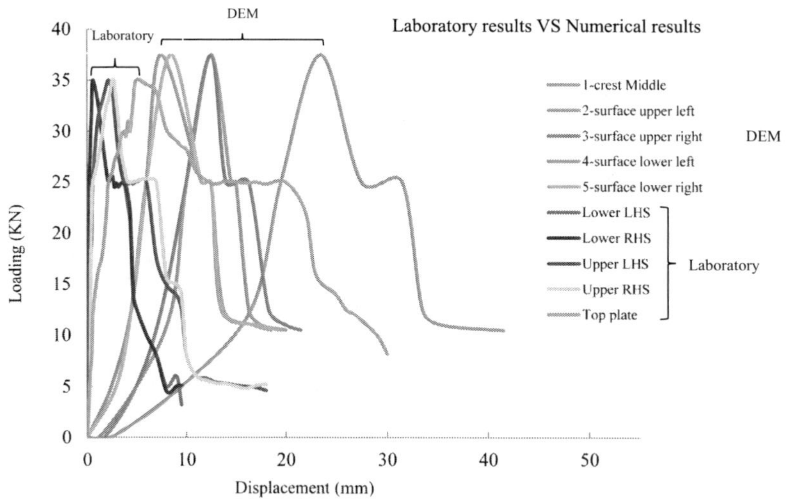

Fig. (4.46). Loading force against displacement curves of the slope at different measuring points from laboratory and numerical results.

In conclusion, the whole slope undergoes two major failures/stress redistribution in the test and numerical modeling when the bearing loading is 25kn at three-quarter of the displacement as well as 10kn at the end of the test. In these two conditions, loading force remains unchanged while displacement increases, and the reason is that the slope reaches critical point and the failure surface is generated with major stress redistribution taken place within the soil mass. The settlement at the lower parts of the slope surface appears earlier than that at the crest, however, the maximum final settlement of 40mm is located at the slope middle crest which is underneath the loading plate and develops larger than the lower positions at the slope surface. In comparison with the laboratory test results in three groups of curves from the left hand side of Fig. (**4.46**), also as shown in Fig. (**4.37**), the trend of three curves are qualitatively similar, and the final failure loading of 35kn from the laboratory test was also close to the value 37kn from the dem analysis. Dem can hence give a relative good qualitative assessment about the slope failure process up to the initiation of the failures.

For the post-failure results, the matching between the DEM model and the laboratory test is less satisfactory quantitatively. Gardiner and Tordesillas (2005) have given some discussion about the treatment for the loss of contact which is not considered in the present study due to the lack of suitable material parameters. In theory, better matching may be achieved by choosing other types of contact rule or other micro-parameters, and the micro-parameters may even be changed after the maximum load has been reached. The authors have not carried out such numerical calibration for this study, and such post-failure parameters/contact rules are difficult to be assessed unless the corresponding laboratory tests are carried out which are practically not possible for normal engineering problems. Even though the matching is not good quantitatively, the load displacement relation from DEM analysis and the model test results are similar qualitatively. Based on the comparison as mentioned above, the results of numerical modeling are considered as effective and acceptable for engineering application.

CONCLUSION

Out of the various numerical methods, the finite element and distinct element methods are the most important in geotechnical engineering. Unlike the upper and lower approaches as illustrated in chapter 2, these methods can provide the complete stress/strain relation as well as the ultimate condition at the expense of much greater computations. While the finite elements are now commonly adopted by the engineers for the solution of variety of engineering problems, the distinct element method is still limited to research purpose at the moment. It is also unlikely that the distinct element method will be adopted for routine application within the near future, nevertheless, distinct element analysis does provide some new insight about the large displacement condition which are sometimes useful but is not available from the finite element analysis. The results as given in this chapter have clearly illustrated the advantages of the distinct element method, however, the limitations of this method must also be considered by the engineers and researchers.

ACKNOWLEDGEMENTS

Declared none.

REFERENCES

Barber J.R., Elasticity, London: Springer, 2010. Available at: http://www.springer.com/us/book/9789048138081.

Chang M. (2002), A 3D slope stability analysis method assuming parallel lines of intersection and differential straining of block contacts, Canadian Geotechnical Journal, 39(4), 799–811.

Chen R.H., Chameau J.L. (1982), Three-dimensional limit equilibrium analysis of slopes, Geotechnique, 32(1), 31 –40.

Cheng Y. P., Nakata Y. and Bolton M. D. (2003), Distinct element simulation of crushable soil. Geotechnique 53, No. 7, 633–641.

Cheng Y.M., Liu H.T., Wei W.B. and Au S.K. (2005), Location of critical three- dimensional non-spherical failure surface by NURBS functions and ellipsoid with applications to highway slopes, Computers and Geotechnics, 32(6), 387–399.

Cheng Y.M. and Yip C.J. (2007), Three-dimensional asymmetrical slope stability analysis – Extension of Bishop's, Janbu's, and Morgenstern–Price's techniques, Journal of Geotechnical and Geoenvironmental Engineering, 133(12), 1544–1555.

Cheng Y.M., Liu Z. N., Song W.D., Au S.K. (2009), Laboratory test and particle flow simulation of silos problem with nonhomogeneous materials, Journal of Geotechnical and Geoenvironmental Engineering, ASCE, 135:1754-1761.

Cheng Y.M., Chau K.T., Xiao L.J. and Li N. (2010), Flow pattern for silo with two layers of materials with single and double openings, Journal of Geotechnical and Geoenvironmental Engineering, ASCE, 136(9):1278:1286.

Cheng Y.M. and Lau C.K. (2014), Soil Slope stability analysis and stabilization – new methods and insights, 2nd edition, Spon Press.

Crouch S.L. and Starfield A.M. (1983), Boundary element methods in solid mechanics, S. L. Crouch and A. M. Starfield, George Allen & Unwin, London.

Cundall P. A. (1971), A computer model for simulating progressive, large-scale movements in blocky rock systems. In: Proceedings of the International Symposium on Rock Mechanics. Nancy, France: [s.n.], 1971: 129–136.

Cundall P.A. (1978), Ball – A computer program to model granular medium using the distinct element method, Technical note TN-LN-13, Advanced Technology Group, Dames and Moore , London, 129-163.

Cundall P.A. and Strack O.D.L. (1979), A discrete model for granular assemblies, Geotechnique 29(1):47-65.

Cundall P.A. (1980) UDEC-A generalized distinct element program for modelling jointed rock, Final Tech. Rep. Eur. Res. Office (US Army Contract DAJA37-79-C-0548).

Cundall P.A., Hart R.D. (1985) Development of generalized 2-D and 3-D distinct element programs for modeling jointed rock, Misc. Paper SL-85-1, US Army Corps of Engineers.

Cundall, P.A., (1987), Distinct Element Models of Rock and Soil structure, in Analytical and Computational Methods in Engineering Rock Mechanics, Brown, E.T. ed., London, George Allen and Unwin, 129-163.

Cundall P.A., Hart D.H. (1992) Numerical modelling of discontinua, Eng. Comput.9:101-11.

Goodman, R. E. (1976), Methods of geological engineering in discontinuous rocks. West Publishing Company, San Francisco, CA, USA.

Hovland H.J. (1977), Three-Dimensional Slope Stability analysis method, Journal of the Geotechnical Engineering Division, ASCE, 103, 971-986.

Huang C.C. and Tsai C.C. (2000), New method for 3D and asymmetrical slope stability analysis, Journal of Geotechnical and Geoenvironmental Engineering, ASCE, 126(10), 917–927.

Iwashita K. and Oda M. (1998), Rolling resistance at contacts in simulation of shear band development by DEM. ASCE J Eng Mech 124(3):285–292

Jing L.R., Stephansson O. and Erling N. (1993), Study of rock joints under cyclic loading conditions, Rock Mechanics and Rock Engineering, Vol. 26, No. 3, 1993, p. 215-232

Jiang, M.J., Yu, H.S., and Harris, D., 2005, A novel discrete model for granular material incorporating rolling resistance: Computers and Geotechnics, v. 32, p. 340-357

Hertz H. (1881), Ueber die Berührung fester elastischer Körper. J. Reine Angew. Math. 92, 156–171.

Huang C.C., Tsai C.C. and Chen Y.H. (2002), Generalized method for three- dimensional slope stability analysis, Journal of Geotechnical and Geoenvironmental Engineering, ASCE, 128(10), 836–848

Lam L. and Fredlund D.G. (1993), A general limit equilibrium model for three-dimensional slope stability analysis, Canadian Geotechnical Journal, 30(6), 905–919.

Itasca Consulting Group Inc. (1995), PFC3D 1.0, User Guide.

Itasca Consulting Group Inc. (2004), PFC3D 3.1, User Guide.

Rao S.S., The finite element method in engineering, London: Butterworth Heinemann, 2011.

Rassam, D.W., Williams, D.J. (1999). 3-dimensinal effects on slope stability of high waste rock dumps. International Journal of Surface Mining, 13:19-24.

Ugai K. (1988), 3-D slope stability analysis by slice methods, Proceedings of the 6th International Conference on Numerical Methods in Geomechanics, Innsbruck, Austria, pp. 1369–1374.

Ugai K. and Leshchinsky D. (1995), Three-dimensional limit equilibrium and finite element analysis: a comparison of results. Soils and Foundations, 35(4), 1–7.

Zhang X. 1988. Three-dimensional stability analysis of concave slopes in plan view. Journal of Geotechnical Engineering 114(6): 658-671

Zienkiewicz O.C., Taylor R.L., Zhu J.Z. (2011), The finite element method: Its basis and fundamentals, 6th edition, London: Butterworth Heinemann.

CHAPTER 5

Numerical Techniques for Solution of Stability Problems

Abstract: Analytical solutions to most of the geotechnical problems are not available unless the geometry under consideration is highly simplified. For practical purposes, numerical method is indispensable. In this chapter, the numerical methods for some of the stability methods as discussed in previous chapters are elaborated.

Keywords: Discretization, Limit analysis, Limit equilibrium, Optimisation, Work done.

In the past few decades, great strides have been made in the computational methods of stability analysis. This chapter briefly discusses three main numerical-based methods for performing geotechnical stability analysis, namely:

(a) limit equilibrium,
(b) the displacement finite element method and
(c) lower- and upper-bound finite element limit analysis

5.1. LIMIT EQUILIBRIUM

Generally speaking, limit equilibrium methods assume that failure occurs across a predefined slip surface. It is a commonly used method for stability analysis in geotechnical context due to its simplicity, but the method may also be inadequate to analyse complex failure mechanisms not well defined by a failure surface (for instance progressive creep, soil liquefaction and brittle fracture). Sloan (2013) summarizes some shortcomings of the limit equilibrium technique as follows:

Y. M. Cheng, H. Wong, C. J. Leo & C. K. Lau

(a) The resulting stresses do not satisfy equilibrium at every point in the domain.
(b) There is no simple means of checking the accuracy of the solution.
(c) It is hard to incorporate anisotropy and inhomogeneity.
(d) It is difficult to generalise the procedure from two to three dimensions.

One of the most common applications of the limit equilibrium method (LEM) in geotechnical engineering is for slope stability analysis. Several limit equilibrium methods have been developed for slope stability analysis and many are covered in standard geotechnical textbooks. These include the widely used ones: Bishop and Simplified Bishop method of slices (1955), Ordinary or Fellenius method of slices (1936), Spencer, Janbu (1973), and Morgenstern and Price (1965) methods. Broadly speaking, all limit equilibrium methods presuppose a failure surface and define the factor of safety, *F*, as the factor by which the shear strength of the soil would have to be factored down to bring the slope into a state of limiting equilibrium, that is to the limit of stability so that it is on the verge of failure. Thus, the factor of safety is given as,

$$F = \frac{\text{shear strength of soil}}{\text{shear stress at limiting equilibrium}} = \frac{\tau_{\text{ult}}}{\tau} \quad \textbf{(5.1)}$$

that is,

$$\tau = \frac{\tau_{\text{ult}}}{F} = \frac{c+\sigma_n \tan\phi}{F} \quad \textbf{(5.2)}$$

where c is the cohesion, ϕ is the internal friction angle and σ_n the normal stress. It is customary to apply a common "F" for both c and ϕ although it is not necessary the case. As the number of equations in limit equilibrium is less than the number of unknowns in slope analysis, the problem is in fact statically indeterminate. Various assumptions are then made in the various methods to render the problem determinate. Fellenius and Simplified Bishop methods make assumptions regarding the horizontal or vertical force equilibrium, while Bishop, Spencer, Janbu, and Morgenstern and Price methods satisfy all conditions of equilibrium.

As Duncan (1996) pointed out, the advent of computers has dramatically changed the computational approach for analysing slope stability. One of the most consequential changes is in respect of numerical optimisation techniques for determining the critical slip surface, as discussed below.

5.1.1. Critical Slip Surface

A fundamental part of slope stability analysis is to determine the slip surface that has the lowest, therefore the most critical, slip surface be it a circular or a non-circular surface. The trial slip surface is generally a pre-defined shape consisting of straight line segments or smooth curve or both. Search methods employing variational calculus (Baker and Gaber, 1978), dynamic programming (Baker, 1980), alternating variable methods (Celestino and Duncan, 1981), Monte Carlo technique (Greco, 1996) and genetic algorithm (Goh, 1999) are then used to identify the critical surface. The problem can be posed as illustrated in the following example (*e.g.* Sun *et al.,* 2008; Malkawi, 2001).

Fig. (**5.1**) shows a trial slip surface represented by *n* nodal points $(x_1, y_1), (x_2, y_2), \cdots (x_n, y_n)$ and *n*-1 segments. For sake of minimizing the number of variables, the nodal points are constrained to the same spacing along the abscissa so that,

$$x_i - x_{i-1} = \Delta x = \frac{1}{(n-1)}(x_n - x_1) \qquad i = 2,\cdots, n-1 \qquad \textbf{(5.3)}$$

and the coordinates of the nodal point *1* and *n* on the slip surface can also be defined using the surface profile $y = h_0(x)$ (see Fig. **5.1**), namely:

$$y_1 = h_0(x_1); \qquad y_n = h_0(x_n) \qquad \textbf{(5.4)}$$

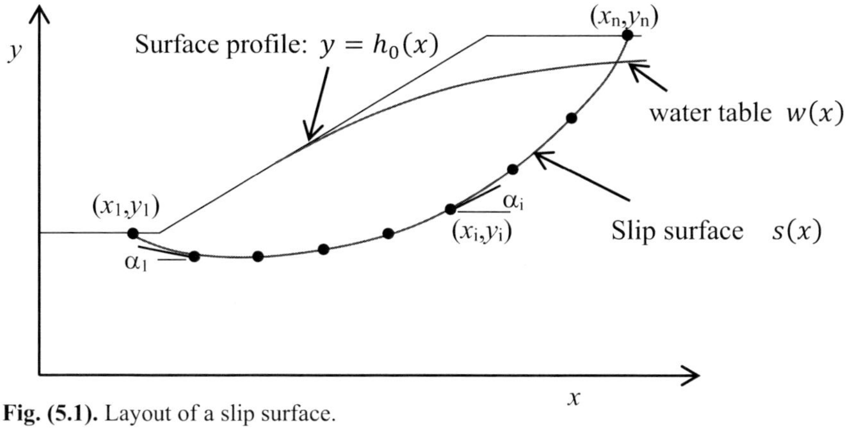

Fig. (5.1). Layout of a slip surface.

Noting that once x_1, x_n , are defined, then y_1, y_n, the abscissas $x_2, x_2, \cdots x_{n-1}$are also defined as a result of equations (5.3) and (5.4). Thus the trial slip surface $s(x)$ can now be mathematically represented by an *n*-dimensional array (Sun *et al.*, 2008),

$$S = [x_1, y_2, y_3, \cdots, y_{n-1}, x_n]^T \tag{5.5}$$

The objective function for locating the critical slip surface, which has the minimum factor of safety can be stated as,

$$\min F(S) \tag{5.5}$$

subject to the following constraints:

$$0 < s(x_i) < h_0(x_i) \tag{5.6}$$

$$x_i - x_{i-1} = \Delta x = \frac{1}{(n-1)}(x_n - x_1) \qquad i = 2,\cdots,n-1 \tag{5.7}$$

In order to generate a kinematically admissible slip surface (Malkawi *et al.*, 2001), it is assumed to concave upwards, such that:

$$\alpha_1 < \alpha_2 < \alpha_3 \cdots\cdots \alpha_i \cdots < \alpha_{n-1} \quad \textbf{(5.8)}$$

where α_i is the inclination of segment *I* (see Fig. **5.1**).

5.2. DISPLACEMENT FINITE-ELEMENT ANALYSIS

Finite element analysis has advanced significantly both in terms of computing resources and computational techniques. At the very heart of present geotechnical finite element programs is the availability of constitutive models which have been developed to deal with complex behaviour of geomaterials in response to loading and material inhomogeneity. Constitutive models and numerical techniques have been introduced to capture nonlinearity, visco-elastoplasticity, work hardening and softening, creep of material in context of coupled processes (thermo-mechanical, poro-mechanical, thermo-poro-mechanical), associative and non-associative flow rules, small and large deformations, soil-structure interaction and dynamics. Conventional displacement based finite element is typically employed in two ways to analyse stability under static loading (*e.g.* Sloan, 2013),

(a) One approach is the so-called strength reduction method (SRM) where the shear strength of the materials is reduced successively by the strength reduction factor (SRF) until equilibrium can no longer be achieved, which is indicated by the deformations at the control points becoming unacceptably large or the finite element solutions do not converge (*e.g.* Zienkiewicz *et al.*, 1975; Dawson *et al.*, 1999; Griffiths & Lane, 1999). The factor of safety obtained in this way is in terms of the strength of the materials. Commercial finite element codes have widely adopted this approach, serving as a powerful alternative to traditional limit equilibrium methods. Cheng *et al.*, (2007) summarised the main advantages of the SRM as follows: (i) the critical failure surface is found automatically from the shear strain arising from the application of gravity loads and the reduction of shear strength; (ii) it requires no assumption on the interslice shear force distribution; (iii) it is applicable to many complex conditions and can give information such as stresses, movements, and pore pressures which are not possible with the LEM. One of the main disadvantages of the SRM is the long solution time required to set up the computer model and to perform the analysis. However, with the development of computer hardware and software, the SRM can now be performed within a reasonable time span suitable for routine analysis and design. Other limitations of the SRM include the choice of an appropriate constitutive model and parameters, boundary conditions and the definition of the failure condition/failure surface.

(b) The second approach is to apply loading incrementally until the displacement field indicate imminent collapse and obtain a factor of safety in terms of loading rather than strength (*e.g.* Sloan, 1979, 1981; Toh & Sloan, 1980; Sloan & Randolph, 1982; De Borst & Vermeer, 1984). This approach produces instability as collapse nears if loading is by prescribing forces instead of displacements. Special numerical techniques to treat parametrized nonlinear algebraic equations of the finite element method are required to deal with the instability especially if prescribed forces loading were employed. Broadly speaking two types of incremental solution techniques are used:

  (i) Purely incremental methods, also known as predictor-only methods

  (ii) Corrective methods, also known as predictor-corrector methods

Nonlinear finite element analysis which arises in solution of stability problems is much more difficult to deal with numerically than the linear case.

Determinate linear system of equations has a unique solution and the computational effort to solve the system can be determined fairly accurately. A nonlinear system on the other hand has possibly a significant number of solutions, of which only the ones with physical meaning should be the main focus without bothering with the others. Underlying this thinking is the concept of incremental analysis as mentioned above, which is used to trace the behaviour of the system by advancing small increments from an initial starting point. Each incremental solution is used as a starting point for the next solution search.

Nonlinearities are tagged as "smooth" and "rough". Smooth nonlinearities are characterised by continuous, path-independent nonlinear relations at the local level, for instance:

(i) *Nonlinear elasticity relationships*. Stresses are nonlinear but reversible functions of strains.
(ii) *Well-behaved nonlinear effects* arising from strain-displacement compatibility relations.
(iii) *External forces* are smooth nonlinear functions of displacements.

Initially the incremental analysis for these problems was dominated by purely incremental methods. However the concern for consistent linearization has prompted investigators to propose rigorous forms of the "residual" out of-balance forces and tangent stiffness matrix, leading to implementation of incremental steps with corrective iterations. Conventional and modified Newton methods were used for correction. These were further extended through restricted step (safeguarded Newton) and, variants of the conjugate-gradient and quasi-Newton methods. Oden (2006) presented a comprehensive review of this early work.

Solution techniques to deal with smooth nonlinearities are in a fairly satisfactory state. Incremental-iterative methods implemented with incremental control strategies (*e.g.* state, arclength, global hyperelliptic and local hyperelliptic) are already employed to enhance robustness and accuracy, although further refinements in the area of traversing bifurcation points can be expected. For rough nonlinearities, though, case-by-case handling is still necessary in view of the lack of general theories and implementation procedures. Separation or nesting of nonlinearities, when applicable, can lead to significant gains in efficiency and reliability, but at the cost of programming complexity and problem-dependent implementations.

5.3. FINITE ELEMENT LIMIT ANALYSIS

Formulation and development of finite element methods within limit analysis framework have increasingly gained traction in recent years (Sloan, 2013). Finite element limit analysis provides rigorous lower and upper bound solutions on the factor of safety of stability problems (*e.g.* Sloan, 1988, 1989; Sloan & Kleeman, 1995; Lyamin & Sloan, 2002a, 2002b; Krabbenhoft *et al.*, 2005). The lower bound solution is derived from optimisation of an objective function of statically admissible stress field, while the corresponding upper bound solution is deduced from kinematically admissible velocity field. It has the usual advantages of finite element methods in being able to successfully deal with complexities in geometry, material layering and inhomogeneity, loading and boundary conditions, in addition to discontinuities and interfaces. Finite-element limit analysis is particularly powerful when both upper- and lower-bound estimates are calculated so that the true collapse load (for the idealised material) is bracketed from above and below. The difference between the two bounds then provides an exact measure of the error in the solution, and can be used to refine the meshes until a suitably accurate estimate of the collapse load is found. A brief description of the concepts of lower- and upper-bound approaches for two dimensional cases is given below.

5.4. LOWER BOUND FINITE ELEMENTS

Seminal work was carried out by Lysmer (1970) and Sloan (1988) on the finite element estimation of lower bound solutions. The following is used to illustrate the technique used by Sloan (1988, 2013) to formulate the lower bound problem based on linear elements. Lyamin and Sloan (2002a) discussed why there are compelling reasons for choosing linear finite elements and not higher order finite elements for lower bound computations. Here, a 3-noded triangular element as shown in Fig. (**5.2**) was employed in the finite element formulation wherein the stress field under plane strain conditions is allowed to vary linearly:

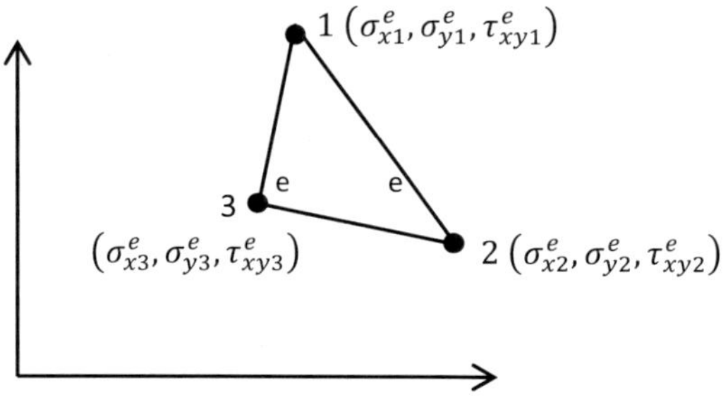

Fig. (5.2). Lower bound linear stress element in 2-dimensions.

$$\sigma_x = \sum_{i=1}^{3} N_i\, \sigma^e_{xi};\ \sigma_y = \sum_{i=1}^{3} N_i \sigma^e_{yi};\ \tau_{xy} = \sum_{i=1}^{3} N_i \tau^e_{xyi};\ \boldsymbol{\sigma} = \boldsymbol{N}\boldsymbol{\sigma}^e \quad \textbf{(5.9)}$$

where $\sigma^e_{xi}, \sigma^e_{yi}, \tau^e_{xyi}$ are the nodal stresses of element "*e*", and N_i the classical shape functions of a 3-noded triangular element;

$\boldsymbol{\sigma} = \{\sigma_x, \sigma_y, \tau_{xy}\}^T$ is the vector of the stress field;

$\boldsymbol{\sigma^e} = \{\sigma^e_{x1}, \sigma^e_{y1}, \tau^e_{xy1}, \sigma^e_{x2}, \sigma^e_{y2}, \tau^e_{xy2}, \sigma^e_{x3}, \sigma^e_{y3}, \tau^e_{xy3}\}^T$ is the vector of elemental nodal stresses and,

$$\boldsymbol{N} = \begin{bmatrix} N_1 & 0 & 0 & N_2 & 0 & 0 & N_3 & 0 & 0 \\ 0 & N_1 & 0 & 0 & N_2 & 0 & 0 & N_3 & 0 \\ 0 & 0 & N_1 & 0 & 0 & N_2 & 0 & 0 & N_3 \end{bmatrix} \quad \textbf{(5.10)}$$

In this case more than one node may share the same co-ordinates. Moreover, each node within an element may have different stresses than the same node of an adjoining element, that is, statically admissible stress discontinuity is allowed at the inter-element boundary. It follows that if there are n triangular elements, there will be $3n$ nodes and $9n$ unknown stresses.

5.4.1. Continuum and Discontinuity Equilibrium

To ensure static admissibility the stress field must satisfy equilibrium equations within the continuum and at any discontinuity. Within the continuum, the governing equations of equilibrium may be written as,

$$[\partial]^T \boldsymbol{\sigma} = \boldsymbol{f}_b \tag{5.11a}$$

$$[\partial] = \begin{bmatrix} \frac{\partial}{\partial x} & 0 \\ 0 & \frac{\partial}{\partial y} \\ \frac{\partial}{\partial y} & \frac{\partial}{\partial x} \end{bmatrix} \quad ; \quad \boldsymbol{f}_b = \begin{Bmatrix} b_x \\ b_y \end{Bmatrix} \tag{5.11b}$$

where b_x, b_y are the body forces. In the simplest and most common case in practice, $b_x = 0$, $b_y = \gamma$ where γ is the unit weight of the material. Substituting (5.9) into (5.10) and (5.11) for the triangular element "e" leads to,

$$\left[\boldsymbol{B}^e_{equil}\right]\{\boldsymbol{\sigma}^e\} = \{\boldsymbol{f}^e_b\} \tag{5.12a}$$

where $\left[\boldsymbol{B}^e_{equil}\right] = [\boldsymbol{B}_1^T \quad \boldsymbol{B}_2^T \quad \boldsymbol{B}_3^T]$, $\{\boldsymbol{f}^e_b\} = \{0, \gamma^e\}^T$ denotes the vector of the body forces for the element and it can be readily shown that,

$$\boldsymbol{B}_i = \frac{1}{2A^e}\begin{bmatrix} b_i & 0 \\ 0 & c_i \\ c_i & b_i \end{bmatrix} \tag{5.12b}$$

where A^e is the area of the element and b_i, c_i are parameters of the shape function N_i (i = 1,2,3) which are essentially functions of the nodal coordinates. A statistically admissible discontinuity must also satisfy equilibrium of the normal and shear stresses at the inter-element edge although the tangential stress is allowed to be discontinuous.

Now, the plane normal and shear stresses can be expressed in terms of the Cartesian stresses as shown in Fig. (**5.3**) as:

$$\begin{Bmatrix} \sigma_n \\ \tau \end{Bmatrix} = \begin{bmatrix} \sin^2\theta & \cos^2\theta & -\sin2\theta \\ -\frac{1}{2}\sin2\theta & \frac{1}{2}\sin2\theta & \cos2\theta \end{bmatrix} \begin{Bmatrix} \sigma_x \\ \sigma_y \\ \tau_{xy} \end{Bmatrix} = \boldsymbol{T\sigma} \tag{5.13}$$

Using the approach of Pastor and Turgeman (1976), Sloan (1988) and Lyamin and Sloan (2002a), and invoking equilibrium along edge "*s*" (Fig. **5.4**), gives

$$\begin{Bmatrix} \sigma_n^a \\ \tau^a \end{Bmatrix} = \begin{Bmatrix} \sigma_n^b \\ \tau^b \end{Bmatrix} \tag{5.14}$$

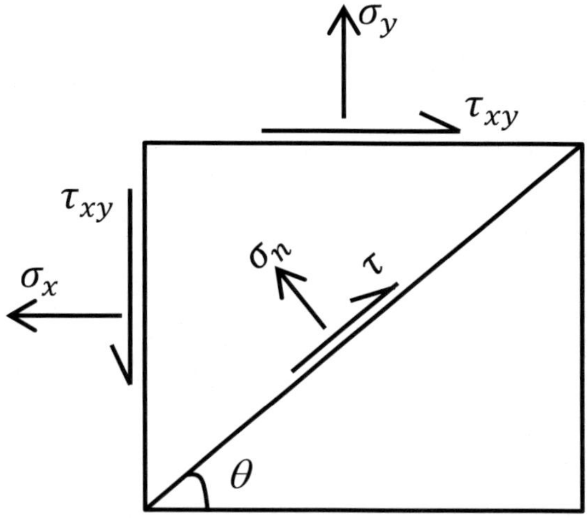

Fig. (5.3). Coordinate transform of stresses.

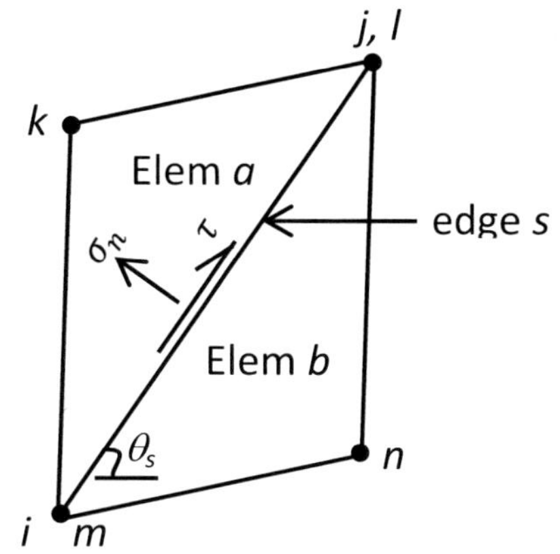

Fig. (5.4). Stress discontinuity between adjacent triangles.

which must also apply to nodes *i,m* and nodes *j,l* of edge "*s*". Hence using (5.13) and (5.14) gives,

$$\begin{bmatrix} \sin^2\theta_s & \cos^2\theta_s & -\sin 2\theta_s \\ -\frac{1}{2}\sin 2\theta_s & \frac{1}{2}\sin 2\theta_s & \cos 2\theta_s \end{bmatrix} \begin{Bmatrix} \sigma_{xi}^a \\ \sigma_{yi}^a \\ \tau_{xyi}^a \end{Bmatrix} = \boldsymbol{T}^s \boldsymbol{\sigma}_i^a =$$

$$\begin{bmatrix} \sin^2\theta_s & \cos^2\theta_s & -\sin 2\theta_s \\ -\frac{1}{2}\sin 2\theta_s & \frac{1}{2}\sin 2\theta_s & \cos 2\theta_s \end{bmatrix} \begin{Bmatrix} \sigma_{xm}^b \\ \sigma_{ym}^b \\ \tau_{xym}^b \end{Bmatrix} = \boldsymbol{T}^s \boldsymbol{\sigma}_m^b \qquad \textbf{(5.15a)}$$

$$\begin{bmatrix} \sin^2\theta_s & \cos^2\theta_s & -\sin 2\theta_s \\ -\frac{1}{2}\sin 2\theta_s & \frac{1}{2}\sin 2\theta_s & \cos 2\theta_s \end{bmatrix} \begin{Bmatrix} \sigma_{xj}^a \\ \sigma_{yj}^a \\ \tau_{xyj}^a \end{Bmatrix} = \boldsymbol{T}^s \boldsymbol{\sigma}_j^a =$$

$$\begin{bmatrix} \sin^2\theta_s & \cos^2\theta_s & -\sin 2\theta_s \\ -\frac{1}{2}\sin 2\theta_s & \frac{1}{2}\sin 2\theta_s & \cos 2\theta_s \end{bmatrix} \begin{Bmatrix} \sigma_{xl}^b \\ \sigma_{yl}^b \\ \tau_{xyl}^b \end{Bmatrix} = \boldsymbol{T}^s \boldsymbol{\sigma}_l^b \qquad \textbf{(5.15b)}$$

After combining equations (5.15a) and (5.15b), the equilibrium equations for edge "*s*" can be written as,

$$\begin{bmatrix} \boldsymbol{T}^s & -\boldsymbol{T}^s & 0 & 0 \\ 0 & 0 & \boldsymbol{T}^s & -\boldsymbol{T}^s \end{bmatrix} \begin{Bmatrix} \boldsymbol{\sigma}_i^a \\ \boldsymbol{\sigma}_m^b \\ \boldsymbol{\sigma}_j^a \\ \boldsymbol{\sigma}_l^b \end{Bmatrix} = \begin{Bmatrix} 0 \\ 0 \\ 0 \\ 0 \end{Bmatrix} \qquad \textbf{(5.16a)}$$

or in compact form as,

$$\left[\boldsymbol{T}_{\text{equil}}^s\right]\{\boldsymbol{\sigma}^s\} = \{\boldsymbol{0}\} \qquad \textbf{(5.16b)}$$

Equation (5.16) shows that the equilibrium condition generates four equality constraints on the nodal stresses for each statically admissible discontinuity along an element edge. An alternative approach to produce the constraints for the discontinuity is presented by Lyamin *et al.,* (2005a), using the continuum equilibrium equation (5.11).

Equilibrium equation (5.12a) must hold true for any element including for interface elements D_1 and D_2 in Fig. (**5.5**), even as $\delta \rightarrow 0$. Referring to element D_1, it can be shown that $b_i = y_m - y_j$, $b_m = y_j - y_i$, $b_j = y_i - y_m$, $c_i = x_j - x_m$, $c_m = x_i - x_j$, $c_j = x_m - x_i$, and as $\delta \rightarrow 0$: $x_i \rightarrow x_m$, $y_i \rightarrow y_m$ so that: $\boldsymbol{B}_m^T \rightarrow -\boldsymbol{B}_i^T; \boldsymbol{B}_j^T \rightarrow \boldsymbol{0}$. Defining $\overline{\boldsymbol{B}}_i^T = A^{D_1}\boldsymbol{B}_i^T$ (where A^{D_1} is the area of interface element D_1) and rearranging equation (5.12a) gives,

$$[\overline{\boldsymbol{B}}_i^T \quad -\overline{\boldsymbol{B}}_i^T \quad \boldsymbol{0}]\{\boldsymbol{\sigma}_i^{D_1} \quad \boldsymbol{\sigma}_m^{D_1} \quad \boldsymbol{\sigma}_j^{D_1}\}^T + A^{D_1}\boldsymbol{f}_{\boldsymbol{b}}^{D_1} = \boldsymbol{0} \qquad \textbf{(5.17)}$$

which leads to:

$\overline{\boldsymbol{B}}_i^T\boldsymbol{\sigma}_i^{D_1} = \overline{\boldsymbol{B}}_i^T\boldsymbol{\sigma}_m^{D_1}$ since $A^{D_1}\boldsymbol{f}_{\boldsymbol{b}}^{D_1} \rightarrow 0$ as $\delta \rightarrow 0$. Repeating the development for element D_2 yields $\overline{\boldsymbol{B}}_j^T\boldsymbol{\sigma}_j^{D_2} = \overline{\boldsymbol{B}}_j^T\boldsymbol{\sigma}_l^{D_2}$ so that all four discontinuous equilibrium conditions are satisfied.

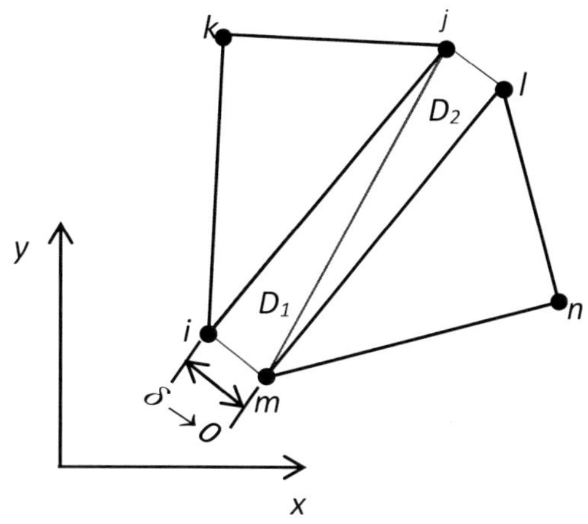

Fig. (5.5). Stress discontinuities at inter element boundary.

5.4.2. Boundary Conditions

The boundary conditions must be prescribed to satisfy the surface tractions $\boldsymbol{q} = \{q_n, t\}^T$ (Fig. **5.6**). These must match the stress components which may vary linearly along the edge Γ^e, namely,

$$\sigma_{ni}^e = q_{ni}; \quad \sigma_{nj}^e = q_{nj}; \qquad \tau_i^e = t_i; \qquad \tau_j^e = t_j \tag{5.18}$$

Using equation (5.13), the four boundary constraints may thus be expressed as,

$$[\boldsymbol{T}_{\text{bound}}^{\Gamma}]\{\boldsymbol{\sigma}^{\Gamma}\} = \{\boldsymbol{q}^{\Gamma}\} \tag{5.19}$$

where, $[\boldsymbol{T}_{\text{bound}}^{\Gamma}] = \begin{bmatrix} \boldsymbol{T}^{\Gamma} & \mathbf{0} \\ \mathbf{0} & \boldsymbol{T}^{\Gamma} \end{bmatrix}$, $\{\boldsymbol{\sigma}^{\Gamma}\} = \begin{Bmatrix} \boldsymbol{\sigma}_i^{\Gamma} \\ \boldsymbol{\sigma}_j^{\Gamma} \end{Bmatrix}$, $\{\boldsymbol{q}^{\Gamma}\} = \begin{Bmatrix} \boldsymbol{q}_i \\ \boldsymbol{q}_j \end{Bmatrix}$, $\{\boldsymbol{q}_i\} = \begin{Bmatrix} q_{ni} \\ t_i \end{Bmatrix}$

and noting that the matrix T^{Γ} has been defined in (5.15a) or (5.15b).

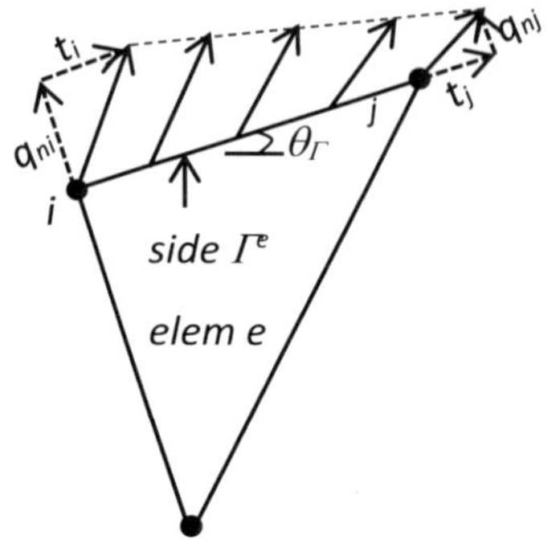

Fig. (5.6). Prescribed boundary tractions.

5.4.3. Yield Conditions, Linear Programming

To satisfy the conditions of the lower bound theorem, the stresses must not violate the yield constraint within the domain, namely $f(\boldsymbol{\sigma}) \leq 0$. For convex $f(\boldsymbol{\sigma})$, this constraint can be satisfied by requiring compliance at each node of the triangular element (Sloan, 2013), that is: $f(\boldsymbol{\sigma}_i^e) \leq 0$, ($i$=1 to 3). In the simpler case, the lower bound theorem may be formulated as a linear programming approach (especially for problems in two-dimensions), which requires the yield criterion to be linearized in terms of the stresses (*e.g.* Sloan, 1988). The linearized yield function surface must lie inside the actual yield surface in stress space. Initial lower bound finite element methods were based on linear triangular elements using linear programming (*e.g.* Anderheggen and Knopfel, 1972; Pastor, 1978; Bottero *et al.,* 1980). The linearization of the yield criterion will give rise to a large number of inequalities, hence the approach may not be appropriate for problems in three dimensions. The linear formulations thus result in very sparse and unsymmetric constraint equations which could only be exploited to solve small problems in the early years due to a lack of efficient algorithms. Sloan (1998) introduced the steepest-edge search to tackle this issue which enables practical two-dimensional geotechnical stability problems of practical interest to be solved (*e.g.* Sloan and Assadi, 1991, 1992), Ukritchon *et al.*, 2003).

The yield constraints are imposed at each node *i* and written as,

$$[\boldsymbol{Y}^{i}_{\text{yield}}]\{\boldsymbol{\sigma}_i\} \leq \{\boldsymbol{b}^{i}_{\text{yield}}\} \quad \textbf{(5.20)}$$

As mentioned earlier, in order to formulate the lower bound theorem as a linear programming problem, the yield constraints are required to be linearized. Using this approach Sloan (1988) as shown in Fig. (**5.7**), for instance, linearized the Mohr-Coulomb yield criterion for plane conditions in terms of shear strength parameters (c,ϕ) and Cartesian stresses as,

$$F = X^2 + Y^2 - R^2 = 0 \quad \textbf{(5.21)}$$

where,

$$X = (\sigma_x - \sigma_y); \quad = 2\tau_{xy}; \quad R = 2c\cdot\cos\phi - (\sigma_x + \sigma_y)\sin\phi \quad \textbf{(5.22)}$$

by approximating it as a inscribing yield polygon of *m* vertices and *m* sides. The coordinates of the kth vertex are thus given by:

$$X_k = R\cos(\pi\,(2k-1)/m); \qquad Y_k = R\sin(\pi\,(2k-1)/m) \quad \textbf{(5.23)}$$

To keep a stress point inside or on the yield polygon, it must satisfy the inequality,

$$(X_{k+1} - X)(Y_k - Y) - (X_k - X)(Y_{k+1} - Y) \leq 0; \;\; k = 1,2,\cdots m \quad \textbf{(5.24)}$$

So after substituting equation (5.22) and (5.23) into (5.24), it may be shown that the linearized yield criterion is:

$$A_k\sigma_x + B_k\sigma_y + C_k\tau_{xy} \leq D; \qquad k = 1,2,\cdots m \quad \textbf{(5.25)}$$

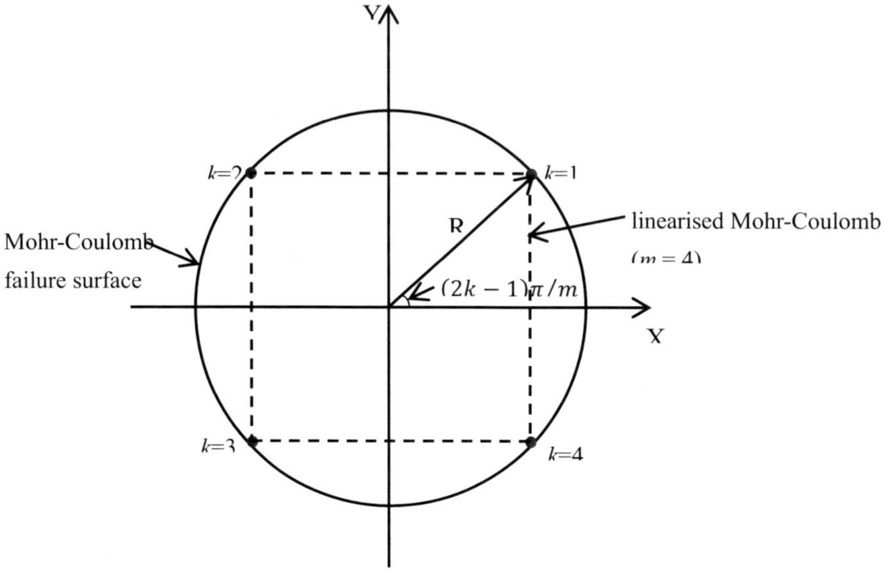

Fig. (5.7). Linearised Mohr-Coulomb failure surface, lower bound approach (after Sloan, 1988) for m =4.

where,

$A_k = \cos(2\pi k/m) + \sin\phi\cos(\pi/m)$; $B_k = \sin\phi\cos(\pi/m) - \cos(2\pi k/m)$

$C_k = 2\sin(2\pi k/m)$; $D = 2c\cos\phi\cos(\pi/m)$ **(5.26)**

The yield constraints are obtained by imposing Equation (5.25) at each node i of each of the mesh. Using Equation (5.25), the matrices in Equation (5.20) at node i, can be expressed in terms of the linearized yield criterion as:

$$\left[\boldsymbol{Y}_{\text{yield}}^{i}\right] = \begin{bmatrix} A_1 & A_2 & \cdots & A_k & \cdots & A_m \\ B_1 & B_2 & \cdots & B_k & \cdots & B_m \\ C_1 & C_2 & \cdots & C_k & \cdots & C_m \end{bmatrix}^T ; \{\boldsymbol{\sigma}_i\} = \{\sigma_{xi} \quad \sigma_{yi} \quad \tau_{xyi}\}^T \quad \textbf{(5.27)}$$

$$\left\{\boldsymbol{b}_{\text{yield}}^{i}\right\} = \{2c_i\cos\phi\cos(\pi/m) \quad 2c_i\cos\phi\cos(\pi/m) \quad \cdots \quad 2c_i\cos\phi\cos(\pi/m)\}^T$$

Equation (5.27) generates m constraint inequalities at each node i. For each triangle, the nodal stresses are furthermore subjected to three inequality constraints defined by Equation (5.20).

5.4.4. Yield Conditions, Nonlinear Programming

Nonlinear formulations result in a non-linear optimisation problem which produces a more modest number of inequality constraints (*e.g.* Zouain *et al.*, 1993; Lyamin and Sloan, 2002a; Krabbenhoft and Damkilde, 2003) compared to the linearized version described above. The non-linear yield constraints can be expressed generally for each node i as,

$$f(\boldsymbol{\sigma}_i) \leq 0 \quad \textbf{(5.28)}$$

5.4.4.1 Objective Function

The exact form of the objective function varies from case to case and will be illustrated here by way of a bearing capacity problem. In this particular case, the objective function is the collapse load to be maximised and it typically takes the form of,

$$Q = t_\Delta \int_{\Gamma^q} q_n \, d\Gamma \quad \textbf{(5.29)}$$

where Q is the collapse load, q_n is the normal pressure from loading applied over the part of boundary Γ^q and t_Δ is the out of plane thickness. The objective function may, however, be generalised to include the resultant of $\boldsymbol{q}$ (*e.g.* $|\boldsymbol{q}|$, see Fig. **5.6**), instead of q_n, where appropriate. Given that the stresses in the 3-noded triangular element are supposed to vary linearly, the contribution Q^e of collapse load from edge Γ^e of element e is (see Fig. **5.6**):

$$Q^e = \frac{t_\Delta L^e}{2}\left(q_{ni} + q_{nj}\right) \quad \textbf{(5.30)}$$

where L^e is the length of edge Γ^e. Equation (5.19) is invoked to transform the normal pressures q_{ni}, q_{nj} to the elemental stresses so that (5.22) gives,

$$Q^e = \{\boldsymbol{c}^{\Gamma^e}\}^T \{\boldsymbol{\sigma}^e\} \tag{5.31}$$

in which,

$$\{\boldsymbol{c}^{\Gamma^e}\}^T =$$

$$t_\Delta L^e \{\sin^2\theta_\Gamma \quad \cos^2\theta_\Gamma \quad -\sin 2\theta_\Gamma \quad \sin^2\theta_\Gamma \quad \cos^2\theta_\Gamma \quad -\sin 2\theta_\Gamma\} \tag{5.32}$$

$$\{\boldsymbol{\sigma}^e\}^T = \{\sigma_{xi}^e, \sigma_{yi}^e, \tau_{xyi}^e, \sigma_{xj}^e, \sigma_{yj}^e, \tau_{xyj}^e\} \tag{5.33}$$

5.4.5. Lower Bound Optimisation Problem

The equality and inequality constraints equations are assembled from all contributions to give:

$$[\boldsymbol{B}_{\text{equil}}] = \sum_{e=1}^{M} \boldsymbol{B}_{\text{equil}}^e \,; \qquad \{\boldsymbol{b}_1\} = \sum_{e=1}^{M} \{\boldsymbol{f}_b^e\} \tag{5.34a}$$

$$[\boldsymbol{T}^{s+\Gamma}] = \sum_{i=1}^{D} [\boldsymbol{T}_{\text{equil}}^s] + \sum_{i=1}^{S} [\boldsymbol{T}_{\text{bound}}^\Gamma]; \quad \{\boldsymbol{b}_2\} = \sum_{i=1}^{S} \{\boldsymbol{q}^\Gamma\} \tag{5.34b}$$

$$[\boldsymbol{Y}_{\text{yield}}] = \sum_{i=1}^{N} [\boldsymbol{Y}_{\text{yield}}^i]; \quad \{\boldsymbol{b}_{yield}\} = \sum_{i=1}^{N} \{\boldsymbol{b}_{\text{yield}}^i\} \tag{5.34c}$$

$$\{\boldsymbol{c}\} = \sum_{i=1}^{P} \{\boldsymbol{c}^{\Gamma^e}\} \tag{5.34d}$$

where M is the total number of elements, N is the total number of nodes, D is the total number of discontinuities and S is the total number of boundary edges with prescribed tractions, P is the total number of boundary edges over which the normal stresses are to be optimized. The lower bound solution may be found by solving the following linear programming problem:

Maximise

$$\{c\}^T\{\boldsymbol{\sigma}\} \quad \textbf{(5.35a)}$$

subject to,

$$[\boldsymbol{Y}_{\text{yield}}]\{\boldsymbol{\sigma}\} \leq \{\boldsymbol{b}_{yield}\} \quad \text{yield constraints} \quad \textbf{(5.35b)}$$

$$[\boldsymbol{B}_{\text{equil}}]\{\boldsymbol{\sigma}\} = \{\boldsymbol{b}_1\} \quad \text{continuum equilibrium} \quad \textbf{(5.35c)}$$

$$[\boldsymbol{T}^{s+\Gamma}]\{\boldsymbol{\sigma}\} = \{\boldsymbol{b}_2\} \quad \text{discontinuity equilibrium and stress boundary conditions} \quad \textbf{(5.35d)}$$

For nonlinear programming, the yield constraints in Equation (5.35) is replaced by the nonlinear yield criterion

$$f(\sigma_i) \leq 0, \quad \text{for each node } i \quad \textbf{(5.35e)}$$

In some cases, additional inequalities are required to be applied to the unknown variables aside from the nonlinear yield criterion. Equation (5.35e) can be generalised, if required, to include such inequalities which are usually linear (Lyamin and Sloan, 2002a):

$$f_j(\sigma_i) \leq 0, \quad j = 1 \text{ to } J, \quad \text{for each node } i \quad \textbf{(5.35f)}$$

5.4.6. Solution Procedures

5.4.6.1 Linear Programming

The theory and implementation of linear programming for optimization of the linear objective function subject to linear equality and linear inequality constraints in Equation 5.35 (a-d) may be found in Best and Ritter (1985). The optimisation problem in Equation 5.35 (a-d) can be expressed in canonical form required by the revised simplex method (Sloan, 1988) as follows:

Minimize

$$\{\boldsymbol{c}^*\}^T\{\boldsymbol{\sigma}^*\} \quad \textbf{(5.36a)}$$

subject to,

$$[\boldsymbol{B}^*]\{\boldsymbol{\sigma}^*\} = \{\boldsymbol{b}^*\} \quad \textbf{(5.36b)}$$

$$\{\boldsymbol{\sigma}^*\} \geq \{\boldsymbol{0}\} \quad \textbf{(5.36c)}$$

where,

$$\{\boldsymbol{c}^*\}^T = \{\{-\boldsymbol{c}\}^T \quad \{\boldsymbol{c}\}^T \quad \{\boldsymbol{0}\}^T\}; \quad \{\boldsymbol{\sigma}^*\}^T = \{\{\boldsymbol{\sigma}^+\}^T \quad \{\boldsymbol{\sigma}^-\}^T \quad \{\boldsymbol{s}\}^T\} \quad \textbf{(5.35d)}$$

$$[\boldsymbol{B}^*] = \begin{bmatrix} [\boldsymbol{Y}_{\text{yield}}] & -[\boldsymbol{Y}_{\text{yield}}] & [\boldsymbol{I}] \\ [\boldsymbol{B}] & -[\boldsymbol{B}] & [\boldsymbol{0}] \end{bmatrix}; \quad \{\boldsymbol{b}^*\}^T = \left\{\{\boldsymbol{b}_{yield}\}^T \quad \{\boldsymbol{b}\}^T\right\} \quad \textbf{(5.36e)}$$

$$[\boldsymbol{B}] = \begin{bmatrix} [\boldsymbol{B}_{\text{equil}}] \\ [\boldsymbol{T}^{s+\Gamma}] \end{bmatrix}; \quad \{\boldsymbol{b}\} = \begin{Bmatrix} \boldsymbol{b}_1 \\ \boldsymbol{b}_2 \end{Bmatrix} \quad \textbf{(5.36f)}$$

$$\{\boldsymbol{\sigma}\} = \{\boldsymbol{\sigma}^1\} - \{\boldsymbol{\sigma}^2\}; \quad \{\boldsymbol{\sigma}^1\}, \{\boldsymbol{\sigma}^2\}, \{\boldsymbol{s}\} \geq \{\boldsymbol{0}\} \quad \textbf{(5.36g)}$$

In the canonical form above, $\{\boldsymbol{s}\}$ is the vector of slack variables and $[\boldsymbol{I}]$ is the identity matrix. The vectors $\{\boldsymbol{\sigma}^+\}, \{\boldsymbol{\sigma}^-\}$ arise because the stresses are unrestricted in sign. Instead of the primal problem defined by Equation 5.36 (a-c), Sloan (1988) explained that due to the lower bound theorem typically generating more rows than columns in the basis matrix of the equation, the revised simplex algorithm is better suited to deal with the problem in its dual form:

Minimize

$$\{\boldsymbol{b}_{yield}\}^T\{\boldsymbol{\Sigma}^1\} + \{\boldsymbol{b}\}^T\{\boldsymbol{\Sigma}^2\} \quad \textbf{(5.37a)}$$

subject to,

$$\left[[\boldsymbol{Y}_{\text{yield}}]\right]^T \{\boldsymbol{\Sigma}^1\} + [\boldsymbol{B}]^T\{\boldsymbol{\Sigma}^2\} = \{\boldsymbol{c}\} \quad \textbf{(5.37b)}$$

$$\{\boldsymbol{\Sigma}^1\} \geq \{\boldsymbol{0}\} \quad \textbf{(5.37c)}$$

which may be expressed in canonical form required by the revised simplex method as,

Minimize

$$\{\boldsymbol{c}^*\}^T\{\boldsymbol{\Sigma}^*\} \quad \textbf{(5.38a)}$$

subject to,

$$[\boldsymbol{B}^*]\{\boldsymbol{\Sigma}^*\} = \{\boldsymbol{b}^*\} \quad \textbf{(5.38b)}$$

$$\{\boldsymbol{\Sigma}^*\} \geq \{\boldsymbol{0}\} \quad \textbf{(5.38c)}$$

where,

$$\{\boldsymbol{c}^*\}^T = \left\{\{\boldsymbol{b}_{yield}\}^T \ \ \{\boldsymbol{b}\}^T - \{\boldsymbol{b}\}^T\right\}; \{\boldsymbol{\Sigma}^*\}^T = \left\{\{\boldsymbol{\Sigma}^1\}^T \ \ \{\boldsymbol{\Sigma}^a\}^T \ \ \{\boldsymbol{\Sigma}^b\}^T\right\} \quad \textbf{(5.38d)}$$

$$[\boldsymbol{B}^*] = \left[[\boldsymbol{Y}_{\text{yield}}]^T \quad [\boldsymbol{B}]^T \quad -[\boldsymbol{B}]^T\right]; \quad \{\boldsymbol{b}^*\} = \{\boldsymbol{c}\} \quad \textbf{(5.38e)}$$

$$\{\boldsymbol{\Sigma}^2\} = \{\boldsymbol{\Sigma}^a\} - \{\boldsymbol{\Sigma}^b\} \quad \textbf{(5.38f)}$$

5.4.6.2 Nonlinear Programming

The need to linearise the yield function can be averted by using the feasible point algorithm based on original work of Zouain *et al.,* (1993) and Herskovits (1986, 1998) and modified by Lyamin and Sloan (2002a). The method developed by Lyamin and Sloan (2002a) is a two-stage quasi-newton algorithm that typically requires less than about 50 iterations, regardless of the problem size. One powerful feature of this method is that because it does not require linearization of the yield function, the approach may also be used for 3 dimensional problems for a wide range of smooth yield functions. For yield functions with sharp vertices (*e.g.* Tresca and Mohr-Coulomb), this approach requires the vertices to be smoothened to give C^1 continuity and to obtain good convergence. Other alternative approaches are the use of second-order cone programming and semi-definite programming algorithms for two- and three-dimensional cases respectively (Ciria, 2004; Makrodimopoulos and Martin, 2006; Krabbenhoft *et al.*, 2007).

5.5. UPPER BOUND FNITE ELEMENTS

The upper bound finite element formulations of Sloan (1989) and Lyamin and Sloan (2002b), which deal with the kinetically admissible velocity field $\dot{\boldsymbol{u}} = \{\dot{u}_x, \dot{u}_y, \dot{u}_z\}^T$ are discussed here. To be kinetically admissible, the velocity field is required to satisfy strain rate and velocity compatibility, flow rule and velocity boundary conditions. The upper bound approach then is to minimize the internal power dissipation,

$$\dot{U}^{int} = \int_\Omega \boldsymbol{\sigma}^T \dot{\boldsymbol{\varepsilon}}^p \, d\Omega \qquad \textbf{(5.39a)}$$

less the rate of work done by ***fixed*** external loads,

$$\dot{W}_t^{ext} = \int_{\Gamma_t} \boldsymbol{t}^T \dot{\boldsymbol{u}} \, d\Gamma + \int_\Omega \boldsymbol{g}^T \dot{\boldsymbol{u}} \, d\Omega \qquad \textbf{(5.39b)}$$

where $\boldsymbol{\sigma}$ is the stress vector, $\dot{\boldsymbol{\varepsilon}}^p = \{\dot{\varepsilon}_x^p, \dot{\varepsilon}_y^p, \dot{\varepsilon}_z^p, \dot{\gamma}_{xy}^p, \dot{\gamma}_{yz}^p, \dot{\gamma}_{zx}^p\}$ generally in 3-dimensions or more specifically $\dot{\boldsymbol{\varepsilon}}^p = \{\dot{\varepsilon}_x^p, \dot{\varepsilon}_y^p, \dot{\gamma}_{xy}^p\}$ in two-dimensions, are the plastic strains, $\boldsymbol{t}$ is the vector of fixed tractions and $\boldsymbol{g}$ the fixed body forces, Γ the boundary and Ω the domain. The rate of work done due to external tractions $\boldsymbol{q}$ over the part of boundary Γ_q or some form of body forces $\boldsymbol{f}_b$ over Ω is given by,

$$\dot{W}_q^{ext} = \int_{\Gamma_q} \boldsymbol{q}^T \dot{\boldsymbol{u}}\, d\Gamma + \int_{\Omega} \boldsymbol{f}_b^T \dot{\boldsymbol{u}}\, d\Omega \qquad \textbf{(5.40)}$$

According to the upper bound theorem, the external loads which are determined by equating $\dot{W}_q^{ext}$ to $\left(\dot{U}^{int} - \dot{W}_t^{ext}\right)$ will be higher than or equal to the actual limit load. Thus the upper bound on the limit load is found by optimizing $\left(\dot{U}^{int} - \dot{W}_t^{ext}\right)$ in terms of either the external forces due to $\boldsymbol{q}$ or $\boldsymbol{f}_b$.

Adopting the linear elements suggested by Lyamin and Sloan (2002b) and Sloan (2013), the velocity field $\dot{\boldsymbol{u}}$ is assumed to vary linearly over the element while the stress field remains constant within the element. Here, both the velocity field $\dot{\boldsymbol{u}}$ and the stress field $\boldsymbol{\sigma}$ are included as the unknown variables. Sloan (2013) explained that the inclusion of the stress in the formulation presents an avenue to develop a mesh refinement strategy for minimising the gap between the lower and upper bound approaches. For the linear triangular element shown in Fig. (**5.8**), each node *i* is connected with unknown velocities$\{\dot{u}_{xi}, \dot{u}_{yi}\}^T$ and each element *e* with unknown element stresses $\{\sigma^e{}_x\, \sigma^e{}_y\, \tau^e{}_{xy}\}$ and non-negative plastic multiplier rate λ. The velocity field in the element can be expressed as:

$$\dot{u}_x = \sum_{i=1}^{3} N_i\, \dot{u}_{xi}\ ;\quad \dot{u}_y = \sum_{i=1}^{3} N_i\, \dot{u}_{yi} \qquad \textbf{(5.41)}$$

where N_i, $i = 1,2,3$ are the shape functions of the linear triangular element.

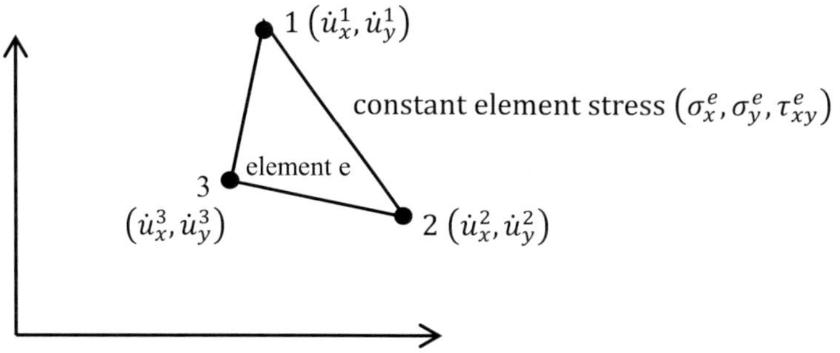

Fig. 5.8. Upper bound linear element in 2-dimensions.

Moreover, the unknown velocities must also satisfy the associated flow rule and consistency requirement. The plastic strain rates $\dot{\boldsymbol{\varepsilon}}^p = \{\dot{\varepsilon}_x^p, \dot{\varepsilon}_y^p, \dot{\gamma}_{xy}^p\}^T$ which is taken as constant for each element, is given by using the associated flow rule,

$$\dot{\varepsilon}_x^p = \partial \dot{u}_x / \partial x = \dot{\lambda}\, \partial F / \partial \sigma_x \qquad \textbf{(5.42a)}$$

$$\dot{\varepsilon}_y^p = \partial \dot{u}_y / \partial y = \dot{\lambda}\, \partial F / \partial \sigma_y \qquad \textbf{(5.42b)}$$

$$\dot{\gamma}_{xy}^p = \partial \dot{u}_x / \partial y + \partial \dot{u}_y / \partial x = \dot{\lambda}\, \partial F / \partial \tau_{xy} \qquad \textbf{(5.42c)}$$

where $\dot{\lambda} \geq 0$ and F is the yield function. Like in the lower bound finite element limit analysis, the upper bound approach may be formulated as a linear (Sloan, 1989) or nonlinear (Lyamin and Sloan, 2002b) optimisation problem. The corresponding linearized Mohr-Coulomb yield function for the upper bound approach is illustrated in Fig. (**5.9**) below, which is a polygon circumscribing the Mohr-Coulomb yield surface. The coordinates of the kth vertex is then given by:

$$X_k = R\cos((2k-1)\pi/m)/\cos(\pi/m)\,;\; Y_k = R\sin((2k-1)\pi/m)/\cos(\pi/m) \quad \textbf{(5.43)}$$

where R and the Mohr-Coulomb yield surface for plane conditions are given in Equations (5.22) and (5.21) respectively. To keep a stress point inside or on the external yield polygon, requires that

$$(Y_k - Y_{k+1})(\sigma_x - \sigma_y) + 2(X_{k+1} - X_k)\tau_{xy} + (X_k Y_{k+1} - X_{k+1}Y_k) \le 0;\; k = 1,2,\cdots m \quad \textbf{(5.44)}$$

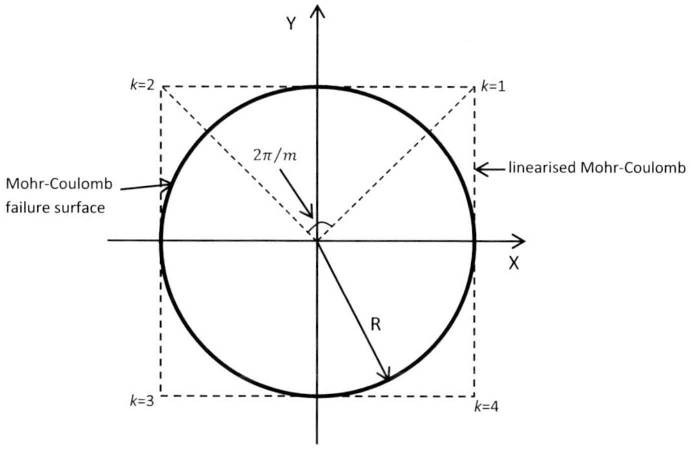

Fig. (5.9). Linearised Mohr-Coulomb failure surface, upper bound approach (after Sloan, 1989) for m =4.

After substituting equation (5.43) into (5.44), the linearized yield criterion is (Sloan, 1989):

$$F_k(\boldsymbol{\sigma}) = A_k\sigma_x + B_k\sigma_y + C_k\tau_{xy} - D = 0; \qquad k = 1,2,\cdots m \quad \textbf{(5.45)}$$

where,

$$A_k = \cos(2\pi k/m) + \sin\phi; \qquad B_k = \sin\phi - \cos(2\pi k/m)$$

$$C_k = 2\sin(2\pi k/m); \qquad D = 2c\cos\phi \quad \textbf{(5.46)}$$

Substituting the linearized Mohr-Coulomb yield criterion (Equation 5.45) into Equation (5.42), the plastic strain rates can be expressed as:

$$\dot{\varepsilon}_x^p = \partial \dot{u}_x/\partial x = \sum_{k=1}^{m} \dot{\lambda}_k A_k \quad \textbf{(5.47a)}$$

$$\dot{\varepsilon}_y^p = \partial \dot{u}_y/\partial y = \sum_{k=1}^{m} \dot{\lambda}_k B_k \quad \textbf{(5.47b)}$$

$$\dot{\gamma}_{xy}^p = \partial \dot{u}_x/\partial y + \partial \dot{u}_y/\partial x = \sum_{k=1}^{m} \dot{\lambda}_k C_k \quad \textbf{(5.47c)}$$

with $\dot{\lambda}_k \geq 0$. Applying the Equation (5.41), the plastic strain rates can be expressed as:

$$\{\dot{\boldsymbol{\varepsilon}}^p\} = [\boldsymbol{B}_{vel}^e]\{\dot{\boldsymbol{u}}^e\} \quad \textbf{(5.48a)}$$

$$[\boldsymbol{B}_{vel}^e] = [\boldsymbol{B}_1 \quad \boldsymbol{B}_2 \quad \boldsymbol{B}_3] \quad \textbf{(5.48b)}$$

$$\{\dot{\boldsymbol{u}}^e\} = \{\dot{u}_{x1}^e \quad \dot{u}_{y1}^e \quad \dot{u}_{x2}^e \quad \dot{u}_{y2}^e \quad \dot{u}_{x3}^e \quad \dot{u}_{y3}^e\}^T \quad \textbf{(5.48c)}$$

Equation (5.47) and (5.48) are then applied to obtain the linearised flow rule constraints for the element:

$$[\boldsymbol{B}_{vel}^e]\{\dot{\boldsymbol{u}}^e\} + [\boldsymbol{B}_{flow}^e]\{\dot{\boldsymbol{\lambda}}^e\} = \{\boldsymbol{0}\} \quad \textbf{(5.49a)}$$

where following the linearization in Equation (5.47),

$$[\boldsymbol{B}_{flow}^e] = \begin{bmatrix} -A_1 & -A_2 & \cdots & -A_m \\ -B_1 & -B_2 & \cdots & -B_m \\ -C_1 & -C_2 & \cdots & -C_m \end{bmatrix} \quad \textbf{(5.49b)}$$

$$\{\dot{\boldsymbol{\lambda}}^e\} = \{\dot{\lambda}_1^e \quad \dot{\lambda}_2^e \quad \cdots \quad \dot{\lambda}_m^e\}^T \quad \textbf{(5.49c)}$$

and where $\boldsymbol{B}_i$ (i = 1 to 3) is given in (5.12b). In the nonlinear optimisation case, the yield function and flow rule are not linearised, hence the flow rule is expressed for the element using Equation (5.42) as,

$$\{\dot{\boldsymbol{\varepsilon}}^p\} = \dot{\lambda}^e \nabla F(\boldsymbol{\sigma}^e) = \dot{\lambda}\{\partial F/\partial \sigma_x, \partial F/\partial \sigma_y, \partial F/\partial \tau_{xy}\}^T;\ \dot{\lambda}^e \geq 0,\ \dot{\lambda}^e F(\boldsymbol{\sigma}^e) = 0 \qquad \textbf{(5.50a)}$$

Mulitplying (5.50a) by A^e gives,

$$A^e\{\dot{\boldsymbol{\varepsilon}}^p\} = \dot{\Lambda}^e \nabla F(\boldsymbol{\sigma}^e) = \dot{\Lambda}^e\{\partial F/\partial \sigma_x, \partial F/\partial \sigma_y, \partial F/\partial \tau_{xy}\}^T;\ \dot{\Lambda}^e \geq 0,\ \dot{\Lambda}^e F(\boldsymbol{\sigma}^e) = 0 \qquad \textbf{(5.50b)}$$

where $\dot{\Lambda}^e = A^e \dot{\lambda}^e$. The form of (5.50b) is useful, as will be shown later, in dealing with zero-thickness element at the inter element boundaries. These elements allow velocity jumps to take place at the inter element boundaries. Combining (5.50b) with (5.42) and (5.50) gives for the flow constraints for the nonlinearised case,

$$[\overline{\boldsymbol{B}}^e_{vel}]\{\dot{\boldsymbol{u}}^e\} - \dot{\Lambda}^e \nabla F(\boldsymbol{\sigma}^e) = \{\boldsymbol{0}\};\quad \dot{\Lambda}^e \geq 0,\ \dot{\Lambda}^e F(\boldsymbol{\sigma}^e) = 0 \qquad \textbf{(5.51)}$$

where $[\overline{\boldsymbol{B}}^e_{vel}] = A^e[\boldsymbol{B}^e_{vel}]$. It is also noted that the flow rule imposes 3 equalities and m inequalities on the element unknowns for the linearised case, and 4 equalities and one inequality for the nonlinear case.

5.5.1. Objective Function

The objective function to be minimised is $\left(\dot{U}^{int} - \dot{W}^{ext}_t\right)$, the internal power dissipation less the rate of work done by ***fixed*** external loads, subject to constraints from the flow rule, velocity boundary conditions and the yield functions. It may be noted that unlike the lower bound approach which deals with only the stress unknowns, the upper bound finite elements include velocity and plastic multiplier unknowns in addition to the stress unknowns. Using Equation (5.39), the objective function can be expressed as the sum of the contributions from each element, *e,* that is,

$$\dot{U}^{int} - \dot{W}^{ext}_t = \textstyle\sum_e \{\boldsymbol{\sigma}^T \dot{\boldsymbol{\varepsilon}}^p\}^e A^e - \{\boldsymbol{c}\}^T\{\dot{\boldsymbol{u}}\} = \sum_e \{\{\boldsymbol{\sigma}\}^T[\boldsymbol{B}^e_{vel}]\{\dot{\boldsymbol{u}}^e\}\}^e A^e - \{\boldsymbol{c}\}^T\{\dot{\boldsymbol{u}}\} \qquad \textbf{(5.52)}$$

where,

$$\{\boldsymbol{c}\}^T\{\dot{\boldsymbol{u}}\} = \int_{\Gamma_t} \boldsymbol{t}^T \dot{\boldsymbol{u}}\, d\Gamma + \int_{\Omega} \boldsymbol{g}^T \dot{\boldsymbol{u}}\, d\Omega \qquad \textbf{(5.53)}$$

Since $\boldsymbol{t}$ is the vector of fixed tractions, $\boldsymbol{g}$ is the vector of fixed body forces, $\{\boldsymbol{c}\}$ is a known vector that can be computed from the fixed external loads. Summing up the contributions from all the elements in Equation (5.52) leads to,

$$\dot{U}^{int} - \dot{W}_t^{ext} = \{\boldsymbol{\sigma}\}^T[\overline{\boldsymbol{B}}_{vel}]\{\dot{\boldsymbol{u}}\} - \{\boldsymbol{c}\}^T\{\dot{\boldsymbol{u}}\} \tag{5.54}$$

where,

$$[\overline{\boldsymbol{B}}_{vel}] = \textstyle\sum_e [\boldsymbol{B}_{vel}^e]A^e \tag{5.55}$$

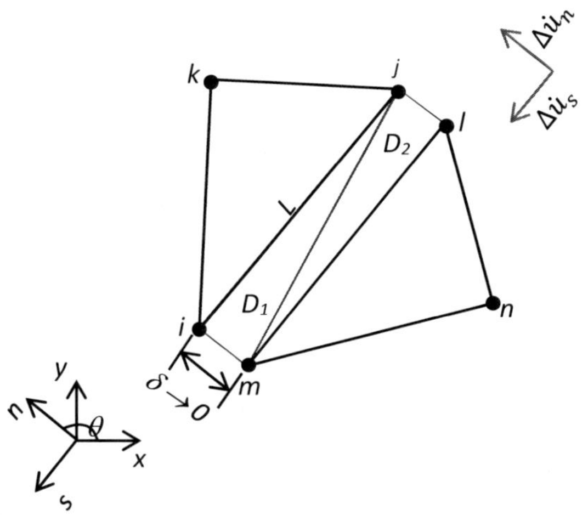

Fig. (5.10). Velocity discontinuities at inter element boundary.

5.5.2. Discontinuity Flow Rule

To allow for velocity discontinuities along inter-element edges, nodes from adjoining elements at the element boundaries may share the same nodal coordinates, yet are unique to each element.

Velocity discontinuities along inter-element boundaries are incorporated using the same approach discussed previously for the lower bound method.

As shown in Fig. (**5.10**), the boundary discontinuity is defined by triangle D_1 and D_2 and width $\delta \to 0$. Following Sloan (2013), using Equation (5.12), (5.48a) and local coordinates (s, n) it can be shown that for triangle D_1, in the limit as $\delta \to 0$,

$$A^{D_1}\varepsilon_s^p = 0 \; ; \; A^{D_1}\varepsilon_n^p = \Delta\dot{u}_{n(im)}L/2; \quad A^{D_1}\gamma_{ns}^p = \Delta\dot{u}_{s(im)}L/2 \qquad \textbf{(5.56a)}$$

where $\varepsilon_s^p, \varepsilon_n^p, \gamma_{ns}^p$ are the local plastic strains, $A^{D_1}$is the area of triangle D_1 , L is the length of the base of D_1, $\Delta\dot{u}_{n(im)} = \dot{u}_{ni} - \dot{u}_{nm}$, $\Delta\dot{u}_{s(im)} = \dot{u}_{si} - \dot{u}_{sm}$. The last two terms are the normal and tangential velocity jumps at nodal pair (i, m). Applying similar procedures to triangle D_2 gives,

$$A^{D_2}\varepsilon_s^p = 0; \; A^{D_2}\varepsilon_n^p = \Delta\dot{u}_{n(jl)}L/2; \; A^{D_1}\gamma_{ns}^p = \Delta\dot{u}_{s(jl)}L/2 \qquad \textbf{(5.56b)}$$

Equations (5.56) shows that velocity jumps are possible in zero thickness continuum elements. For a plane strain case, the velocity jumps can be defined using Equation (5.50b), and normal and shear stress as,

$$\Delta\dot{u}_n = \dot{\Lambda}\,\partial F/\partial\sigma_n; \; \Delta\dot{u}_s = \dot{\Lambda}\,\partial F/\partial\tau \;\; ; \; \dot{\Lambda} \geq 0, \; \dot{\Lambda}F(\sigma_n, \tau_{ns}) = 0 \qquad \textbf{(5.57)}$$

The absolute value of the limiting shear stress τ may be defined in terms of the Mohr-Coulomb criterion as:

$$|\tau\;| = c - \sigma_n\tan\phi \qquad \textbf{(5.58)}$$

where tensile stress is taken as positive, so that the yield surface is:

$$F(\sigma_n, \tau\;) = |\tau\;| + \sigma_n\tan\phi - c = 0 \qquad \textbf{(5.59)}$$

Furthermore, using Equation (5.59) and Koiter's theorem in relation to composite yield surfaces, we define,

$$\Delta\dot{u}_n = (\Delta\dot{u}_n^- + \Delta\dot{u}_n^+) \; ; \qquad \Delta\dot{u}_s = (\Delta\dot{u}_s^- + \Delta\dot{u}_s^+) \qquad \textbf{(5.60)}$$

where,

$$\Delta\dot{u}_n^- = \dot{\Lambda}^- \partial F^-/\partial\sigma_n;\ \Delta\dot{u}_s^- = \dot{\Lambda}^- \partial F^-/\partial\tau\ ;\ \dot{\Lambda}^- \geq 0;\ \dot{\Lambda}^- F^-(\sigma_n, \tau\ ) = 0 \quad \textbf{(5.61a)}$$

$$\Delta\dot{u}_n^+ = \dot{\Lambda}^+ \partial F^+/\partial\sigma_n;\ \Delta\dot{u}_s^+ = \dot{\Lambda}^+ \partial F^+/\partial\tau\ ;\ \dot{\Lambda}^+ \geq 0;\ \dot{\Lambda}^+ F^+(\sigma_n, \tau\ ) = 0 \quad \textbf{(5.61b)}$$

$$F^-(\sigma_n, \tau\ ) = -\tau\ + \sigma_n \tan\phi - c \quad \textbf{(5.61c)}$$

$$F^+(\sigma_n, \tau\ ) = \tau\ + \sigma_n \tan\phi - c \quad \textbf{(5.61d)}$$

Then it follows that,

$$\Delta\dot{u}_n = (\dot{\Lambda}^+ + \dot{\Lambda}^-)\tan\phi\ ;\ \ \Delta\dot{u}_s = (\dot{\Lambda}^+ - \dot{\Lambda}^-) \quad \textbf{(5.62)}$$

Now, the plane normal and shear velocities can be expressed in terms of the Cartesian velocities:

$$\begin{Bmatrix} \Delta\dot{u}_{n(im)} \\ \Delta\dot{u}_{s(im)} \end{Bmatrix} = \begin{bmatrix} \cos\theta & \sin\theta \\ -\sin\theta & \cos\theta \end{bmatrix} \begin{Bmatrix} \Delta\dot{u}_{x(im)} \\ \Delta\dot{u}_{y(im)} \end{Bmatrix} \quad \textbf{(5.63)}$$

which when combined with Equation (5.63) gives for nodal pair (i, m) of element $D^1$and in terms of its elemental normal and shear stresses $\sigma_n^{D_1}, \tau^{D_1}$:

$$\begin{bmatrix} \cos\theta & \sin\theta \\ -\sin\theta & \cos\theta \end{bmatrix} \begin{Bmatrix} \Delta\dot{u}_{x(im)} \\ \Delta\dot{u}_{y(im)} \end{Bmatrix} = \begin{Bmatrix} (\dot{\Lambda}_{im}^+ + \dot{\Lambda}_{im}^-)\tan\phi \\ \dot{\Lambda}_{im}^+ - \dot{\Lambda}_{im}^- \end{Bmatrix} \quad \textbf{(5.64a)}$$

and similarly for nodal pair (j,l) of element D^2 and in terms of its elemental normal and shear stresses $\sigma_n^{D_2}, \tau^{D_2}$:

$$\begin{bmatrix} \cos\theta & \sin\theta \\ -\sin\theta & \cos\theta \end{bmatrix} \begin{Bmatrix} \Delta\dot{u}_{x(jl)} \\ \Delta\dot{u}_{y(jl)} \end{Bmatrix} = \begin{Bmatrix} (\dot{\Lambda}_{jl}^+ + \dot{\Lambda}_{jl}^-)\tan\phi \\ \dot{\Lambda}_{jl}^+ - \dot{\Lambda}_{jl}^- \end{Bmatrix} \quad \textbf{(5.64b)}$$

Moreover, loading unloading conditions require that,

$$\dot{\Lambda}_{im}^{+} \geq 0 \ ; \ \dot{\Lambda}_{im}^{-} \geq 0; \quad \dot{\Lambda}_{jl}^{+} \geq 0 \ ; \ \dot{\Lambda}_{jl}^{-} \geq 0 \qquad \textbf{(5.65)}$$

and,

$$\dot{\Lambda}_{im}^{+} F^{+}\left(\sigma_n^{D_1}, \tau^{D_1}\right) = 0; \quad \dot{\Lambda}_{im}^{-} F^{-}\left(\sigma_n^{D_1}, \tau^{D_1}\right) = 0 \qquad \textbf{(5.66a)}$$

$$\dot{\Lambda}_{jl}^{+} F^{+}\left(\sigma_n^{D_2}, \tau^{D_2}\right) = 0; \quad \dot{\Lambda}_{jl}^{-} F^{-}\left(\sigma_n^{D_2}, \tau^{D_2}\right) = 0 \qquad \textbf{(5.66b)}$$

The constraints imposed by Equations (6.64), (5.65) an (5.66) must be applied at the inter element nodes together with the element flow rule to model velocity discontinuity.

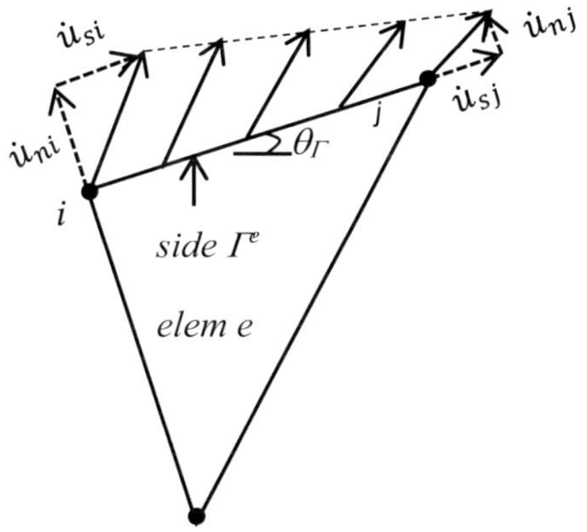

Fig. (5.11). Prescribed boundary velocities.

5.5.3. Prescribed Boundary Velocities

The kinematically admissible velocity field must also satisfy prescribed velocities along the boundaries and these boundary conditions can be compiled into a matrix of the form,

$$[\boldsymbol{I}]\{\dot{\boldsymbol{u}}\}|_{\Gamma_v} = \{\boldsymbol{v}\} \quad \textbf{(5.67)}$$

where $[\boldsymbol{I}]$ is the identity matrix, Γ_v denotes the boundaries with prescribed velocities and $\{\boldsymbol{v}\}$ is the vector of prescribed velocities. If prescribed velocities are defined in terms of the local normal and tangential components along the boundary edge, as shown in Fig. (**5.11**), then they can be easily transformed to Cartesian velocities using the following transformation relationship:

$$\begin{Bmatrix} \dot{u}_{ni} \\ \dot{u}_{si} \end{Bmatrix} = \begin{bmatrix} \cos\theta_\Gamma & \sin\theta_\Gamma \\ -\sin\theta_\Gamma & \cos\theta_\Gamma \end{bmatrix} \begin{Bmatrix} \dot{u}_{xi} \\ \Delta\dot{u}_{yi} \end{Bmatrix} \quad \textbf{(5.68)}$$

5.5.4. Loadings

Two typical types of boundary loadings may be imposed. To model loadings caused by a rigid structure, *e.g.* footing, the boundary velocities defined by Equations (5.67) and (5.68) are applied, except in this case the velocities are assumed. For instance, to model edge *ij* in (Fig. **5.11**) as a rigid boundary loading, the normal velocities are set as:

$$\dot{u}_{ni} = \dot{u}_{nj} = -v_0 \quad \textbf{(5.69)}$$

where v_0 is some assumed constant. Its value is not an issue as v_0 will cancel out when using Equation (5.52)-(5.54). In the tangential direction, $\dot{u}_{si} = \dot{u}_{sj} = 0$ for a "rough" contact, while for a "smooth" contact, the tangential velocities are unrestrained. When the loading is an applied uniform normal pressure q_0 over the part of boundary which is Γ_q, for instance in the case of a flexible footing, then the imposed constraint is,

$$\int_{\Gamma_q} \dot{u}_n \, d\Gamma = v_0 \quad \textbf{(5.70)}$$

where v_0 is the prescribed material velocity across the boundary, often taken as one. Discretising Equation (5.70) in terms of Cartesian coordinates using Equation (5.68) and along edges of linear elements gives,

$$\sum_{ext\ press\ boundary} \frac{1}{2} L_{ij} \{(\dot{u}_{xi} + \dot{u}_{xj})\cos\theta_\Gamma + (\dot{u}_{yi} + \dot{u}_{yj})\sin\theta_\Gamma\} = v_0 \qquad \textbf{(5.71)}$$

Where L_{ij} is the length of the edge ij. Substituting (5.71) into Equation (5.40), allows the uniform pressure to be directly minimised.

A third type of loading constraint is associated with the body force, such as when the unit weight of the material is to be optimised by taking,

$$\int_\Omega \dot{u}_y \, d\Omega \approx \frac{1}{3} \sum_{e=1}^{M} A^e (\dot{u}_{yi} + \dot{u}_{yj} + \dot{u}_{yk}) = -v_0 \qquad \textbf{(5.72)}$$

where v_0 is a constant, again typically taken as one. The integral is discretised as the sum of the contributions from each element and the integrand is expressed in terms of the nodal vertical velocities $(\dot{u}_{yi}, \dot{u}_{yj}, \dot{u}_{yk})$ of the element.

The loading and velocity constraint conditions represented by Equation (5.67), (5.68), (5.69), (5.71) and (5.72) can be assembled and expressed in general form as:

$$[\boldsymbol{B}]\{\dot{\boldsymbol{u}}\} = \{\boldsymbol{b}\} \qquad \textbf{(5.73)}$$

5.5.5. Upper Bound Optimisation Problem

5.5.5.1 Non-Linear Programming

After assembling the contributions from the entire mesh, the upper bound solution is found by solving the following linear programming problem:
Minimise

$$\{\boldsymbol{\sigma}\}^T[\overline{\boldsymbol{B}}_{vel}]\{\dot{\boldsymbol{u}}\} - \{\boldsymbol{c}\}^T\{\dot{\boldsymbol{u}}\} \quad \textbf{(5.74)}$$

subject to,

$[\overline{\boldsymbol{B}}_{vel}^e]\{\dot{\boldsymbol{u}}^e\} = \dot{\Lambda}^e \nabla F(\boldsymbol{\sigma}^e)$ (for each element e)

$\dot{\Lambda}^e \geq 0$ (for each element e)

$\dot{\Lambda}^e F(\boldsymbol{\sigma}^e) = 0$ (for each element e)

$F(\boldsymbol{\sigma}^e) \leq 0$ (for each element e)

$[\boldsymbol{B}]\{\dot{\boldsymbol{u}}\} = \{\boldsymbol{b}\}$

where $\{\boldsymbol{\sigma}\}$ is a global vector of unknown stresses, $\{\dot{\boldsymbol{u}}\}$ is the global vector of unknown nodal velocities, $[\boldsymbol{B}]$ is the matrix of equality constraint coefficients, and $\{\boldsymbol{b}\}$ is a known vector of coefficients. The solution to (5.74) can be found efficiently using the two-stage quasi-Newton solver proposed by Lyamin and Sloan (2002b).

5.6. APPLICATIONS

5.6.1. Finite Element Lower Bound Limit Analysis Using Linear Programming

The following example of a smooth rigid strip footing resting on infinite half space of undrained purely cohesive soil is given in Sloan (1988) as shown in Fig. (**5.12**).

The exact collapse pressure of a smooth rigid footing under undrained conditions is given by Prandtl (Prandtl, 1920; Shield, 1954) as,

$$q_f = N_c c_u = (2 + \pi)c_u = 5.14c_u \quad \textbf{(5.75)}$$

where c_u is the undrained cohesion of the soil. The lower bound solutions for this problem are computed using linear programming for varying *m*, the number of sides of the polygon used to approximate the yield surface (which in this case is a Tresca yield function since $\phi_u = 0$ under undrained conditions); *g*, the number of rows in the constraint matrix; *h*, the number of columns in the constraint matrix; *i*, the total number of iterations for active set algorithm; *t*, the CPU time required to solve the problem in a VAX 8550, as shown in Table **5.1** below.

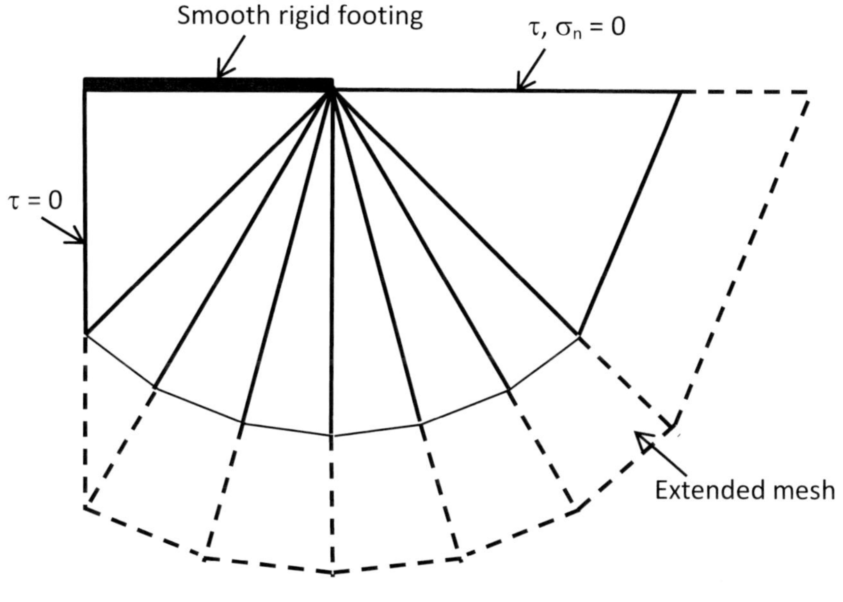

Fig. (5.12). Mesh for smooth rigid footing on purely cohesive soil ($\phi_u = 0$) (after Sloan, 1988).

Sloan (1988) reported that for *m* = 24 and 48, the lower bound estimate of the collapse pressure is within one percent of the exact solution. With *m* > 12, sufficient accuracy has been obtained for the undrained loading case. The extended mesh yielded a solution which is very similar to the smaller mesh and converges to the exact solution. This demonstrates that the solutions in Table **5.1** are valid lower bounds for the infinite half-space.

Table 5.1. Lower bound solutions for a smooth rigid footing on purely cohesive soil (after Sloan, 1988).

m	*g*	*h*	*i*	*t* (sec)	N_c
6	72	195	40	0.5	4.72
12	72	339	58	0.7	5.03
24	72	627	85	1.4	5.06
48	72	1203	125	2.9	5.07
48*	72	1203	210	4.2	5.08

*denotes the case with extended mesh.

5.6.2. Finite Element Upper Bound Limit Analysis Using Linear Programming

The finite element upper bound limit analysis of the same problem for a smooth rigid footing resting on infinite half-space under undrained conditions was presented by Sloan (1989). An active set method for solving large, sparse linear programming problems based on the algorithm by Best and Ritter (1985) is used to compute the solutions. The coarse and fine meshes used in the analysis are shown in Fig. (**5.13**).

The results corresponding to the upper bound analysis are shown in Table **5.2.** The upper bound solutions were found to be quite insensitive to mesh refinement for this problem, as evidenced by the % error in N_c. For $m \geq 12$, the approximation of the yield surface is fairly insensitive to the "*m*" linearization refinement. For given mesh, the computation time is approximately directly proportional to m used in the linearization, whereas for a given m, the computation time increases approximately as the square of the number of elements. Sloan (1989) noted that the special arrangement using 4 triangles to form a quadrilateral with central node allows the incompressibility condition $\dot{\varepsilon}_x + \dot{\varepsilon}_y = 0$ to be satisfied, which may not be achieved if an arbitrary arrangement was employed.

Table 5.2. Upper bound solutions for a smooth rigid footing on purely cohesive soil (after Sloan, 1989).

Mesh	M	g	h	I	t **(sec)**	N_c	**% error in** N_c
Coarse	6	329	709	228	12	5.63	9.5
Fine	6	1227	2757	1243	241	5.50	7.0
Coarse	12	329	1285	326	20	5.44	5.8
Fine	12	1227	5061	1729	389	5.31	3.3
Coarse	24	329	2437	537	38	5.39	4.9
Fine	24	1227	9669	2582	732	5.27	2.5

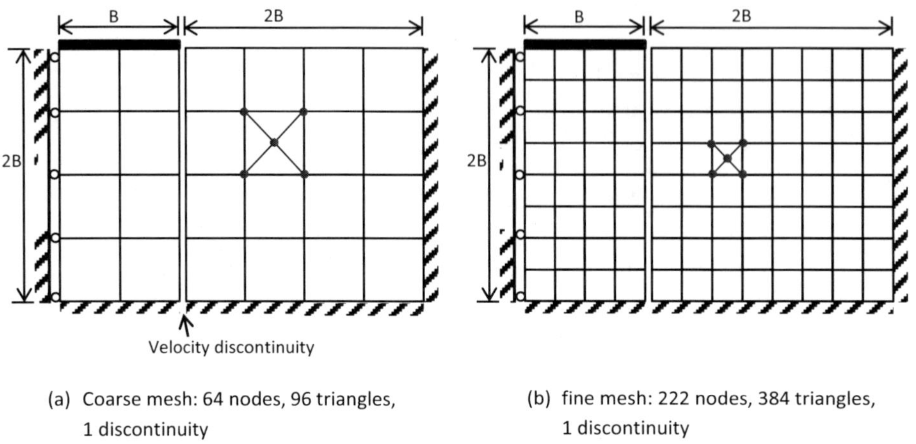

Fig. (5.13). Coase and fine mesh for smooth rigid footing on purely cohesive soil (Upper bound solutions) (after Sloan, 1989).

The plastic zones and the corresponding velocity field for the fine mesh are shown in Figs. (**5.14a** and **5.14b**) respectively. The plastic region resembles the one predicted by Prandtl's mechanism. The mesh for the upper bound approach is generally designed based on similar principles used for the classic displacement finite element method. That is, finer elements are concentrated where the strain gradients are expected to be steepest.

The size of the mesh need only be large enough to accommodate the plastic region; while mesh smaller than this will produce a valid upper bound solution, the value will be too high. Moreover, while judicious specification of the locations of any discontinuities will naturally help in better prediction of the true velocity field, the upper bound solution is in fact not highly sensitive to the location of the velocity discontinuities.

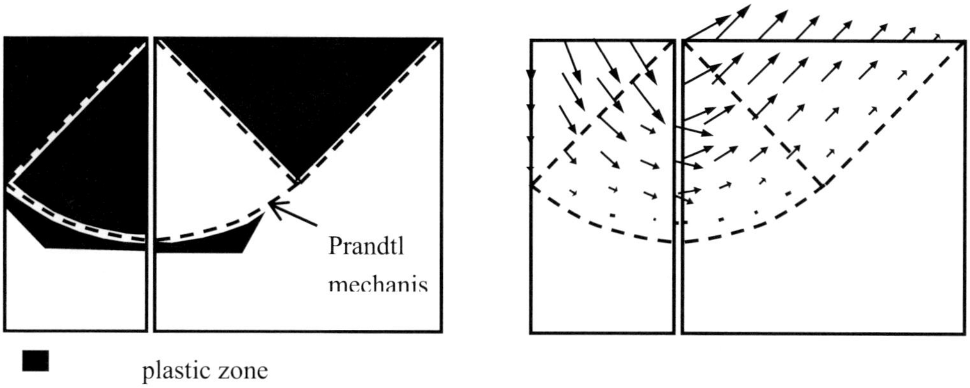

(a) Plastic zones for smooth rigid footing on (b) Velocity field for smooth rigid footing on purely cohesive soil on purely cohesive soil

Fig. (5.14). Development of plastic zones and velocity fields (after Sloan, 1989).

5.6.3 Finite Element Lower Bound Limit Analysis Using Nonlinear Programming

This example of the lower bound solution for a rigid strip footing on infinite halfspace cohesive-frictional soil is given by Lyamin and Sloan (2002a). The exact solution for the special case of weightless soil and no surchrage elsewhere on the surface is given by Prandtl (1920) as,

$$(q/c')_f = \{\exp(\pi \tan \phi') \tan^2(45 + \phi'/2) - 1\} \cot \phi' \quad \textbf{(5.76)}$$

where q, c' and ϕ' are the footing pressure, effective cohesion and effective friction angle respectively. Collapse $q/c' = 46.15$ for $\phi' = 35°$. Coarse, medium and fine meshes were used to compute the lower bound collapse pressure. Fig. (**5.15**) shows the medium mesh and the makeup of each of the meshes. The following soil properties have been adopted for the analysis: $c' = 1$; $\phi' = 35°$; $\gamma = 0$(for weightless soil). The results of the lower bound finite element limit analysis using non-linear programming are shown in Table **5.2**. Lyamin and Sloan (2002) compared their solutions with that of Sloan (1988) using linear programming (with active set method as described previously, a simplex method from IBM optimizations solutions library (OSL) and the interior-point method from the OSL). The non-linear two-stage algorithm (NLP) performed very well in demonstrating fast convergence to the optimum solution, which importantly does not result in rapid increase in the required number of iterations with respect to problem size. It was reported that for the coarsest mesh, the non-linear approach computed nearly 3 times faster than the active set linear programming method while giving a lower bound collapse pressure which is 2 percent better.

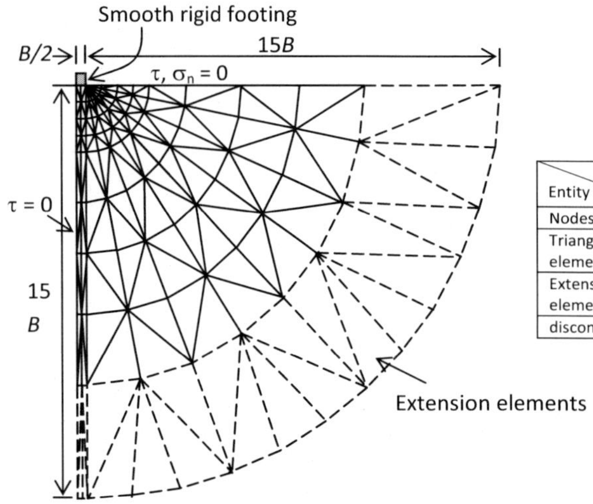

Mesh / Entity	Coarse	Medium	Fine
Nodes	228	712	1452
Triangular elements	66	220	452
Extension elements	8	14	26
discontinuities	101	335	697

Fig. (5.15). Lower bound mesh for rigid strip and smooth footing (after Lyamin and Sloan, 2002a).

With the fine mesh, the performance of the non-linear programming approach is even more superior, achieving a lower bound result which is only 1.2% lower than the exact solution while requiring only 5 seconds of CPU time. Compared to the active set method, this is a reduction by a factor of 57.

Moreover for the non-linear programming method, the number of iterations remains essentially constant, while the CPU time increases approximately linearly with respect to the mesh size as shown in Table **5.3**. It enjoys greater comparative advantage over the linear programming approach (especially the simplex and active set methods) in terms of computational time, the larger the problem size. The results also show that the OSL simplex method fared worse than the other two linear programming techniques in requiring more considerably more iterations and CPU time. The best performing linear programming technique is the OSL interior point method, which kept the number of iterations largely independent of problem size. However, compared to the non-linear programming technique, it required significantly longer CPU time while achieving less accuracy (as reflected in the % error in q/c').

Table 5.3. Lower bound solutions for smooth rigid strip footing.

Mesh	**Linear Programming**						**Non-Linear Programming**	
	Simplex (*m*=24)		**Active Set (*m*=24)**		**Interior-Point (*m*=24)**		**Lyamin and Sloan(2002a) (ε= 0.01)**	
	q/c' (error %)	CPU (s) (iterations)	q/c' (error %)	CPU (s) (iterations)	q/c' (error %)	CPU (s) (iterations)	q/c' (error %)	CPU (s) (iterations)
Coarse	37.791 (-18.1)	4.23 (926)	37.791 (-18.1)	1.71 (327)	37.791 (-18.1)	3.52 (27)	38.685 (-16.2)	0.60 (29)
Medium	43.032 (-6.7)	77.8 (4019)	43.032 (-6.7)	33.8 (1928)	43.032 (-6.7)	15.9 (42)	44.246 (-4.1)	2.3 (30)
Fine	44.064 (-4.5)	473 (10234)	44.064 (-4.5)	283 (5554)	44.063 (-4.5)	25.2 (29)	45.568 (-1.2)	5.0 (29)

m = number of sides in linearized yield surface

ε = convergence tolerance

CONCLUSION

Except for simple problems where the geometry are regular without complicated loadings, analytical solutions are not available for many stability problems. This chapter has introduced the concept where a problem can be discretised and solved in a systematic manner. The procedures and the mathematical techniques for the solution of the discretised analysis are demonstrated with case studies, and the numerical results are demonstrated to be consistent with the classical solutions for simple cases. The numerical approaches however have the advantage that they are applicable over varieties of problems, with irregular geometry and complicated loadings.

ACKNOWLEDGEMENTS

Declared none.

REFERENCES

Anderheggen, E. and Knopfel, H. (1972), Finite element limit analysis using linear programming, Int. J. Solids Structures, 8(12), pp1413-1431.

Baker, R. and Gaber, M., (1978). Theoretical analysis of the stability of slopes, Geotechnique, 28(4), pp395-411.

Baker, R. (1980). Determination of the critical slip surface in slope stability computations. International Journal Numerical and Analytical Methods in Geomechanics, 4(4), pp333-359.

Best, M.J. and Ritter, K. (1985), Linear programming: active set analysis and computer programs, Prentice-Hall, Englewood Cliffs.

Bishop A.W. (1955), The use of the slip circle in the stability analysis of earth slopes, Geotechnique, 5(1), 7–17.

Bottero, A., Negre, R., Pastor, J. and Turgeman, S. (1980), Finite element method and limit analysis theory for soil mechanics problems, Comput. Methods Appl. Mech. Engng, 22(1), pp131-140.

Celestino, T.B. and Duncan, J.M. (1981). Simplified search for non-circular slip surfaces, Proc. 10th Int. Conf. SMFE, Stockholm, Sweden, pp391-394.

Cheng, Y.M., Lansivarra, T. and Wei, W.B., (2007), Two-dimensional slope stability analysis by limit equilibrium and strength reduction methods, Computers and Geotechnics, 34, pp137-150.

Ciria, H., (2004), Computation of upper and lower bounds in limit analysis using second-order cone programming and mesh adaptivity, M.S. thesis, Massachusetts Institute of Technology, Cambridge, MA 02139.

Dawson, E.M., Roth, W.H. and Drescher, A. (199). Slope stability by strength reduction. Geotechnique, 49(6), pp835-840.

De Borst, R. and Vermeer, P.A., (1984), Possibilities and limitations of finite elements for collapse. Geotechnique, 34(2), pp199-210.

Duncan, M.J., (1996), State of the art: limit equilibrium and finite-element analysis of slopes, Journal of Geotechnical Engineering, 122(7), pp577-596.

Goh, A.T.C. (1999). Genetic algorithm search for critical slip surface in multiple-wedge stability analysis, Canadian Geotech. J., 36(2), pp382-391.

Greco, V.R. (1996), Efficient Monte Carlo technique for locating critical slip surface, J. Geotechnical Engineering, 122(7), pp517-525.

Griffiths, D.V. and Lane, P.A. (1999). Slope stability analysis by finite elements. Geotechnique, 49(3), pp387-403.

Herskovits, J., (1986), A two-stage feasible directions algorithm for nonlinearly constrained optimization, Mathematical Programming, 36, pp19-38.

Herskovits J. (1998), Feasible direction interior-point technique for nonlinear optimization, Journal of optimization theory and applications 99 (1), 121-146.

Janbu N. (1973), Slope stability computations, in: R.C. Hirschfield and S.J. Poulos (eds.), Embankment-Dam Engineering, pp. 47–86, John Wiley.

Krabbenhoft, K. and Damkilde, L. (2003), A general nonlinear optimization problem algorithm for lower bound limit analysis, Int. J. Numer. Methods Engng, 56(2), pp165-184.

Krabbenhoft, K., Lyamin, A.V., Hjaij, M. and Sloan, S.W., 2005, A new discontinuous upper bound limit analysis formulation, Int. J. Numer. Methods Engineering, 63(7), pp1069-1083.

Krabbenhoft, K., Lyamin, A.V. and Sloan, S.W. (2007), Formulation and solution of some plasticity problems as conic programs, Int. J. Solids Struct., 44(5), pp1533-1549.

Lyamin, A.V. and Sloan, S.W. (2002a), Lower bound limit analysis using non-linear programming, International Journal of Numerical Methods for Engineering, 55(5), pp573-611.

Lyamin, A.V. and Sloan, S.W. (2002b), Upper bound limit analysis using non-linear programming, International Journal of Numerical Methods in Engineering, 26(2), pp181-216.

Lyamin, A.V., Krabbenhoft, K., Abbo, A.J. and Sloan, S.W. (2005a), General approach for modelling discontinuities in limit analysis, Proc. 11th int. conf. Int. Assoc. Computer Methods and Advances in Geomechanics, Torino, 1, pp95-102.

Lyamin, A.V., Sloan, S.W., Krabbenhoft, K. and Hjiaj, M. (2005b), Lower bound limit analysis with adaptive remeshing, International Journal of Numerical Methods in Eng, 63(14), pp1961-1974.

Lysmer, J. (1970, Limit analysis of plane problems in soil mechanics, J. Soil Mech. Found. Div. ASCE, 96, No. SM4, pp1311-1334.

Makrodimopoulos, A. and Martin, C.M., (2006), Lower bound limit analysis of cohesive-frictional materials using second-order cone programming, Int. J. Numer. Methods Engng, 66(4), pp604-634.

Malkawi, A., Hassan, W. and Sarma, S. (2001), Global search method for locating general slip surface using Monte Carlo techniques, Journal of Geotechnical and Geoenvironmental Engineering, 127(8), pp688-698.

Oden, J.T. (2006), Finite elements of nonlinear continua, reprint, Dover Pubs, Inc. Mineola, NY.

Pastor J. and Turgeman S. (1976), Mise en ceuvre numerique de methods de l'Analyse limite pour les materiaux de von Mises et de Coulomb standards en deformation plane, Mech. Res. Comm. 3, 469-476.

Pastor, J. (1978), Analyse limite: determination numerique des solution statiques completes: application au talus vertical, J. Mecanique Appl.,2(2), pp167-196 (in French).

Prandtl, L., 1920, Uber Die Haerte Plastischer Korper, Nachrichten Von Der Koeniglichen Gesellschaft Der Wissenschaften Zu Geottingen, Mathematisch-physikalische Klasse, pp. 74-85.

Shield, R. T., 1954, "Plastic Potential Theory and the Prandtl Bearing Capacity Solution", Journal of Applied Mechanics, Vol. 21, No.2, June, pp. 193-194.

Sloan, S.W. (1979). Numerical prediction of collapse loads using finite element methods. MPhil thesis, University of Cambridge, UK.

Sloan, S.W. (1981). Numerical analysis of incompressible and plastic solids using finite elements. PhD thesis, University of Cambridge, UK.

Sloan, S.W. and Randolph, M.F. (1982), Numerical prediction of collapse loads using finite element methods. Int. Journal of Numerical Analy. Methods Geomech, 6(1), pp47-76.

Sloan, S.W. (1988), Lower bound limit analysis using finite elements and linear programming, International Journal for Numerical and Analytical Methods in Geomechanics, 12(1), pp61-77.

Sloan, S.W., (1989), Upper bound limit analysis using finite elements and linear programming, International Journal for Numerical and Analytical Methods in Geomechanics, 13(3), pp262-283.

Sloan, S.W. and Assadi, A. (1991), Undrained stability of a square tunnel in a soil whose strength increases linearly with depth, Computer and Geotechnics, 12(4), pp321-346.

Sloan, S.W. and Assadi, A., (1992), The stability of tunnels in soft ground, Proceedings Peter Wroth memorial symposium on predictive soil mechanics, Oxford, pp644-663.

Sloan, S.W. and Kleeman, P.W. (1995), Upper bound limit analysis using discontinuous velocity fields. Computer Methods in Applied Mechanics and Engineering, 127, pp293-314.

Sun J., Li, J. and Liu, Q., 2008, Search for critical slip surface in slope stability analysis by spline-based GA method', Journal of Geotechnical and Geoenvironmental Engineering, 134(2), pp252-256.

Turgeman, S. and Pastor, J., (1982), Limit analysis; a linear formulation of the kinematic approach for axisymmetric mechanic problems, Int. J. Numer. Analy. Methods Geomech., 6(1), pp109-128.

Toh, C.T. and Sloan, S.W. (1980). Finite element analysis of isotropic and anisotropic cohesive soils with a view to correctly predicting impending collapse. Int. J. Numerical Analy. Methods in Geomech., 4(1), pp1-23.

Ukritchon, B., Whittle, A.J. and Sloan, S.W. (2003), Undrained stability of braced excavations in clay, J. Geotech. Geoenviron. Div. ASCE, 129(8), pp738-755.

Zienkiewicz, O.C., Humpheson, C. and Lewis, R.W. (1975). Associated and non-associated visco-plasticity and plasticity in soil mechanics. Geotechnique, 25(4), pp671-689.

Zouain N., Herskovits, J., Borges, L.A. and Feijoo, R.A., (1993), An iterative algorithm for limit analysis with nonlinear yield functions, International Journal of Solids and Structures, 30(10), pp1397-1417.

CHAPTER 6

Applications of Plasticity Theory and Limit Analysis to the Bearing Capacity of Shallow Foundations

Abstract: In this chapter, the plasticity theory (slip line analysis) and limit analysis will be applied to the bearing capacity problems for illustration. Through these applications, the readers will be able to strengthen their understanding on the various types of stability analysis methods.

Keywords: Axisymmetry, Bearing capacity, Footings/foundations, Geotechnical centrifuge modelling, Method of characteristics, Numerical modelling, Physical modelling, Plane strain, Plasticity, Sands, Shear strength, Silts, Stress analysis.

6.1. OUTLINE

The method of characteristics which has been introduced in chapter 3 is used to establish consistent factors for the vertical bearing capacity of circular and strip footings on soil which satisfies the linear (c, Ø) Mohr-Coulomb strength criterion. This method of solution avoids the assumption of arbitrary slip surfaces, and failure zones are automatically generated within which equilibrium and plastic yield are simultaneously satisfied for the given boundary stresses. Although similar solutions have previously been published for circular footings, their application has been hindered by errors and confusions over terminology which are now resolved and is explained by the method of solution in this chapter. For bearing capacity problems, it is well-known that Terzaghi's superposition of bearing capacity terms containing N_q, N_γ, and N_c is both

Y. M. Cheng, H. Wong, C. J. Leo & C. K. Lau

safe and sufficiently accurate for circular footings as well as for strip footings. The values to be adopted are tabulated as functions of Ø for ease of application. It is also found that the differences between the factors applicable to circular and strip footings far exceed the empirical shape factors commonly in use. Some new shape factors are suggested that can better represent the relationship between the limiting equilibrium of circular and strip foundations. Some current shape factors attempt to relate the axisymmetric (triaxial) and plane strain soil parameters. This approach cannot succeed, as the relationship between strength parameters depends also strongly on the relative density. The new bearing factors which are proposed facilitate a more rational approach in which the soil parameters appropriate to the geometry can first be determined and then used to find the appropriate bearing capacity factors.

The solutions so far were based either on a straight (c, Ø) envelope or simply on a constant angle of shearing, Ø. For granular soils, sec Ø usually varies linearly with the logarithm of mean effective stress. A new method of calculation permits Ø to vary throughout the stress field as an arbitrary (or empirical) function of stress. This method is verified for both plane-strain and axisymmetric conditions by forcing a variation in sec Ø equivalent to generating a constant-cohesion envelope, for which solutions already exist. The variable- Ø analysis is used to demonstrate the highly significant effect of stress variation around and beneath a footing. Finally, it is shown that an equivalent constant value $Ø_m$ can be derived empirically, using the new solutions to identify an equivalent mean effective stress p_m. However, only the variable- Ø solution can simultaneously capture the bearing capacity and the geometry of the bearing mechanism.

This approach is validated here for the case of model circular footings on dense beds of silica sand and silica silt. The models were tested at 1 **g** with surcharge to explore the N_q behaviour, and in a centrifuge to determine self-weight effects for N_γ. It is shown that triaxial Ø values expressed as a function of the logarithm of p can be used to predict model bearing capacities within a deviation in Ø of 2°.

It should be noted that part of this chapter has previously been published in Canadian Geotechnical Journal, 1993, 30(6): Geotechnique, 2011, 61(8): 627-638 and Geotechnique, 2011, 61(8): 634-650.

6.2. INTRODUCTION

The derivation of bearing capacity for foundations for frictional soil requires the relation between the strength parameters and the effective stresses, and then on the use of bearing capacity factors. This chapter will establish the approach of modelling the strength envelope by a simple constant- Ø relation; which is later expanded into a more general (c, Ø) envelope. The objective is to derive the corresponding estimates of the bearing capacity factors in both plane and axisymmetric load cases under a more realistic formulation.

The method of characteristics proposed by Sokolovskii (1960) is used here. This method assumes that limiting stresses have been reached at every point inside the solution domain, and the plastic equilibrium is determined with the applied load. The requirement of a trial slip surface as adopted in the limit equilibrium analysis by Terzaghi (1943) and Meyerhof (1951) can then be avoided. There are various doubts regarding the method of characteristics which include:

(i) The difficulty in the assigning boundary conditions, particularly where the mobilisation of the tangential friction should be such as to oppose the relative motion and when the kinematics of plastic soil strain is itself uncertain.
(ii) The difficulty in accepting the assertion that certain zones (*e.g.* in

the far field or in a wedge/cone 'trapped' beneath the centre of a footing) are not in a limiting state simply because the characteristics have not been extended that far.

(iii) It is also possible that yield is not fully achieved within a small region bounded by the major failure mass. The use of yield criteria for the whole solution has been questioned by many researchers.

There are various numerical methods such as the finite element/boundary element/distinct element technique which can account for kinematics and equilibrium everywhere, and can be used to derive bearing capacities in close agreement with simpler solutions, albeit with more computing effort (Griffiths, 1982) under some cases. The author has also found that sometimes it is not easy to reproduce the plasticity solution by the use of commercial programs, and the use of method of characteristic appear to be superior in the simple cases. The limiting equilibrium following Sokolovskii (1960) is more strenuous than that following Terzaghi (1943) and is also not computation intensive. The objective of this chapter is not to offer empirical evidence, however, but to discuss the extension of the method of characteristics to axisymmetric (circular or, approximately square) footings. The characteristic solutions have been found for axisymmetric footings by Cox *et al.,* (1961) and Cox (1962) which can provide estimates which are rather different from the application of commonly used shape factors as appear in many geotechnical design codes. Solutions by the method of characteristics should logically form the basis of engineering judgement for circular footings.

Triaxial test results by De Beer (1965b), shown in Fig. (**6.1**), have clearly shown that scc Ø is not a material constant for granular soils, but is extremely sensitive to the mean stress level. For the purpose of this chapter, all stresses are effective stresses unless otherwise stated. The vertical stress in the supporting soil ranges from the bearing pressure σ_f directly under the footing to the overburden pressure σ_o on either side (see Fig. **6.2**). The stress can vary by a factor of up to 1,000 for high Ø values in constant- Ø analysis. The contact stress distribution under the footing is also highly non-uniform. The stress tends to be highest at the centre and eases off rapidly towards the edge, especially for shallow circular footings. Stress variations therefore cause $Ø_m$ to vary significantly from point to point.

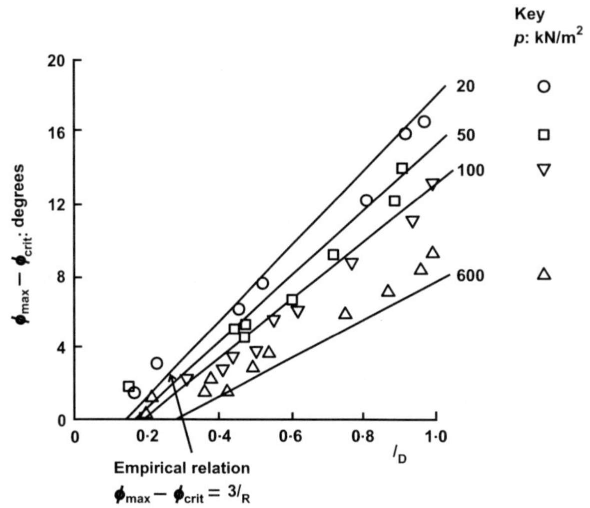

Fig. (6.1). Triaxial test data for Berlin sand (after De Beer, 1965a).

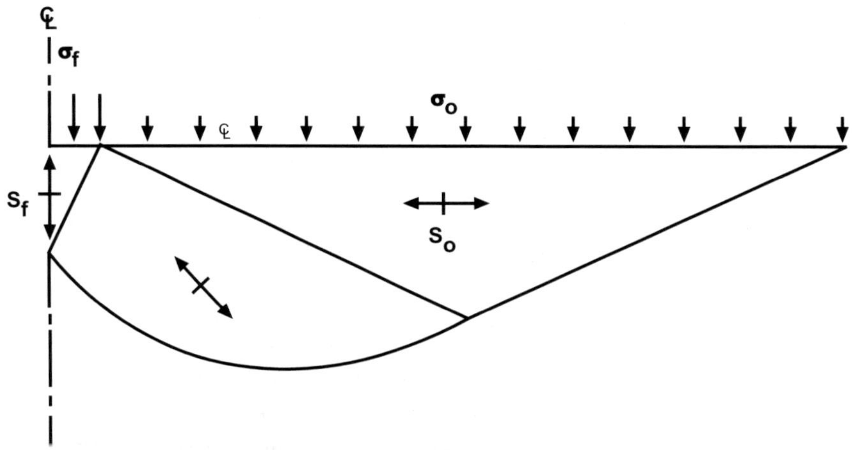

Fig. (6.2). Principal stress directions beneath a collapsing strip footing on weightless soil.

Constant- Ø analysis ignores the 7° change in $Ø_{max}$ as a result of the tenfold change in confining pressure observable in Fig. (**6.1**), and typical for a dense sand. Such analysis is therefore incapable of accurately representing the true situation. Ueno *et al.,* (1998) reported good agreement with test data for both strip and circular footings in their non-linear finite-element analysis, after incorporating the confining- stress-dependent shear strength. Kumar and Khatri (2008) reported similar results for smooth strip footings.

A new form of analysis that takes the pressure effects on Ø into account has been investigated by Lau and Bolton (2011a). Their work will be reported below. Essentially, the calculations will use the Ø value corresponding to the locally prevailing stress level.

Bolton's (1986) work on the strength and dilatancy of sands accounts for the variation in secant angle of Ø rather than attempting to define the Mohr-Coulomb envelope (see Fig. **6.3a**). Attempts to fit a straight c- Ø envelope (Kutter *et al.*, 1988) proved successful when the envelope was fitted over the appropriate stress range. Fig. (**6.3b**) demonstrates, however, that for a stress range of 0 to 10 MPa that involves crushing of particles at the high end of the stress range, no simple envelope could be selected (Vesic and Clough, 1968).

In their introduction to foundation problems, Terzaghi and Peck (1948) developed the classical bearing capacity equation, and defined bearing capacity factors N_q, N_γ, and N_c in terms of a linear strength envelope $\tau_{max} = c + \sigma \tan Ø$. They advised that peak strength parameters should be used for c and Ø only when the supporting soil would fail in 'general shear'. They postulated that, otherwise, shear would occur beneath, but not around, the foundation, and proposed that in this case of 'local shear' the peak strength should be reduced by a factor of 1.5 to achieve satisfactory designs. This idea was taken up by Vesic (1963). However, subsequent investigators have found difficulty in separating various effects that Terzaghi and Peck were unable to discriminate between. These include: a non-linear peak strength envelope; progressive failure due to strain localisation and softening to critical states; corrections for depth/width ratio by including an allowance for increase of depth (*i.e.* settlement) during loading; anisotropy; and an allowance for footing shape between a long strip (plane strain) and a circular pad (axisymmetric), which itself must comprise partly a statement regarding appropriate strength parameters (plane strain or triaxial) and partly a recognition that the state of limiting equilibrium is influenced by 3D effects, arching and so on.

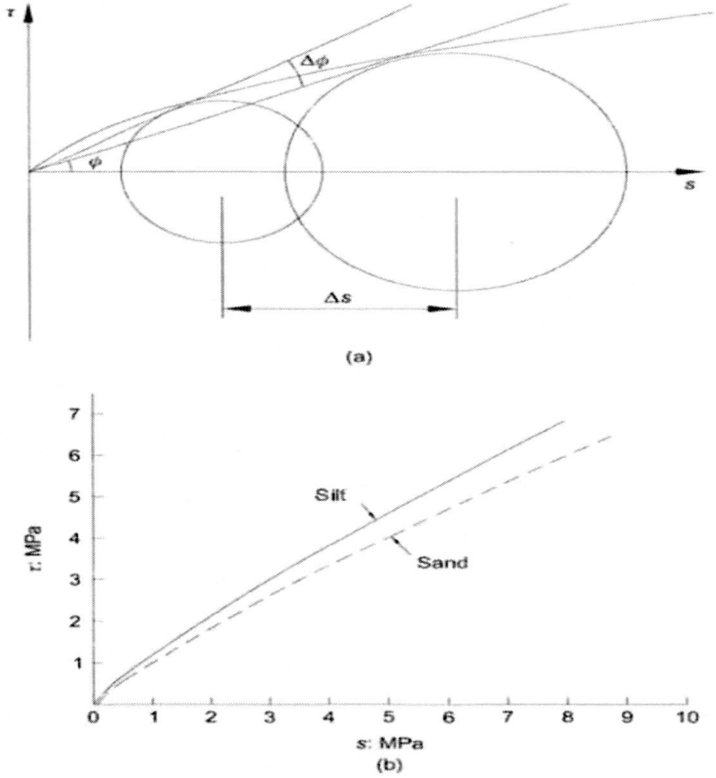

Fig. (6.3). (**a**) Typical shear strength for sands; (**b**) shear strength for two granular materials.

De Beer (1965a) reported triaxial tests on sand in which the strength envelope was non-linear, and characterised by a sec Ø value that increased linearly with the relative density I_D and reduced linearly with the logarithm of mean stress p. Bolton (1986) confirmed that these observations were a general feature of the peak strength of sands, and suggested an empirical relation between Ø, I_D and log p that took account of the greater strength observed in plane strain compared with axisymmetric (triaxial) strain. De Beer (1965b) developed an understanding of the possible scale effect on footings in terms of a reduction in sec Ø with p: the bearing capacity factors for large footings carrying high stresses should logically be smaller than those for small footings.

The work of De Beer shows that it is essential to re-create correct stress levels in model tests if correct behaviour is to be observed. Ovesen (1975) explained the use a geotechnical centrifuge to achieve full-scale stresses in reduced-scale models of foundations. Ovesen (1979) went on to show that the effect of reducing footing width B while retaining particle size d is negligible, at least when $B/d > 30$.

The non-uniform approach to peak strength, and especially the possibility of progressive failure due to strain concentration in soils with a brittle peak strength, was investigated by Muhs (1965) in footing tests at 1 **g**. Yamaguchi *et al.*, (1976, 1977) performed centrifuge tests on model footings in which soil strains were measured around the footing, leading to an understanding of the progressive mobilisation of strength in different regions. Non-uniformity of strength, due not to the strain distribution, but to anisotropy leading to reduced Ø values on slip surfaces parallel to soil bedding, was considered by Tatsuoka *et al.*, (1991). Fig. (**33**) shows, in sketches, the variations of Ø that might concern a practising engineer who wishes to use some bearing capacity equation to predict the plastic indentation of footings on sand.

In the investigation that follows it has been assumed, following Corte (1989) and Cerato and Lutenegger (2007), that stress level effects in Fig. (**6.4c**) are paramount in the bearing capacity problem. The chapter will start by discussing the use of plasticity theory to predict the bearing capacity of footings where Ø varies as in Fig. (**6.4c**), and explains how an equivalent constant Ø might be deduced. Other influences in Fig. (**6.4b** and **6.4d**) will be discussed in the light of the experimental findings. There is some interest shown from the oil exploration industry in shallow foundation design principles. This is a result of the on-going usage of spudcan foundation for offshore oil rigs (LeBlanc, 1981). Like interpretation of penetrometer data, the main point of interest for spudcan design has always been the evaluation of failure loads. This is because spudcans have to be pushed into the seabed to a predetermined depth with ballast on the rig during preloading before unloading the rig to its normal working conditions. This is partly a safeguard against accidental overloading and scouring of the seabed by under-currents, which may undermine their foundations during operation. As spudcans can normally reach 20 m in diameter (Randolph *et al.*, 2005), scale effects therefore remain a problem to be resolved.

The methodology adopted for this experimental validation exercise (Lau and Bolton, 2011b) is as follows.

(a) It was decided to acquire two granular materials that differed in nominal diameter by a factor of 50, but which were otherwise practically indistinguishable in all other grain characteristics. The idea was that they could be used as model materials for each other.

(b) Soil models of different sizes were constructed using the same material for both 1 **g** and centrifuge tests, in order to investigate the relative particle size effects. It has been common practice in Cambridge (Schofield, 1980) to use models of models. The idea is model a centrifuge model with another centrifuge model at a different scale in order to check their internal consistency. The major disadvantage is that the differences in scale between two such centrifuge models are not generally larger than seven.

(c) Soil models were constructed of a particular footing bearing on each of the two model soils of different particle size. The footing: particle size ratio could then be varied over a wide range, while most of the existing laboratory apparatus and equipment could be utilised without much adaptation. This approach inevitably introduced some complications as a result of changing the grain size. In particular, any absolute particle size effects had to be accounted for when interpreting the experimental results.

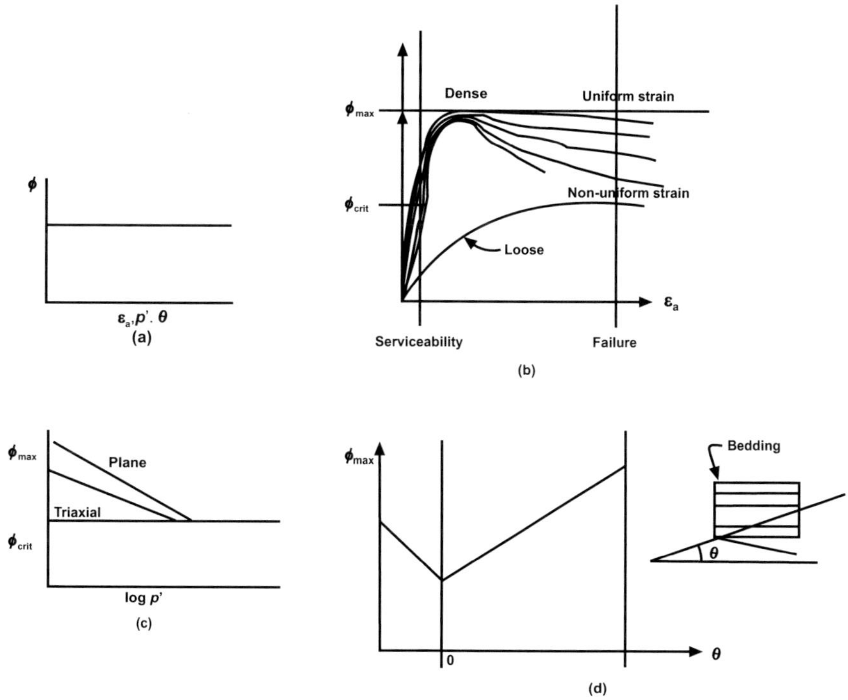

Fig. (6.4). The variations of Ø that might concern a practicing engineer: (**a**) ideal soil: bearing capacity equation; (**b**) $\varnothing = f(\varepsilon_a)$; (**c**) $\varnothing_{max} = f(p')$; (**d**) $\varnothing_{max} = f(\theta)$.

The soil bed under 1 **g** condition and with surcharge is designed to derive its resistance to the advancing punch overwhelmingly from the surcharge term of Terzaghi's equation,

$$\sigma_f = \sigma_o N_q + \frac{1}{2} B\gamma' N_\gamma \quad \textbf{(6.1)}$$

where σ_o is the surcharge, N_q is the bearing capacity factor (surcharge) and N_γ is the bearing capacity factor (self-weight). To achieve this, the surcharge is applied such that the self-weight at 1 **g** when compared with the surcharge is negligible. Under elevated **g** conditions in a geotechnical centrifuge, however, the soil can derive its resistance to the advancement of the punch from its self-weight alone. In the event that settlement is significant, correction should be made to account for the effective increase in overburden equivalent to the surcharge and other geometric effects. By taking advantage of these considerations, it is possible to uncouple the scale effects on N_q and N_γ. The data are then used to validate the theoretical predictions made earlier based on plastic solutions by the method of characteristics, and permitting Ø to vary with p.

6.2.1. The Method of Characteristics

Sokolovskii's (1960) solution which was mainly targeted towards plane strain condition for strip footing was generalised by Shield (1955) for the axisymmetric case with Tresca (cohesive) material. Cox *et al.,* (1961) had obtained axisymmetric solutions for a general (c, Ø) weightless material, and later Cox (1962) had included self-weights into the analysis. Cox had used a very strange substitution, namely $c^* = c + \sigma_o \tan \varnothing$, where σ_o is an arbitrary surcharge (without much theoretical background) defined as the atmospheric pressure (≈ 100 kPa), and a dimensionless parameter $G = \frac{\gamma B}{2c^*}$.

Even though there is some apparent misunderstanding of the physical principle of effective stress, this has not alter the mathematical acceptability of Cox's solution. His result has however simply hindered the use by engineers because of the likelihood of misinterpretation of the terms. Furthermore, Cox had lumped the surcharge (N_q) and self-weight (N_γ) effects together without separate consideration, which avoids Terzaghi's (1943) superposition assumption but makes the results very difficult for the engineers to interpret and use.

Larkin (1968) had produced a more meaningful set of non-dimensional parameters, normalising stresses by dividing by $0.5\gamma B$ and distances by dividing by 0.5B, which then produced results for plane strain condition that differed by a factor of two when comparing with those of Prandtl (1920) which was generally accepted to be exact. These extensions of bearing capacity theory are unnecessarily complex and confusing and restricted to too narrow arange of Ø not adequate for actual application. Although their mathematical formulation for axisymmetric solutions was correct, the engineers are still relying on the use of empirical shape factors. To overcome such situation, the method of characteristics to plane or axisymmetric footings will be recast below in a form that should prove useful in practice.

The following assumptions are fundamental to this approach:

(i) The soil is rigid-plastic without volume change, and the change-of-geometry is neglected, and imperfections due to strain softening and progressive failurc are ignored.

(ii) The Mohr-Coulomb yield criterion is adopted so that its states of plastic equilibrium are consistent with a straight envelope acting as a tangent to the Mohr circle of stress Fig. (**6.5a**). The cohesion intercept can also be dealt with later. The shear strength τ_f is then related to the mean stress s by $\tau_f = s \sin Ø$.

(iii) The intermediate principal stress is not considered in the yield criterion, but it enters into the axisymmetric analysis as the hoop stress σ_θ which influences the radial equilibrium. The Haar and von Karman (1909) hypothesis which states that σ_θ should be equal to one of the two principal stresses lying in the axial plane, *i.e.* either $\sigma_\theta = \sigma_1$ or σ_3 is assumed here. Following the approach by Cox *et al.,* (1961), it is set to be equal to the minor principal stress, *i.e.* as small as possible. Lau (1988) has obtain support from finite element analysis of punch indentation that supports this assumption to be reasonable, which must generally be either true or safe, and it will be shown later that the effect of σ_θ will increase the bearing capacity.

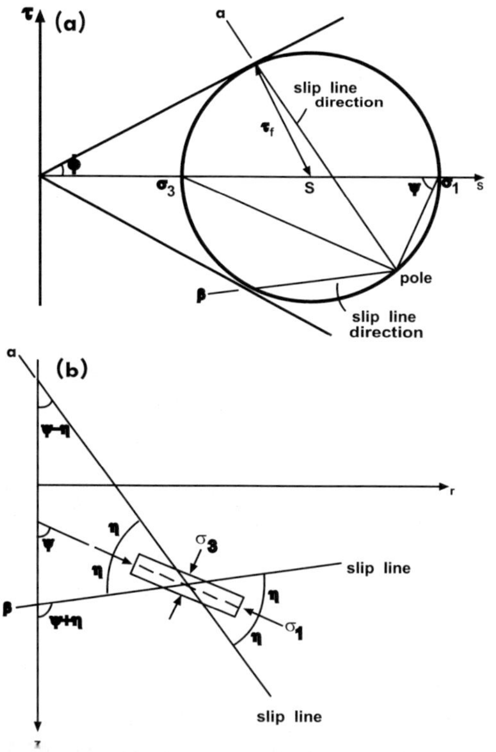

Fig. (6.5). (**a**) Mohr circle under failure condition; (**b**) Sign convention and notation.

The mathematical formulation of axi-symmetry problem can now broadly follow Larkin (1968). The equations of equilibrium for a toroidal element Fig. (**6.6**) can be written in cylindrical co-ordinates r, θ, z as

$$\frac{\partial\sigma_r}{\partial r}+\frac{\partial\tau_{rz}}{\partial z}+\frac{\sigma_r-\sigma_\theta}{r}=0,\qquad \frac{\partial\tau_{rz}}{\partial r}+\frac{\partial\sigma_z}{\partial z}+\frac{\tau_{rz}}{r}=\gamma \tag{6.2}$$

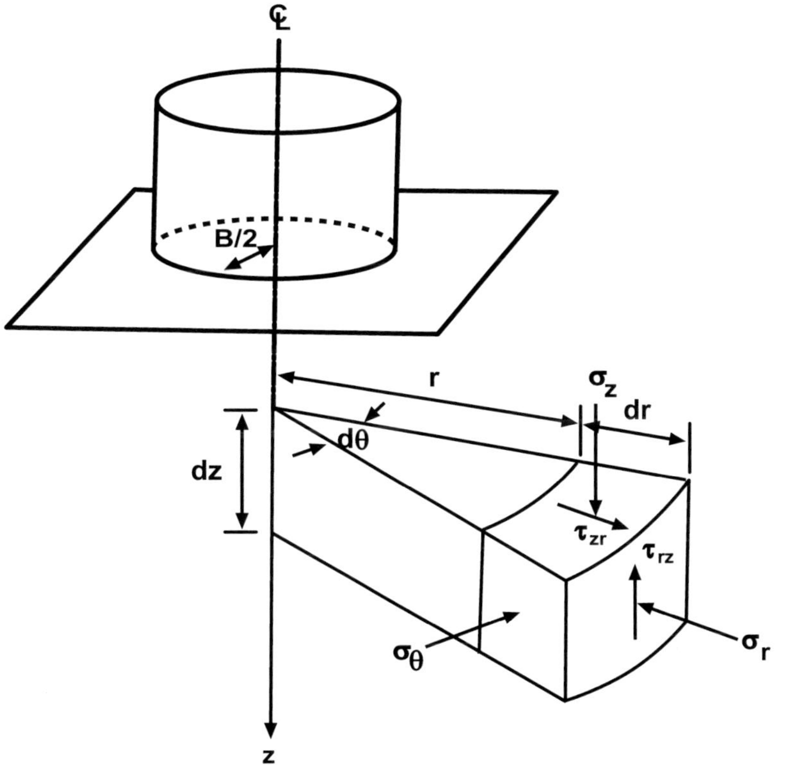

Fig. (6.6). Cylindrical coordinate system and stress components.

The four stress components can be expressed in terms of mean stress s and the inclination φ substended to the z axis by the major principal stress Fig. (**6.6b**):

$\sigma_r = s(1 - \sin\emptyset\cos2\varphi),\quad \sigma_z = s(1 + \sin\emptyset\cos2\varphi),$

$$\tau_{rz} = s\ \sin\emptyset \sin 2\varphi, \qquad \sigma_\theta = \sigma_3 = s(1 - \sin\emptyset) \qquad \textbf{(6.3)}$$

The characteristic lines that will embody the solution are lines α and β on which $\emptyset$ is mobilised. In the literature they are often referred to as slip lines, but their significance is in relation to the equations of equilibrium, not those of displacement. The geometry dictates that their slopes can be written as

$$\frac{dr}{dz} = \tan(\varphi - m\eta) \qquad \text{where } \eta = \frac{\pi}{4} - \frac{\emptyset}{2} \qquad \textbf{(6.4)}$$

and m takes the value +1 for an α line and -1 for a β lines as (shown in Fig. **6.6b**). These equilibrium and yield equations are a set of hyperbolic partial differential equations that reduce to two ordinary differential equations expressing the increase in stress along each characteristic line ($m=\pm1$) in terms of the changing inclination φ and position (r, z):

$$ds\cos\emptyset + m2s\ \sin\emptyset d\varphi + n\frac{s}{r}[\sin\emptyset\cos\emptyset dr + m(\sin^2\emptyset - \sin\emptyset)dz =$$
$$\gamma(-m\ \sin\emptyset dr + \cos\emptyset dz) \qquad \textbf{(6.5)}$$

The factor n has been introduced for convenience; in axisymmetric analysis for circular footings it takes the value +1, whereas it happens that plane solutions for long strip footings are given if n is set to zero.

6.2.2. Method of Computation

Following Larkin (1968) the variables will now be normalised according to a scale length of 0.5B, so we will take

$$\Sigma = \frac{s}{0.5B\gamma}, \qquad R = \frac{r}{0.5B}, \qquad Z = \frac{z}{0.5B} \qquad \textbf{(6.6)}$$

Equation 6.5 can now be written in finite difference form, suitable for the solution of the intersection of an α line, which passed through a known 'point' $(R_1, Z_1, \Sigma_1, \varphi_1)$, and a β line, which passed through another known "point" $(R_2, Z_2, \Sigma_2, \varphi_2)$. Shi (1988) suggested the following substitutions:

$$A = \frac{-2\Sigma_1}{(R+R_1)\cos\emptyset}[\sin\emptyset\cos\emptyset(R-R_1) + (\sin^2\emptyset - \sin\emptyset)(Z-Z_1)] -$$
$$(R-R_1)\tan\emptyset + (Z-Z_1) + \Sigma_1 + 2\Sigma_1\tan\emptyset\varphi_1$$

$$B = \frac{-2\Sigma_2}{(R+R_2)\cos\emptyset}[\sin\emptyset\cos\emptyset(R-R_2) - (\sin^2\emptyset - \sin\emptyset)(Z-Z_2)] +$$
$$(R-R_2)\tan\emptyset + (Z-Z_2) + \Sigma_2 - 2\Sigma_2\tan\emptyset\varphi_2 \quad \textbf{(6.7)}$$

The finite difference equations can now be written as

$$(R-R_1) = (Z-Z_1)\tan(\varphi_1 - \eta), \quad (R-R_2) = (Z-Z_2)\tan(\varphi_2 + \eta) \quad \textbf{(6.8)}$$

$$\Sigma = \frac{A\Sigma_2 + B\Sigma_1}{\Sigma_1 + \Sigma_2}, \quad \varphi = \frac{B-A}{-2\tan\emptyset(\Sigma_1 + \Sigma_2)} \quad \textbf{(6.9)}$$

which are suitable for iteration. If in (Fig. **6.7**), points P and Q are known, and W is to be determined, R_W and Z_W can be found from eq. 6.8 and then φ and Σ can be calculated from eq. 6.9, by putting $\varphi_1 = \varphi_p$ and $\varphi_2 = \varphi_Q$. However, in general, the characteristics are curved and the solution can be improved by updating φ, putting $\varphi_1 = \frac{\varphi + \varphi_P}{2}$ and $\varphi_2 = \frac{\varphi + \varphi_Q}{2}$ and repeating until there is convergence (within some target accuracy) to a stable set of values (Sokolovskii, 1960).

Proceeding from the known boundary condition (normal stress σ_o) on the free surface KS (Fig. **6.8**), the entire stress field within KLMO can be determined. Firstly, a value for R_K is assumed. Secondly, the boundary KS is subdivided into a set of equally spaced points. The solution then marches towards the footing boundary. When the calculation is complete, it should be checked to see if the β characteristic starting from K actually finishes at O. If not, R_K is adjusted according to whether it is too far or too near. The whole calculation process iterates until an acceptable accuracy is achieved (see also chapter 3.2.2).

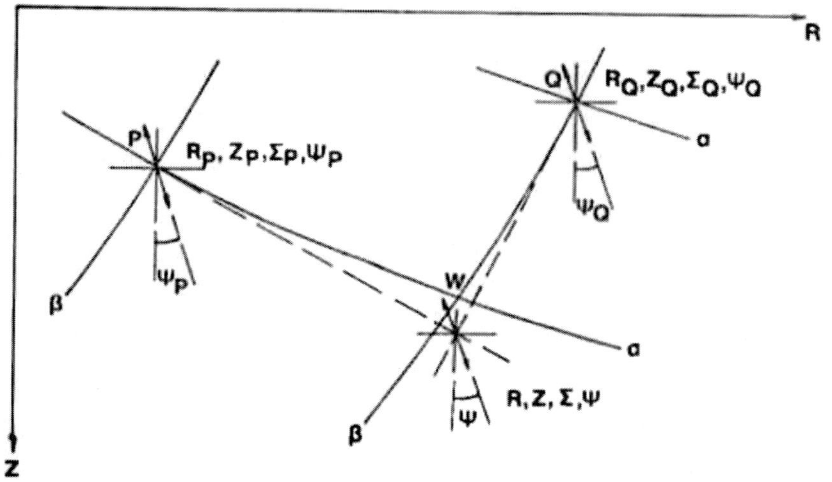

Fig. (6.7). Computation scheme of new point W from known points P and Q.

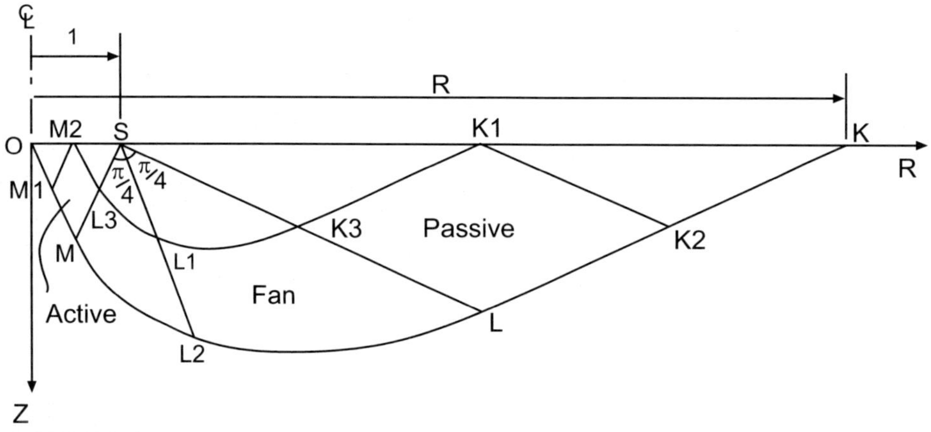

Fig. (6.8). A typical stress characteristic mesh.

A coarse mesh developed for plane strain case with weightless soil as shown in Fig. (**6.8**) will be used to illustrate the marching scheme used in the computer program CONPHI developed based on the above theory. Again, the readers should have no difficult to devise a similar one.

(1) Passive zone – The boundary conditions at S, K1 and K should be known. K2 is found by using K1 and K; K3 is found by S and K1; L is found by K3 and K2. S, K3 and L become the known boundary conditions for the fan zone.

(2) Fan zone – Node S can be viewed as a degenerated β characteristic with unique R and Z, but varying φ and Σ. The first two terms of eq. 6.5 demonstrate that the significant effect in the fan zone is that stress increases exponentially with rotation $\Delta\varphi$ of the β line from K3 to L1, for example, L1 being found from S (at $\Delta\varphi = \pi/4$) and K3. Similarly, L2 is found from L1 and L, L3 from S (at $\Delta\varphi = \pi/2$) and L1, and M from L3 and L2. S, L3 and M now become the known boundary conditions for the active zone.

(3) Active zone – On the footing contact plane, Z is known and the value of φ, being a function of mobilised friction depends on the detailed kinematics, which are a function of dilatancy are not available in this equilibrium solution. A simple extreme case is provided by a frictionless interface for which $\varphi = 0$. The solution for Σ and R at M2 can therefore be found from the condition at L3; then M1 can be found from M2 and M; finally, Σ and R can be found at O using M1 together with the known values of Z and φ at O. O is intended here to home in onto the centreline.

At this stage, all the variables on the footing contact plane are known. The bearing pressure σ_z can be found by multiplying Σ with $0.5B\gamma$ to give s, substituting this, together with φ, into eq. 6.3. The mean bearing pressure under the footing can then be found by numerical integration to give the bearing capacity factors.

6.2.3. Principle of Superposition

The method of characteristics derived above can solve for the bearing capacity of footings with surcharge and soil self-weight. Terzaghi (1943) assumed that the bearing capacity of a foundation on weightless frictional soil due to surcharge σ_o could be calculated and expressed as $\sigma_o N_q$. If the bearing capacity of the foundation due to the self-weight γ alone could be written $0.5\gamma BN_\gamma$, then these components could safely be superposed which will be safe for design purpose. Decomposition of the total bearing capacity into three components has, following Terzaghi, been adopted in practice in nearly all the foundation design codes. The validity and utility of this practice, and its possible extension for circular footings, are now examined.

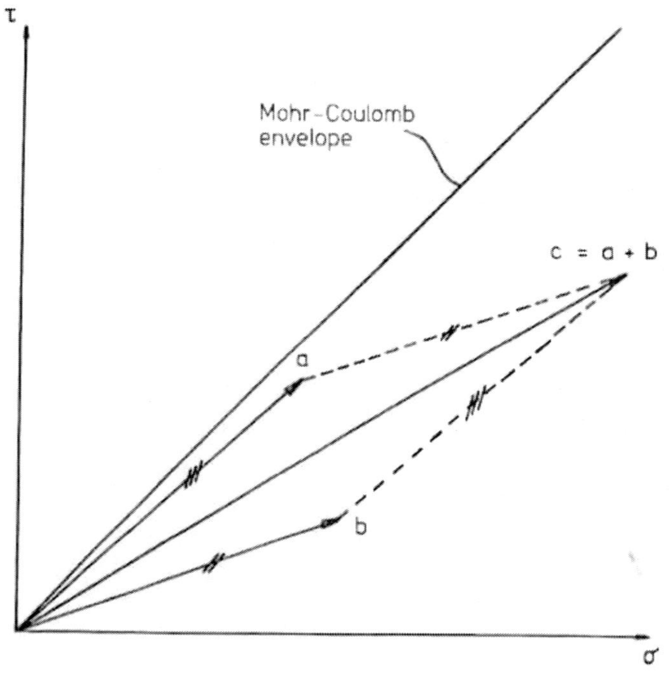

Fig. (6.9). Effects of superposition.

It can easily be demonstrated that superposition must be conservative for materials that obey a linear Mohr-Coulomb envelope with constant $\emptyset$. Suppose, in Fig. (**6.9**), that vector **a** represents the stress on any plane at a point inside a soil body in limiting equilibrium subject to load case A (*e.g.* surcharge plus bearing capacity), while **b** represents the stresses on the same plane due to limiting equilibrium in some other load case B (*e.g.* self-weight plus bearing capacity). If loads A were separately in equilibrium, as were loads B, then the combination C = A + B would also be in equilibrium. Furthermore, the stress **c** would be the vector addition **a** + **b**, so if **a** and **b** separately satisfied Mohr-Coulomb so must **c**. In general, the critical planes mobilisng $\emptyset$ at the point in question will not coincide in the two load cases, and the maximum angle mobilised on any plane in case C will be less than $\emptyset$: hence the conservatism of superposition.

Davis and Booker (1971) had performed rigorous checks on the superposition assumption for the plane strain case and found that it to be a conservative approach. The error was no more than 20% for $\emptyset$ in the range $20^{\circ} - 40^{\circ}$. Cox (1962) had not made such superposition assumption for circular footings and used a dimensionless parameter $G = \frac{0.5B\gamma}{\sigma_o \tan\emptyset}$ to show that solutions depended in a non-linear relation on the relative magnitude of the self-weight term and the surcharge term. It is now proposed to treat σ_o not as atmospheric pressure following Cox, but as an effective surcharge applied to the plane surface of the soil around the foundation. A new dimensionless parameter called the superposition factor $\Omega = \frac{\sigma_o}{0.5B\gamma}$ will be used to demonstrate the relative importance of surcharge and self-weight effects, and the total bearing capacity σ_f will be related to a combined bearing capacity factor $N_{q\gamma} = \frac{\sigma_f}{0.5B\gamma+\sigma_o}$. Fig. (**6.10**) shows the effect of Ω on $N_{q\gamma}$ for smooth circular footings on soil with $\emptyset = 40°$. By converting Cox's data into the new format, it can be seen that the new calculation produces results that are practically identical, although the numerical solution techniques are different. It can also be seen that $N_{q\gamma}$ remains constant for Ω

> 10^3 which corresponds to the limit of N_q. Likewise, $N_{q\gamma}$ remains constant at N_γ for $\Omega < 10^{-3}$. These limits can be used to derive an alternative superposition estimate which are on the safe side with less than 20% error in the vicinity of $\Omega = 1$. Since this result is typical of all those cases that have been checked, and the complexity of the bearing capacity tables is much reduced if only N_q and N_γ are required in each case, the general practicability of Terzaghi's (1943) superposition is hence confirmed.

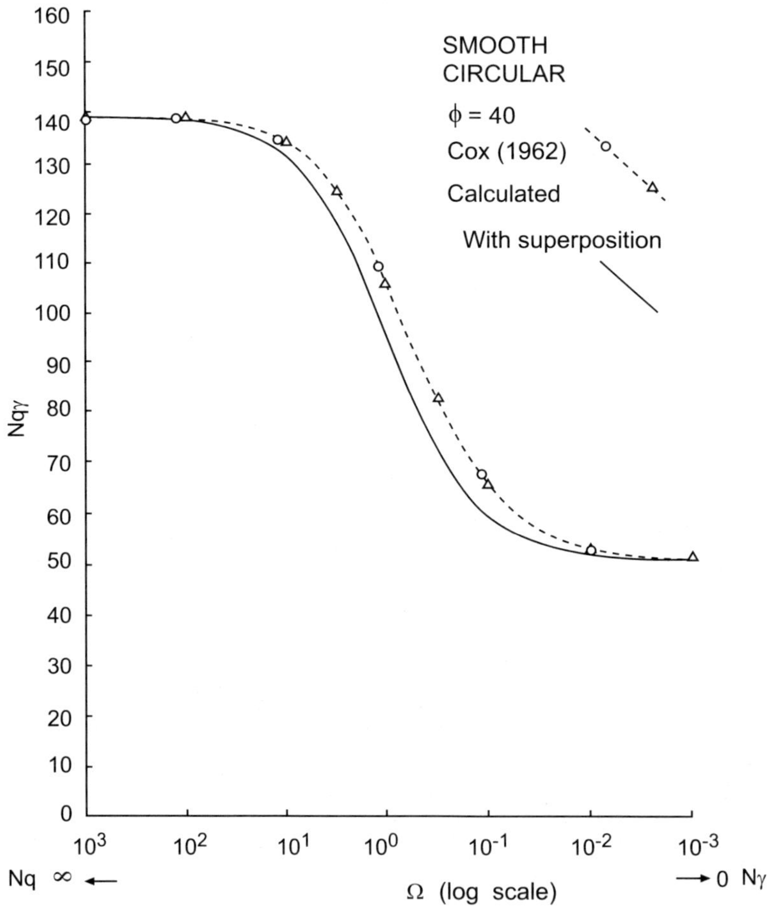

Fig. (6.10). Effects of Ω on $N_{q\gamma}$

An extension of the theory to cover the more general Mohr-Coulomb envelope, $\tau = c + \sigma \tan\emptyset$, can now be made. Here, c and $\emptyset$ can be taken as the envelope parameters on an effective stress diagram that offer a best fit to strength data over the appropriate range of stress. Prandtl's (1920) substitution can be used to find an expression for N_c, leading to the third component of Terzaghi's (1943) bearing capacity c N_c. Define an equivalent surcharge $\sigma_e = c \cot\emptyset$ (see Fig. **6.11**) acting over the whole soil surface including that part that will carry the foundation, causing a hydrostatic stress increase everywhere. The problem may then be seen as a simple constant-$\emptyset$ case with the origin shifted. The total bearing capacity due to cohesion would then be $N_q\ \sigma_e$, of which σ_e was already acting. The *extra* contribution is therefore $(N_q - 1)\sigma_e$. So using superposition we can follow Terzaghi to write

$$\sigma_f = c\, N_c + \sigma_o N_q + 0.5B\gamma N_\gamma \qquad \textbf{(6.10)}$$

finding N_q and N_γ directly for plane or circular footings from the earlier calculations with Ω set to 10^3 and 10^{-3} respectively, and using $N_c = (N_q - 1)\cot\emptyset$.

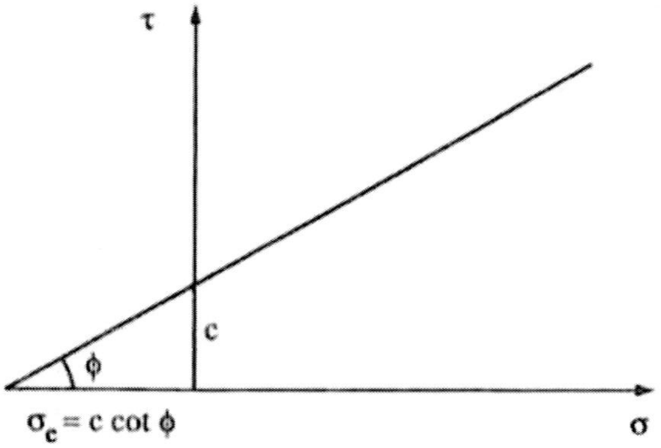

Fig. (6.11). General Mohr-Coulomb envelope.

6.2.4. Calculation of Values for Nq and Nγ for Smooth and Rough Bases

Computation results of N_q and N_γ for smooth, frictionless footings are listed in Table **6.1** as are solutions for rough footings, which are based on an optimistic treatment of friction opposing the spreading of soil beneath the footing. All the analyses were carried out with a mesh consisting of 96 β characteristics and 220 α characteristics. The rotation step in the fan zone was 3°. Figs. (**6.12-6.17**) show the typical stress characteristics and footing pressure distributions for plane strain and axisymmetric cases for both N_q and N_γ at $\emptyset = 40°$. In each case two solutions are shown, a shallow mechanism (after Hill, 1950) consistent with a smooth footing, and a deep mechanism (after Prandtl, 1920) in which it is assumed that friction against the footing stabilises a trapped wedge or cone. Following Meyerhof (1951), it is assumed that the trapped wedge or cone has a base angle $(\frac{\pi}{4}+\frac{\emptyset}{2})$ and mobilises $\emptyset$ on its inclined surface so that it acts as the final characteristic in the marching solution (see the shaded areas in Figs. **6.13b**, **6.15b** and **6.17b**). Each approach satisfies the condition of symmetry, that the principal stress direction is vertical at the centerline of the footing. Indeed, the trapped wedge/cone offers a vertical major principal stress at every point on its surface.

It is well known (Chen 1975, for example) that both shallow and deep mechanisms offer the same solution for N_q in plane strain, so friction on the footing can have no effect, and this is confirmed in Figs. (**6.12** and **6.13**). Although the same has not yet been proved for circular footings, it will be assumed hereafter that the same identity also holds for N_q in axisymmetric strain. On the other hand, it must be anticipated that the self-weight of the larger plastic zone will lead to larger estimates of N_γ for deep mechanism. This is indeed found to be the case (see Figs. **6.14** and **6.15** for example).

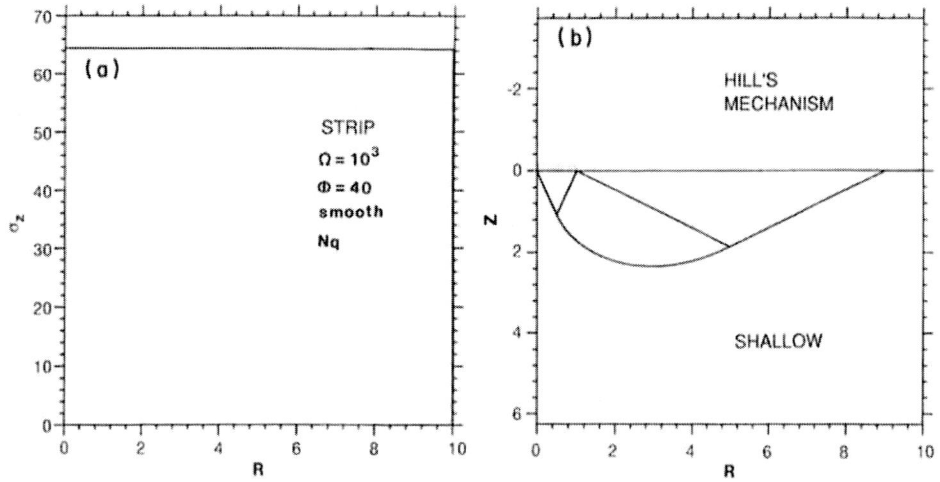

Fig. (6.12). Footing pressure distribution (**a**) and stress characteristics (**b**) for the plane strain case for N_q for a smooth base.

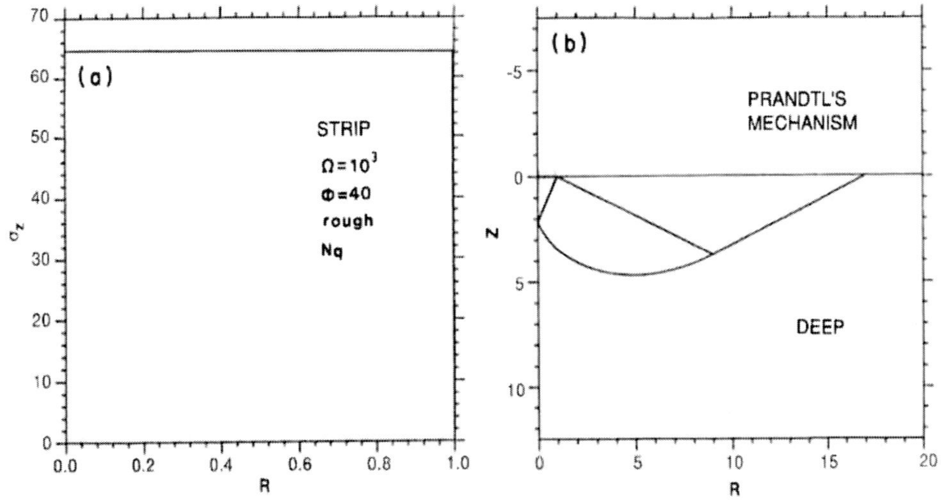

Fig. (6.13). Footing pressure distribution (**a**) and stress characteristics (**b**) for the plane strain case for N_q for a rough base.

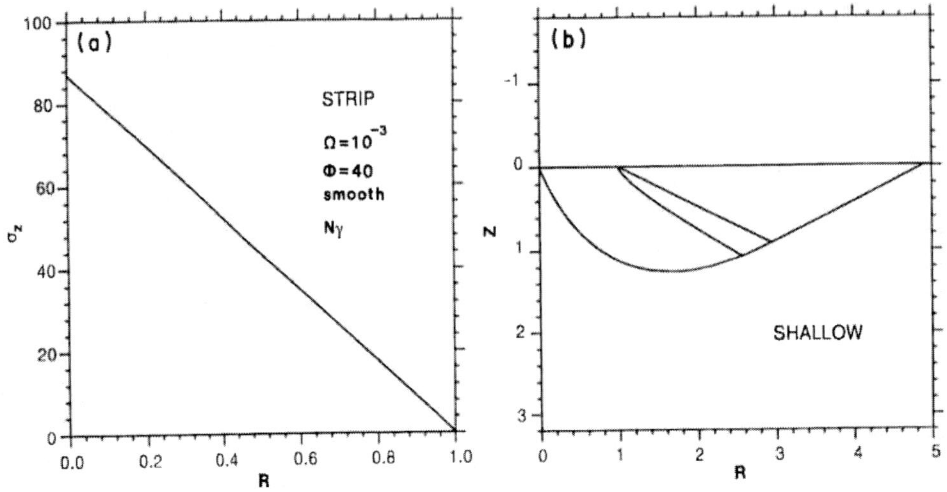

Fig. (6.14). Footing pressure distribution (**a**) and stress characteristics (**b**) for the plane strain case for N_γ for a smooth base.

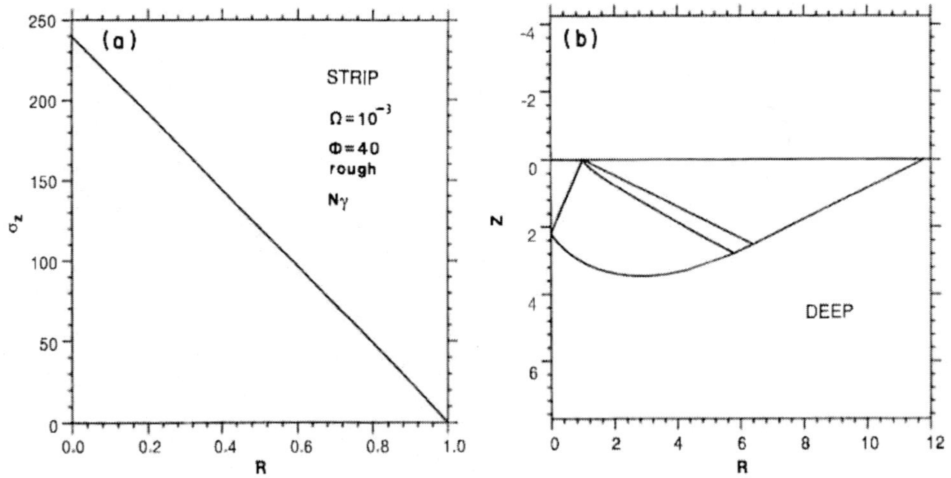

Fig. (6.15). Footing pressure distribution (**a**) and stress characteristics (**b**) for the plane strain case for N_γ for a rough base.

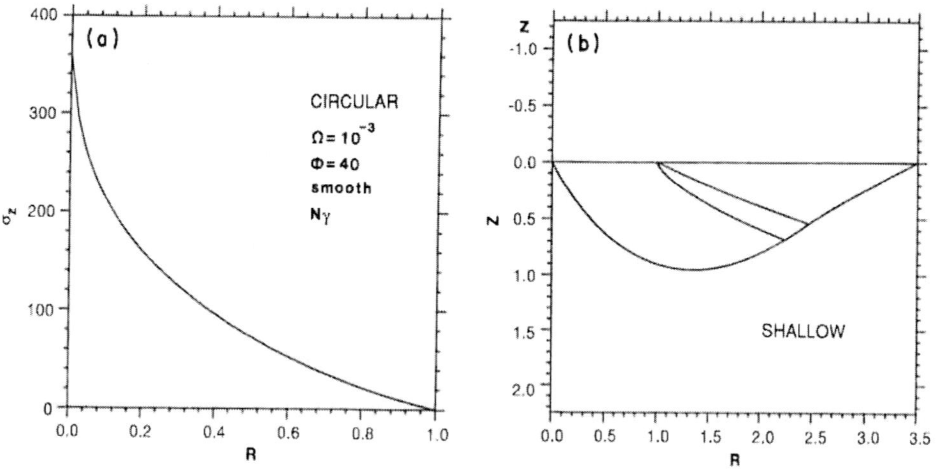

Fig. (6.16). Footing pressure distribution (**a**) and stress characteristics (**b**) for the axisymmetric case for N_γ for a smooth base.

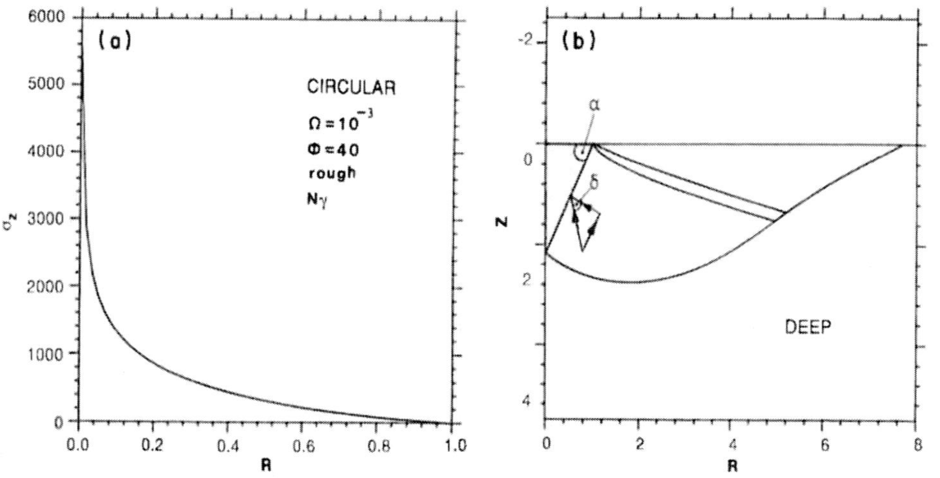

Fig. (6.17). Footing pressure distribution (**a**) and stress characteristics (**b**) for the axisymmetric case for N_γ for a rough base.

The assumption of a trapped zone is tantamount to solidifying the soil beneath the footing: presumably, friction mobilised on the base can do no more, but might do less. Although there is no theoretical justification for the assumption, there is experimental evidence in its favour (for example, Ko and Davidson 1973). It can be seen in Table **6.1** that the effect of allowing for footing roughness in this way increases the value of N_γ by about a factor of 3 for strip footings (equivalent to an increase in $\emptyset$ of about 5° overall) and by a factor of 4 for circular footings (equivalent to an increase in $\emptyset$ of about 7°). The values in columns 2 and 6 of Table **6.1** appropriate to rough strip footings satisfy

$$N_\gamma \approx (N_q - 1)\tan(1.5\emptyset) \quad \textbf{(6.11)}$$

within an equivalent discrepancy on $\emptyset$ of $\pm 2^\circ$ for $\emptyset$ from 30° to 50°. For example, the value of N_γ indicated by eq. 6.11 for $\emptyset = 35^\circ$ would be 42.1 compared with the tabulated value of 51.0 and 42.1 would be the tabulated result for $\emptyset = 33.8^\circ$, which is 1.2° less than requested. The same approximation eq. 6.11 also applies to the relative values in Table **6.1** for rough circular footings. Meyerhof (1961) first suggested an expression such as eq. 6.11, with a factor for strip footings of 1.4 rather than 1.5.

Table 6.1. Bearing capacity factors.

	N_q		N_γ			
ϕ	**Smooth or Rough**		**Smooth**		**Rough**	
(deg)	Strip	Circle	Strip	Circle	Strip	Circle
5	1.57	1.65	0.09	0.06	0.62	0.68
10	2.47	2.80	0.29	0.21	1.71	1.37
15	3.94	4.70	0.71	0.60	3.17	2.83
20	6.40	8.30	1.60	1.30	5.97	6.04
25	10.7	15.2	3.51	3.00	11.6	13.5
30	18.4	29.5	7.74	7.10	23.6	31.9
31	20.6	34.0	9.1	8.6	27.4	38.3
32	23.2	39.0	10.7	10.3	31.8	46.1

(Table 6.1) contd.....

33	26.1	45.0	12.7	12.4	37.1	55.7
34	29.4	52.2	15.0	15.2	43.5	67.6
35	33.3	61.0	17.8	18.2	51.0	82.4
36	38	71	21	22	60	101
37	43	83	25	27	71	124
38	49	99	30	33	85	153
39	56	116	36	40	101	190
40	64	140	44	51	121	238
41	74	166	53	62	145	299
42	85	200	65	78	176	379
43	99	241	79	99	214	480
44	115	295	97	125	262	619
45	135	359	120	160	324	803
46	159	444	150	210	402	1052
47	187	550	188	272	505	1384
48	222	686	237	353	638	1847
49	265	864	302	476	815	2491
50	319	1103	389	621	1052	3403
51	386	1427	505	876	1373	4710
52	470	1854	663	1207	1812	6628

The shallow mechanism clearly offers a safe solution to the problem of vertical bearing capacity, but it will err too far on the safe side in many circumstances. The only collapse mechanisms that can operate in a given situation are those that offer the soil-footing system a kinematically admissible displacement field. The possible displacement fields should take soil dilatancy into account and relate principal directions of compressive stress and strain. Although soil strains are outside the scope of this chapter, Fig. (**6.18**) illustrates some circumstances in which the shallow Hill (1950) mechanism might actually operate for foundations that could appear rough. In these cases the footing can stretch and bend, or split, so that the soil just beneath the footing, and close to the centreline on either side, can acquire relatively large lateral displacements as the footing penetrates the soil.

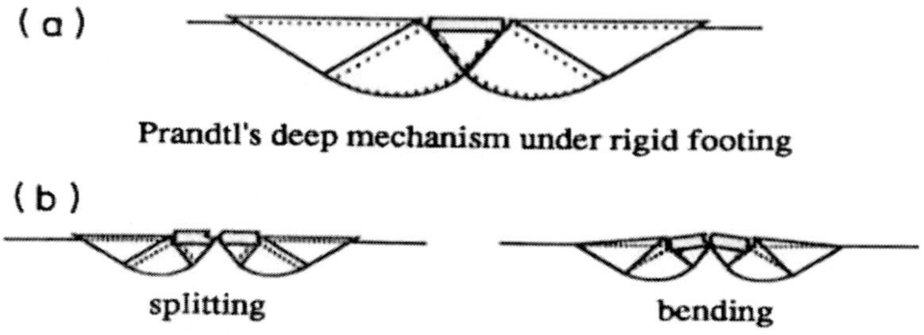

Fig. (6.18). The influence of footing deformability on the soil mechanism. (**a**) Prandtl's (1920) deep mechanism under rigid footing; (**b**) Hill's (1950) shallow mechanism under deformable footing.

6.2.5. Shape Effects

Many semi-empirical shape factors have been suggested for the conversion of values for N_q and N_γ from plane strain to axisymmetric strain. Terzaghi (1943) reduced the N_γ term by a factor of 0.6, but left N_q unaltered. Meyerhof (1963) used a factor $(1 + 0.1\ K_p)$, where K_p is the coefficient of passive earth pressure on both, which enhances the plane factors by 1.3 at $\emptyset = 30^\circ$ and by 1.76 at $\emptyset = 50^\circ$. Other authors suggested intermediate values.

As Meyerhof (1963) points out, the final estimate of bearing capacity must take account of two effects, the greater capacity of circular footings on soil with a given $\emptyset$, and the reduced $\emptyset$ of soil in axisymmetric strain (triaxial tests) compared with plain strain. It is logical to follow Meyerhof's procedure, selecting axisymmetric values directly from Table **1** in this case instead of applying indirect shape factors, and attempt to ensure that an appropriate $\emptyset$ value is used. Careful back-analysis of the punching of circular footings such as spud foundations will reveal whether triaxial strengths are appropriate. The final outcome might be that the ultimate bearing pressure of a circular footing is found to be less than that of a strip footing on the same soil, as Terzaghi (1943) would have predicted.

In detail, it can be seen from Table **6.1** that the shape factors for rough foundations creating deep mechanisms vary from about 1 at low values to about 3.5 at high ∅ values, a range exceeding even the recommendation of Meyerhof (1963). If it is desired to base axisymmetric values on plane values and use some simple factor, it will be found for deep mechanisms that axisymmetric values for N_q and N_γ can best be found by first increasing ∅ by a factor of 1.1, and then finding plane strain values for this enhanced ∅ value. The error expressed as an equivalent discrepancy on ∅ is less than 1° in the range $30^\circ - 50^\circ$. It must be recognised that the enhancement factor of 1.1 is purely to account for load spreading in axisymmetry; it is not determining what ∅ value to enter for the soil. In the case of shallow mechanisms, the plane strain value can first be enhanced by scaling up ∅ by 10%, but the resulting N_γ value must then be divided by 2 to get an acceptable estimate of the axisymmetric capacity according to the values in Table **6.1**.

Coincidentally, a quite different argument is often used to enhance triaxial ∅ values by a factor 1.1 to estimate plane strain values under similar conditions of density and stress. If this were also applied, it would transpire that bearing capacity factors for rough, rigid footings could be selected from plane strain bearing factors using a ∅ value 1.1 times the triaxial value, and irrespective of whether the footing was a circular pad or a strip. Footing shape would not then influence the ultimate bearing pressure.

The first mention of a 10% increase to obtain plane strain strength from triaxial strength was by Bishop (1961) based on some tests on compacted granular soils. For loose sands the increment is very small, or negligible, as was also shown. Bolton (1986) showed that for dense sands under low confining pressures the increment to secant ∅ values could approach 20%, but it was also demonstrated that Mohr-Coulomb envelope was curved when a large range of stress was to be considered. Changes of both c and Ø would therefore be necessary to capture the difference in envelope between plane and triaxial strengths relevant to footings. The fact that the plane strain factor on triaxial Ø could lie between 1.0 and 1.2 means that a careless choice could easily lead to an error in bearing capacities of a factor of 3.

6.3. EXTENDED PRANDTL EQUATION WITH VARIABLE Ø

The Prandtl (1920) solution for a strip footing on weightless frictional soil is based on the linkage between a rotation of principal stress direction and the consequential change in principal stress magnitude. Fig. (**6.19a**) shows a stable boundary XX between two adjacent zones, a and b. Jumping across XX would produce an infinitesimal rotation and shift of principal stresses. However, as the boundary is stable, the two zones must also share a common boundary stress (σ_x, τ_x), as shown in Fig. (**6.19b**).

By the sine rule, we have

$$\frac{ds}{\sin(2d\theta)} = \frac{s\sin\emptyset}{\cos\emptyset}$$

where s is the mean stress in the plane of shear. Therefore

$$\frac{ds}{s} = 2\tan\emptyset d\theta \qquad \textbf{(6.12)}$$

Eq. 6.12 links the shift in the centre of a train of Mohr circles to the rotation of the direction of the major principal stress. By integrating eq. 6.12 over the limits shown in Fig. (**6.20**), we have

$$\int_{s_0}^{s_f} \frac{ds}{s} = \int_0^{\theta} 2\tan\emptyset d\theta$$

$$s_f = s_o e^{2\theta\tan\emptyset}$$

$$\frac{\sigma_f}{1+\sin\emptyset} = \frac{\sigma_o}{1-\sin\emptyset} e^{2\theta\tan\emptyset} \qquad \textbf{(6.13)}$$

Therefore

$$\frac{\sigma_f}{\sigma_o} = \left(\frac{1+\sin\emptyset}{1-\sin\emptyset}\right) e^{2\theta\tan\emptyset} \qquad \textbf{(6.14)}$$

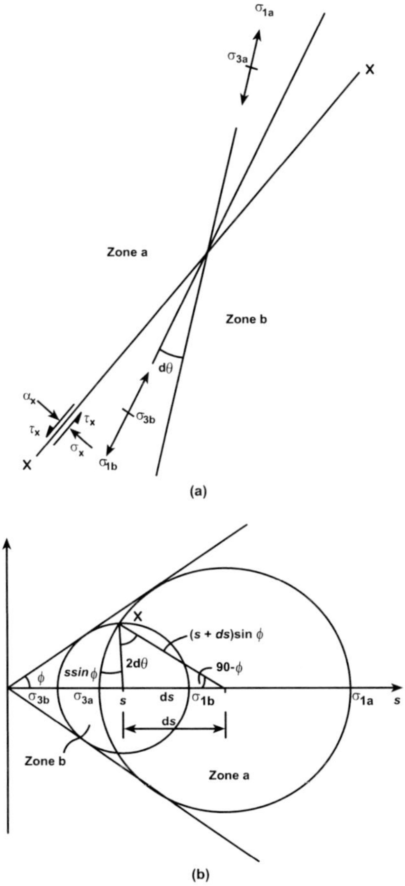

Fig. (6.19). (**a**) Neighbouring zones of limiting friction; (**b**) stresses in equilibrium across boundary XX.

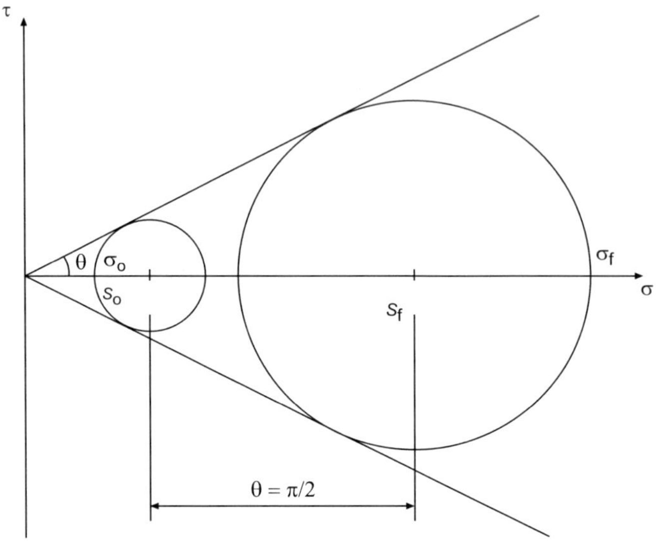

Fig. (6.20). Mohr circles of stress beneath a collapsing strip footing.

For constant $\emptyset$, and with $\theta = \frac{\pi}{2}$, eq. 6.14 becomes the Prandtl equation for bearing capacity factor N_q. The discontinuities can be viewed as stress characteristics (see Fig. **6.21a**). For example, travelling along characteristic $D_1D'_1$ is equivalent to jumping across a series of the other family of characteristics, that is $D_2D'_2$, and *vice versa*. If $\emptyset$ is a function of s, this additional effect should be taken into account in the general stress-rotation equation. Fig. (**6.21b**) shows the geometry of a discontinuity between limiting states (s, $\emptyset$) and (s + ds, $\emptyset + d\emptyset$), where $\frac{d\emptyset}{ds}$ has been taken positive merely for mathematical consistency. By the cosine rule, we have

$$ds^2 = s^2\sin^2\emptyset + (s + ds)^2(\emptyset + d\emptyset) - 2s\sin\emptyset(s + ds)\sin(\emptyset + d\emptyset)\cos 2d\emptyset$$

$$\left(\frac{ds}{s}\right)^2 - 2\left(\frac{ds}{s}\right)\tan\emptyset d\emptyset - (4\tan^2\emptyset d\theta^2 + d\emptyset^2) = 0$$

$$\frac{ds}{s} = \tan\emptyset d\emptyset \pm \sqrt{\sec^2\emptyset d\emptyset^2 + 4\tan^2\emptyset d\theta^2} \qquad \textbf{(6.15)}$$

The general stress-rotation equation can now be established as

$$d\theta = \sqrt{\frac{\left(\frac{ds}{s} - \tan\emptyset d\emptyset\right)^2 - \sec^2\emptyset d\emptyset^2}{4\tan^2\emptyset}} \quad \textbf{(6.16)}$$

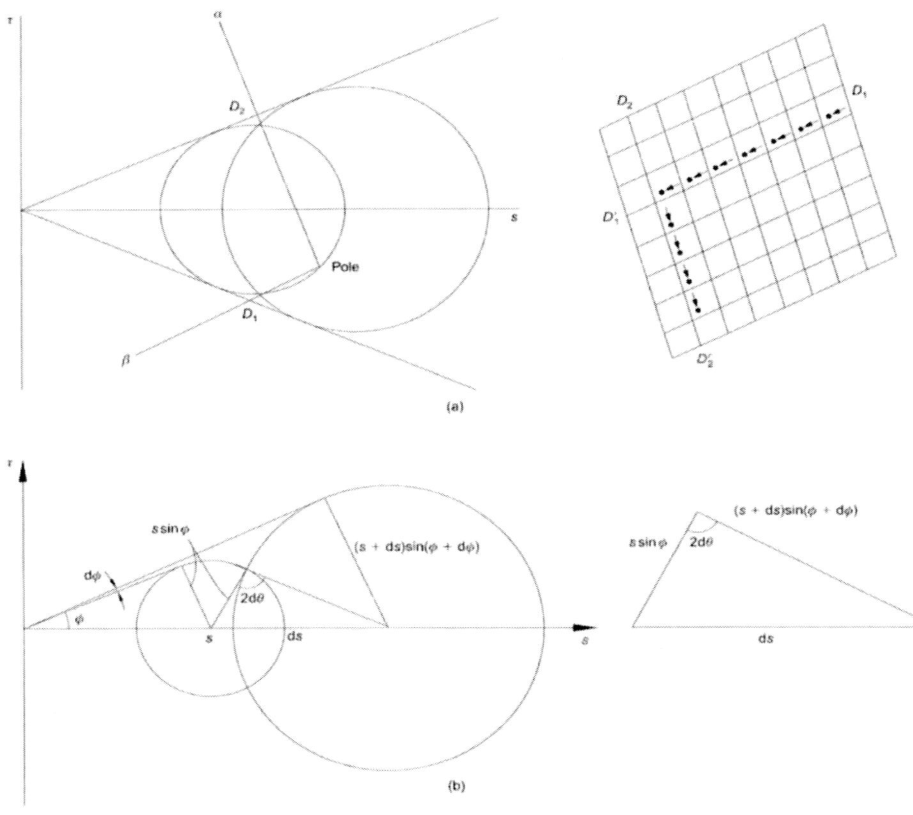

Fig. (6.21). (**a**) Mohr circles of stress across two families of discontinuities; (**b**) effect of varying ϕ on the stress-rotation equation.

6.3.1. Validation of Stress-Rotation Equation

A FORTRAN 77 program, EXPRAN, was written to evaluate N_q for a frictionless strip footing on weightless soil with Ø varying as a function of stress s, based on the theory as just outlined. A flow chart is shown in Fig. (**6.22**) to demonstrate how the calculation is organised. The algorithm commences by calculating Ø in the passive zone (see Fig. **6.23**). Initially, any arbitrary value is assumed, so that the mean stress can be written as $\sigma_o /(1 - \sin Ø)$. A new Ø value can be derived from the functional relationship with s. Iteration is carried out by using the current Ø to calculate the next mean stress with improved accuracy. This process continues until Ø has achieved a target accuracy.

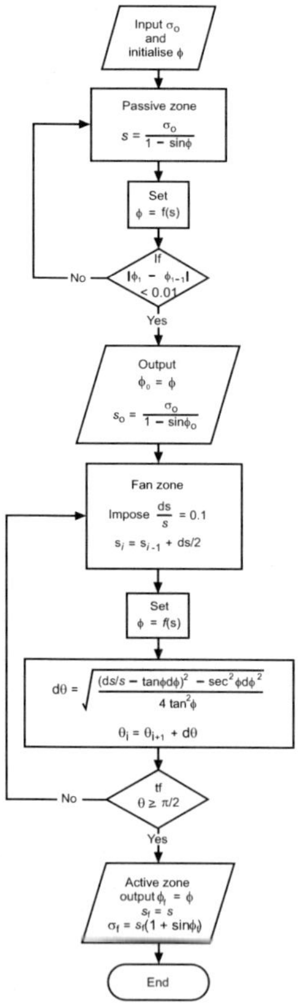

Fig. (6.22). Flow chart for program EXPRAN.

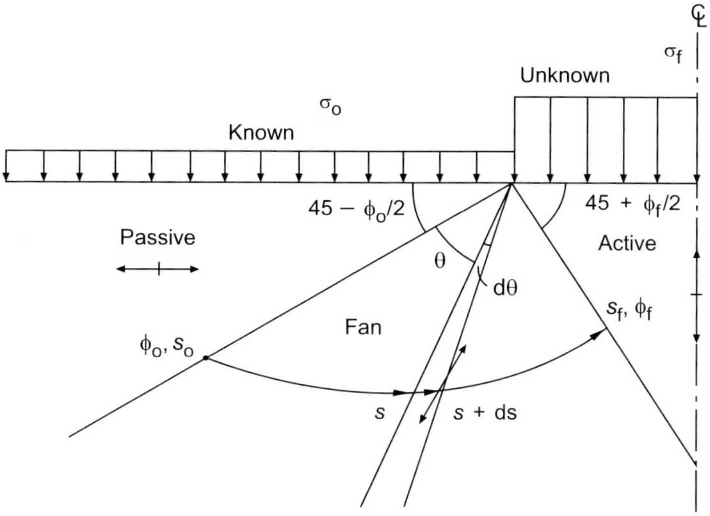

Fig. (6.23). Solution scheme for program EXPRAN.

In the fan zone, a loading step of $ds = 0.1s$ is imposed. Likewise, the current $\emptyset$ is calculated based on the current mean stress. The amount of cumulative rotation of the principal stress direction is updated at the end of each loading step by eq. 6.16. The active zone is deemed reached, and stress ceases to increase, when the rotation of the principal stress direction is greater than or equal to $\pi/2$. Typically, the total rotation over-ran by 0 – 0.04°. The bearing capacity σ_f can now be established as $s_f(1 + \sin\phi_f)$ in the active zone. The bearing capacity factor $N_q = \sigma_f/\sigma_o$ can then be calculated, appropriate to the particular value of σ_o. When σ_o is increased, ø reduces and N_q reduces.

An analysis with purely cohesive material that has a known exact solution (Prandtl, 1920) is selected here as a benchmark to check against the proposed algorithm. The idea is that a purely cohesive material can be viewed as a special case of $\emptyset$ varying with stress (see Fig. **23**)

$$\emptyset = \sin^{-1}\frac{c}{s} \qquad \textbf{(6.17)}$$

so that the shear strength is constant under all pressure conditions. By using eq. 6.17 in Fig. (**6.22**), the algorithm was found to yield $\sigma_f - \sigma_o = 5.145c$, which differs from the exact solution by only 0.07%. Therefore eq. 6.17 in combination with a stepping routine similar to Fig. (**6.22**) has been shown to be suitable for plastic solutions with varying sec ∅, at least in this simple application.

6.3.2. The Method of Characteristic with Variable Ø

For cases with self-weight, analytical solutions do not exist, and the method of characteristics has to be adopted. Graham and Pollock (1972), Graham and Hovan (1986) and Zhu *et al.,* (2001) have incorporated ∅ as a stess-level dependent variable in their analysis by the method of characteristics. Although during computation, ∅ is updated according to the mean stress level, they have used the same governing equations as formulated by Sokolovskii (1960), in which ∅ is treated as a constant. As pointed out by Hill (1950), when ∅ is a function of mean stress, a general set of governing equations that treat ∅ as an additional variable should be formulated instead. Ueno *et al.,* (2001) have treated ∅ as an additional variable, but their method is confined to plane strain strip footings only. Alternatively, geometrical reasoning can be used to establish the effects of varying ∅ on the constant-∅ governing equations.

In order to extend the constant-∅ analysis so that variable-∅ cases can also be included, the geometrical reasoning approach will be adopted. There are two effects on the governing equations when ∅ varies: the first is on the stress-rotation equation and the second is on the geometry of the characteristic lines. Both effects have to be quantified before computation can begin.

Before using the analysis to generate general predictions, it is prudent to check the method against known exact solutions. Two cases for purely cohesive material can be checked, the first for plane strain (Prandtl, 1920), and the second for axisymmetric conditions (Shield, 1955). As before, the purely cohesive soil will be treated in a variable-$\emptyset$ analysis with $\emptyset = \sin^{-1} c/s$. On differentiating, we obtain

$$d\emptyset = -\frac{cds}{s^2\cos\emptyset} \quad \textbf{(6.18)}$$

Inserting eq. 6.17 and eq. 6.18 in eq. 6.16, we obtain

$$\frac{ds}{s} = \pm 2\sin\emptyset d\theta \quad \textbf{(6.19)}$$

Comparing this with 6.12, we can define an operational constant equivalent ø value, $ø_{op}$, such that

$$\tan\emptyset_{op} = \sin\emptyset \qquad \text{or} \qquad \emptyset_{op} = \tan^{-1}(\sin\emptyset) \quad \textbf{(6.19)}$$

A general form of the equation can easily be derived in the same way and written as

$$\tan\emptyset_{op} = \frac{\tan\emptyset}{\sqrt{(1-\tan\emptyset\frac{d\emptyset}{dlns})^2 - \sec^2\emptyset(\frac{d\emptyset}{dlns})^2}} \quad \textbf{(6.20)}$$

While eq. 6.19 or eq. 6.20 describes the value of secant ø and $ø_{op}$ to be used in a stress-rotation equation, it does not deal with the geometry of the characteristic lines themselves. This geometry is related to the tangency of a Mohr circle on the limiting Mohr envelope. A second parameter, $\emptyset_{en}$, must be defined as the instantaneous angle of the tangent to the strength envelope, as shown in Fig. (**24**).

$$\sin\emptyset_{en} = \frac{(s+ds)\sin(\emptyset+d\emptyset)-s\sin\emptyset}{ds}$$

$$= \sin\emptyset + s\frac{d\emptyset}{ds}\cos\emptyset \quad \textbf{(6.21)}$$

Therefore

$$\emptyset_{en} = \sin^{-1}(\sin\emptyset + s\frac{d\emptyset}{ds}\cos\emptyset) \quad \textbf{(6.22)}$$

Substituting eq. 6.18 in eq. 6.22, we have

$$\emptyset_{en} = \sin^{-1}\left(\sin\emptyset - \frac{c}{s}\right) = \sin^{-1}(\sin\emptyset - \sin\emptyset) = 0 \quad \textbf{(6.23)}$$

Although this was obvious for the current 'cohesion' validation, eq. 6.22 will be used in predicting the effect of more general variation of ø with s.

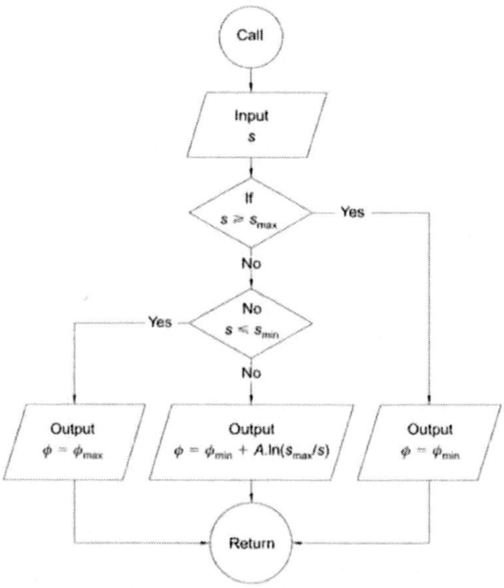

Fig. (6.24). Flow chart for subroutine to set ø.

6.3.3. Ø Against s: Data for Typical Sands

Bolton (1986) shows that

$$\emptyset = \emptyset_{\text{crit}} + \Delta I_{\text{R}} \quad \textbf{(6.24)}$$

fits the data of many quartz sands, where $\emptyset_{\text{crit}}$ is treated as the constant angle mobilised in continuous shearing at constant density, and Δ is found to be 3° in triaxial tests and 5° in plane strain. The dimensionless dilatancy index I_{R} can be written as

$$I_{\text{R}} = I_{\text{D}}\ln\frac{\text{p}_{\text{c}}}{\text{p}} - 1 \quad \textbf{(6.25)}$$

where I_{D} is the initial relative density, p_{c} is an intrinsic parameter related to the crushing of soil grains, and $\text{p} = (\sigma_1 + \sigma_2 + \sigma_3)/3$ is the mean stress at failure. Putting eq. 6.25 in eq. 6.24 and differentiating yields

$$\frac{d\emptyset}{dp} = -\frac{\Delta I_{\text{D}}}{\text{p}} = -\frac{\text{A}}{\text{p}} \quad \textbf{(6.26)}$$

where the new parameter $\text{A} = \Delta I_D$. In the work that follows no attempt has been made to discriminate between the spatial mean stress p at the point and the mean stress $\text{s} = (\sigma_1 + \sigma_3)/2$ in a plane of shearing at the same point.

Substituting eq. 6.26 into the general stress-rotation equation eq. 6.16, we have

$$d\theta^2 = \frac{ds^2}{4s^2\tan^2\emptyset}(1 + 2\tan\emptyset\text{A} - \text{A}^2)$$

$$\frac{ds}{s} = \frac{2\tan\emptyset d\theta}{\sqrt{1+2\text{A}\tan\emptyset-\text{A}^2}} \quad \textbf{(6.27)}$$

Comparing with the stress-rotation equation for constant $\emptyset$

$$\frac{ds}{s} = \pm 2\tan\emptyset_{op} d\emptyset \quad \textbf{(6.28)}$$

as before , we can obtain

$$\tan\emptyset_{op} = \frac{\tan\emptyset}{\sqrt{1 + 2A\tan\emptyset - A^2}}$$

$$\emptyset_{op} = \tan^{-1}\left(\frac{\tan\emptyset}{\sqrt{1+2A\tan\emptyset - A^2}}\right) \quad \textbf{(6.29)}$$

to be used in the stress-rotation equation. Taking A = $5I_D°$ or $5I_D$ π /180 radians, A=0.087 is obtained for the densest sands with $I_D = 1$ in plane strain, which gives $\emptyset_{op} \approx \emptyset - 2^{\circ}$. In all other cases, the difference between $Ø_{op}$ and the local secant Ø value will be smaller than 3°.

Substituting eq. 6.26 into eq. 6.22, we can find the envelope gradient

$$\emptyset_{en} = \sin^{-1}(\sin\emptyset - A\cos\emptyset) \quad \textbf{(6.30)}$$

It is easy to show that the largest deviation between Ø and $Ø_{en}$ for dense soils in plane strain is about 5°. Having established $Ø_{op}$ and $Ø_{en}$, the constant- $\emptyset$ formulation can now be extended to realistic variable- $\emptyset$ analysis for sandy soils. The subroutine in Fig. **(6.24)** was used to calculate $\emptyset$ according to the stress level.

6.3.4. Validation of the Method of Characteristics with Variable Ø

The method of characteristics relies on extending 'characteristic lines', which are lines of incipient shear failure, taking account of the different stress rotation and associated stress increments on two intersecting characteristics, such as QW and PW in Fig. (**6.25**). The four stress components can be expressed in terms of two dependent variables s and φ

$$\sigma_r = s(1 - \sin\emptyset\cos 2\varphi)$$
$$\sigma_z = s(1 + \sin\emptyset\cos 2\varphi)$$
$$\tau_{rz} = s\,\sin\emptyset\sin 2\varphi$$
$$\sigma_\theta = s(1 - \sin\emptyset) = \sigma_3 \quad \textbf{(6.31)}$$

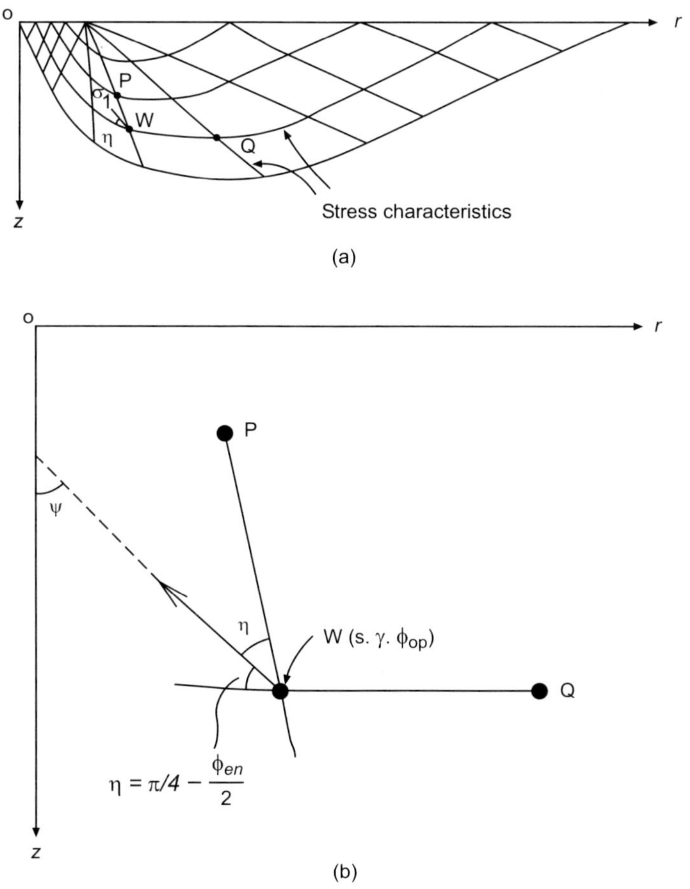

Fig. (6.25). (**a**) A typical stress characteristic mesh; (**b**) computation scheme of new point W from known points P and Q.

where φ is the angle between the major principal stress direction and the z axis. Along two families of stress characteristics, the governing equations become two ordinary differential equations

$$\mathrm{ds}\cos\emptyset + 2\mathrm{s}\sin\emptyset\mathrm{d}\varphi + \mathrm{n}\frac{\mathrm{s}}{\mathrm{r}}[\sin\emptyset\cos\emptyset\mathrm{dr} + (\sin^2\emptyset - \sin\emptyset)\mathrm{dz}]$$

$$= \gamma(-\sin\emptyset\mathrm{dr} + \cos\emptyset\mathrm{dz}) \quad \textbf{(6.32)}$$

along an α characteristic given by

$$\frac{\mathrm{dr}}{\mathrm{dz}} = \tan(\varphi - \eta) \quad \textbf{(6.33)}$$

And

$$\mathrm{ds}\cos\emptyset - 2\mathrm{s}\sin\emptyset\mathrm{d}\varphi + \mathrm{n}\frac{\mathrm{s}}{\mathrm{r}}[\sin\emptyset\cos\emptyset\mathrm{dr} - (\sin^2\emptyset - \sin\emptyset)\mathrm{dz}]$$

$$= \gamma(\sin\emptyset\mathrm{dr} + \cos\emptyset\mathrm{dz}) \quad \textbf{(6.34)}$$

along a β characteristic given by

$$\frac{\mathrm{dr}}{\mathrm{dz}} = \tan(\varphi + \eta) \quad \textbf{(6.35)}$$

where $\eta = \frac{\pi}{4} - \frac{\emptyset}{2}$

For the axisymmetric case, we must take n = 1, and for the plane strain case n = 0 (by inspection).

The finite difference forms of eq. 6.24 and eq. 6.25 can be used to determine s and φ at the point of intersection of α and β characteristics by iteration either from a known boundary condition or from previously computed values at two adjacent points. The four stress components can then be obtained by back substituting s and φ into eq. 6.31. More details about the algorithm can be found elsewhere (Sokolovskii, 1960 and Cox, 1962). The new method additionally permits $\emptyset$ to vary from point to point, so that the geometry of the 'net' varies with $\emptyset_{en}$ and the stress magnitude varies with $\emptyset_{op}$.

A validation exercise for the program VARIPHI was performed for effectively weightless soil, so that it estimated bearing capacity factor N_c by performing an N_q-type analysis by the method of characteristics with varying $\emptyset$. The plane strain case yield $\sigma_f - \sigma_o = 5.14c$, which is practically the same as the exact solution, and similar to the simpler calculation procedure described earlier. The stress characteristics and footing pressure distribution are derived and shown in Fig. (**6.26**). It can be seen that the log spiral in the shear fan has now degenerated into a circular curve, as required.

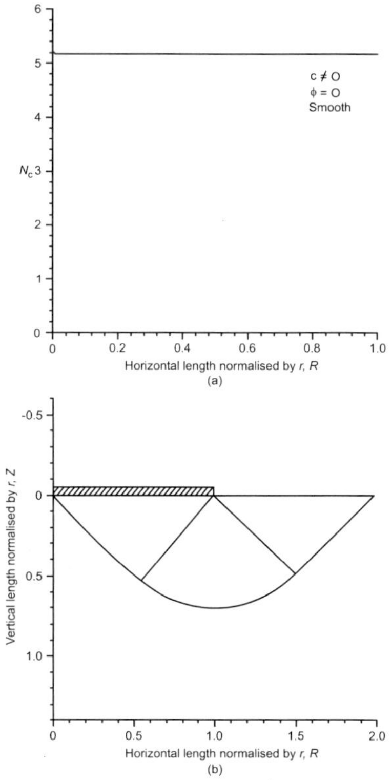

Fig. (6.26). Plane-strain (**a**) footing pressure distribution; (**b**) stress characteristics.

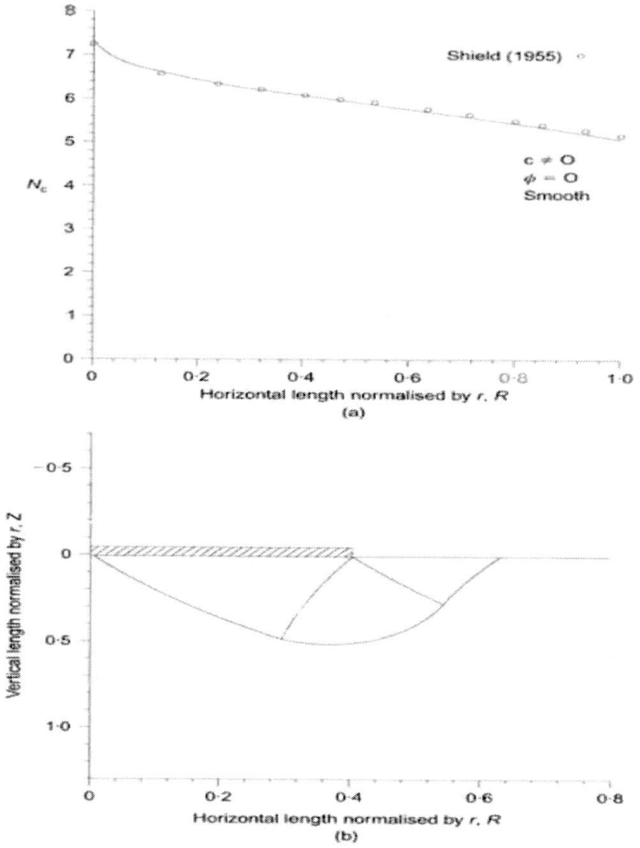

Fig. (6.27). Axisymmetric (**a**) footing pressure distribution; (**b**) stress characteristics.

For the axisymmetric case, the method gives $\sigma_f - \sigma_o = 5.63c$, which differs from Shield's (1955) exact solution of 5.69c by 1%. Fig. (**6.27**) compares the distribution of the calculated footing pressure distribution with Shield's solution, and shows that the present calculation gives a similar footing pressure distribution. It also shows that the characteristics are of the same shape as that calculated by Shield.

These validations have demonstrated that even with a pressure effect on sec ∅ sufficient to suppress shear strength gain entirely, the method of calculation proposed here can yield a very accurate answer. It is therefore believed that this calculation procedure is suitable for application to a wide variety of soils with a general relation of ϕ against s.

6.4. PRACTICAL APPLICATIONS

In this section, prediction of the indentation pressure for a typical jack-up rig will be given as an example. Fig. (**6.28**) shows a circular footing resting on the surface of submerged dense sand with A = 3°, $\emptyset_{crit}$ = 37.5° and γ = 20kN/m^3, so that $\gamma' = 10\text{kN/m}^3$.

The footing, whether dimpled, conical or flat, is assumed to be rough enough to stabilise an active zone of sand beneath the footing with a semi-angle of 28°. The inclined surface of the active cone is assumed to be acted upon by soil in which the major principal stress direction is vertical.

This creates the boundary condition $\delta \approx \phi$ for a wide range of Ø values in the soil near the cone. Any use of the method of characteristics demands this sort of idealization of the relevant boundary conditions.

The results of the analysis are shown in Fig. (**6.28**), for footing diameter (at the sea bed) of 0.4 m. 1.42 m and 5.0 m. In each case, Ø is shown to drop from its maximum value of 57° at shallow depth around the footing to much smaller values at the edge of the active zone – for example, 36° for the 5 m footing. At the same time, the bearing capacity factor N_γ has dropped by more than a factor of 3, from 712 to 207. Here N_γ is simply defined as $\sigma_f/(0.5B\gamma')$ and σ_f is calculated as the average effective stress applied to the footing at the seabed. This calculation ignores disturbance of the sand, which would occur when a real conical footing penetrates a sand bed. It simply offers a 'snapshot' of the limiting equilibrium of a footing wished into place, permitting the size of the footing to reflect the magnitude of the stress, and permitting the soil to mobilise a peak angle of shearing $\emptyset_{max}$, which is corrected for stress level using a secant variation of the type shown in Fig. (**6.1**), equivalent to the curved strength envelope shown in Fig. (**6.3b**).

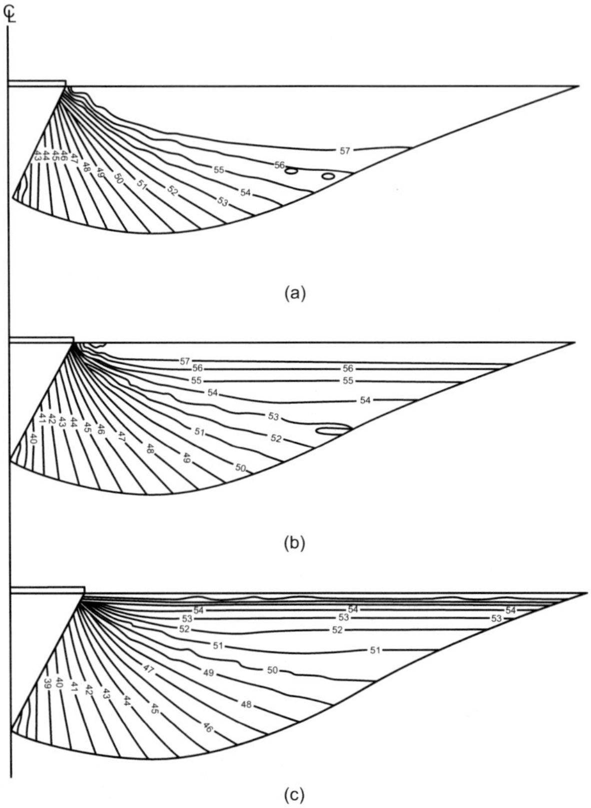

Fig. (6.28). Contour map of varyin ϕ under rough circular footing on sand with no surcharge: (**a**) B=0.4m; (**b**) B=1.42m; (c) B=5.0m.

Effects such as disturbance, or progressive failure, would result in the mobilisation of $Ø_{crit} < Ø_{failure} < Ø_{max}$. Engineers must rely on experimental evidence to guide them on whether allowance must additionally be made for loss of peak strength. In the later part of the chapter, it will be demonstrated that no such allowance is necessary for centrifuge models of flat footings loaded vertically to failure on sand or silt beds.

A program VARPHI which is based on the theory as discussed above is developed by the author, and the readers should have no difficult to devise a similar one which should not be difficult. This program is capable of being applied directly to any problem of the limiting plastic equilibrium of granular soils, if the geometry and boundary conditions are well defined, and the variation of $\emptyset$ and s has been determined. However, a method has been found to generalise the outcome of VARIPHI analysis for a class of problem, leaving the engineer able to use existing conventional constant-$\emptyset$ analysis with a selected mean-mobilisable angle of shearing resistance $\emptyset_m$.

Taking the previous example of a 5 m footing on the surface of dense submerged sand, the N_γ value of 207 would, in a typical constant-$\emptyset$ analysis, imply an equivalent constant value of $\emptyset_m$=39.3°. When this is compared with the contours of $\emptyset_{max}$ shown in Fig. (**6.28**), it seems too close to the lower limit. It must be remembered, however, that variable-$\emptyset$ analysis not only influences the mean mobilised strength; it also alters the geometry of the characteristic net. The apparently low constant equivalent value $\emptyset_m$=39.3°is that value which, when inserted in a constant-$\emptyset$ analysis with a constant-$\emptyset$ geometry, recovers the same value of bearing capacity as a variable-$\emptyset$ analysis. $\emptyset_m$ is of some practical assistance, since it enables us to identify the equivalent mean effective stress p_m at which a triaxial test would recorded $\emptyset_m$.

A number of trial calculations were performed for two different soils at a relative density of 1, namely

(a) Sand: silica, $Ø_{max}$ = 57.5° at s = 10kPa, A = 3°
(b) Silt: silica, $Ø_{max}$ = 57.5° at s = 50kPa, A = 3°

supporting circular footings at the point of plastic indentation. Two calculations were made, one on the basis of weightless soil subject to an effective overburden pressure σ_o (see Table **6.2**), and the other for soil with self-weight and negligible surcharge (see Table **6.3**). In the latter case, a small surcharge (with an equivalent embedment of much less than 1% of the footing width) was required to stabilise the surface film, but a check was made to ensure that this magnitude did not influence the bearing capacity. The nominal surcharge was needed for a purely frictional material to avoid numerical problems due to local yielding of the soil near the surface film.

Table 6.2. Summary of results from the variable-Ø analysis for a circular footing on weightless soil.

σ_o	σ_f: kPa		$Ø_{average}$: degrees		p_m: kPa	
	Sand	Silt	Sand	Silt	Sand	Silt
5	3149	6619	47.6	50.7	270	480
10	4584	9778	46.1	49.5	450	700
25	7645	15673	44.2	47.6	850	1400
50	11370	22735	42.7	46.1	1400	2200
100	17110	33430	41.2	44.6	2300	3700
200	25980	49680	39.6	43.1	3900	6000

Table 6.3. Summary of results from the variable-Ø analysis for a shallow circular footing.

***B*: m**	σ_f: kPa		$Ø_{average}$: degrees		p_m: kPa	
	Sand	Silt	Sand	Silt	Sand	Silt
0.40	1423	3155	44.5	47.4	760	1500
0.90	2161	4670	43.0	45.9	1250	2400
1.42	2735	5943	42.1	45.1	1700	3100

(Table 6.3) contd.....

3.00	4017	8720	40.5	43.7	2900	5000
5.00	5165	11410	39.3	42.8	4200	6900
10.00	7435	16180	37.8	41.3	7200	11000

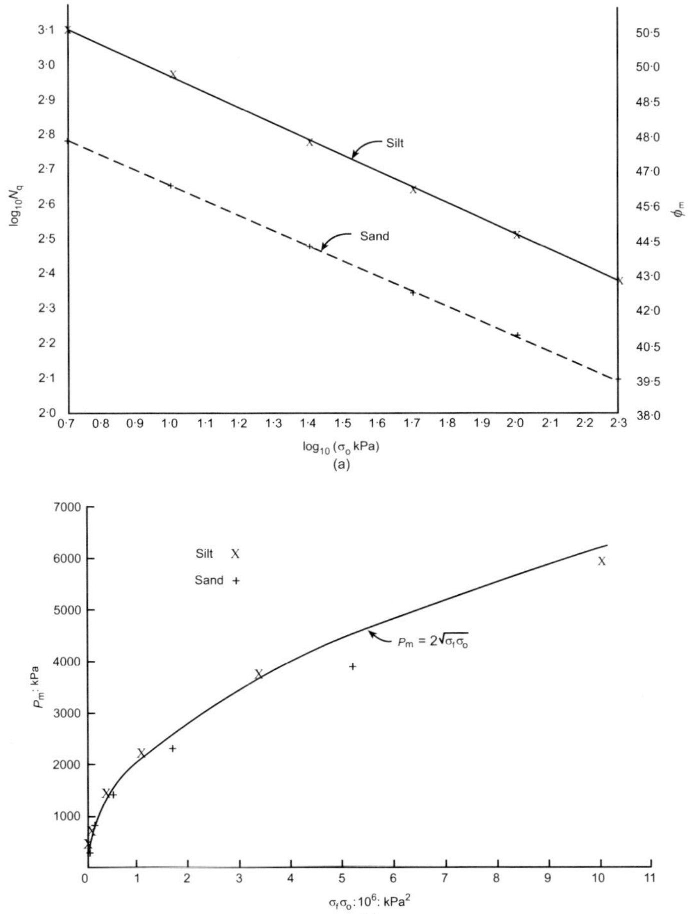

Fig. (6.29). (**a**) $\log_{10}$-$\log_{10}$ plot of N_q and σ_o; (**b**) empirical rule to find the average working stress in the weightless soil supporting a circular surface footing.

The two soils were of different degrees of particle strength, the silt being five times stronger, as represented by its mobilisation of $Ø_{max}$ = 57.5° at s = 50kPa, where as the sand could mobilise 57.5° at s = 10kPa. The typical rate of loss of strength A = 3° invoked per unit increase in the natural logarithm of stress (from eq. 6.20, 6.21 and 6.22) implies that $Ø_{max}$ at s = 50 kPa for the sand would be (57.5° - 3ln5) = 52.7°. The later part of the chapter shows these propositions fit the triaxial data of two soils fitting these descriptions. Either the silt particles may be described as five times stronger, or the silt particles as an aggregate can be described as displaying 5° more shearing resistance than the sand. Each soil loses 7°of Ø per factor 10 increases in mean stress.

In Tables **6.2** and **6.3**, the bearing capacities calculated using VARPHI have been back-analysed to give the equivalent constant value ϕ_m, and this in turn has been processed to find the value of mean stress p_m at which ϕ_m would have been measured in a triaxial test. In Figs. (**6.29a** and **6.30a**), the values of bearing capacity coefficient (N_q or N_γ) are plotted against the relevant resisting stress (σ_o or $0.5B\gamma'$), each on logarithm axes (base 10), and appropriate values of ϕ_m are inserted at intervals on the bearing capacity axis. It can be seen that these two identically dense granular materials may have bearing capacity factors between about 100 and 1,000, depending on the stress magnitudes. The less crushable silt has a bearing capacity about twice that of the sand, but for each soil a tenfold increase in resistive stress (σ_o or $0.5B\gamma'$) produces roughly a threefold reduction in bearing capacity factor. In Figs. (**6.29b** and **6.30b**), the values of p_m associated with ϕ_m are plotted against the harmonic mean of the resisting stress and the bearing capacity σ_f.

It will be seen that the VARIPHI calculations for the two soils can be fitted reasonably well by parabolas, in each of the figures. A reasonable approximation for both soils appears to be

$$p_m = 2\sqrt{\sigma_f \sigma_o} \qquad \text{for } N_q \qquad \textbf{(6.36)}$$

$$p_m = 13\sqrt{\sigma_f 0.5B\gamma} \qquad \text{for } N_\gamma \qquad \textbf{(6.37)}$$

If this value of p_m is used to find $\varnothing_m$ in triaxial tests, the resulting value can be inserted into a constant-$\varnothing$ analysis to derive an appropriate bearing capacity.

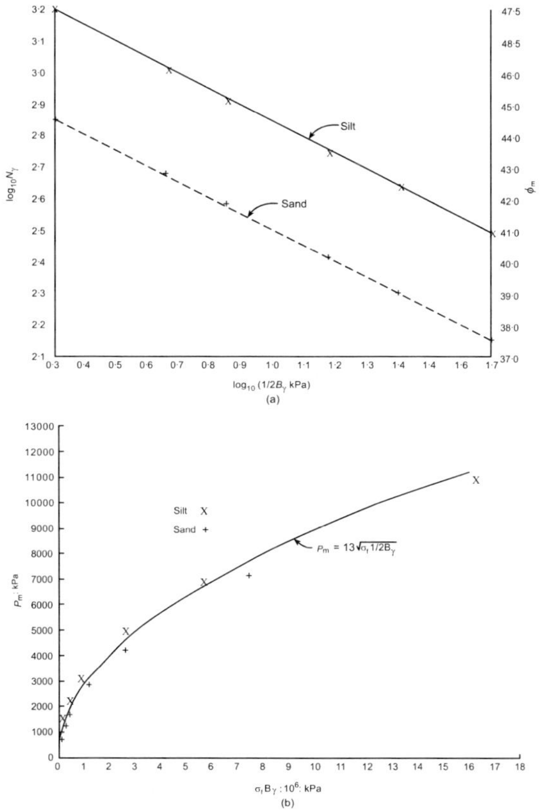

Fig. (6.30). (**a**) $\log_{10}$-$\log_{10}$ plot of N_γ and $\frac{1}{2}B\gamma$; (**b**) empirical rule to find the average working stress in the weightless soil supporting a circular surface footing.

Material Acquisition

Two silica soils, a sand and a silt, which differed in nominal grain diameter by a factor of 50 but were otherwise indistinguishable in terms of shape, grading and mineralogy, were successfully acquired (see Table **6.4** and Fig. **6.31**). These were

(a) Silt: washed silica flour
(b) Sand: 8/40 wires Chatelet flint grit.

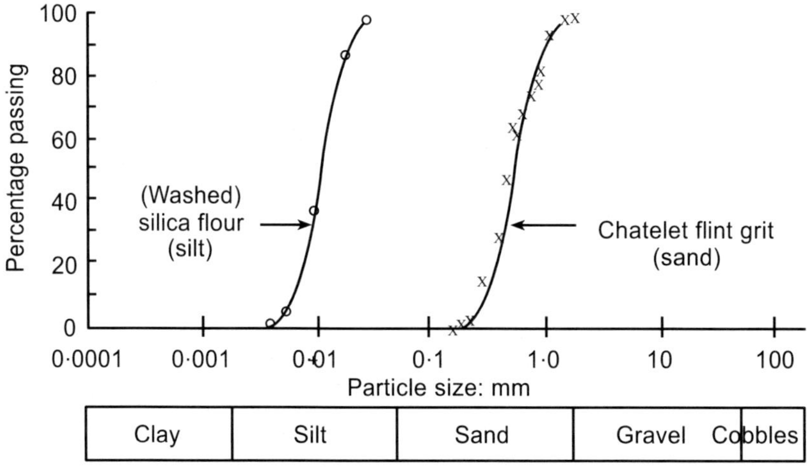

Fig. (6.31). Grading curves of the two granular materials.

Table 6.4. Properties of the two granular materials.

Properties	Silt	Sand
Mineralogical make-up	>99% SiO_2	>97% SiO_2
Specific gravity	2.65	2.65
Roundness	0.40	0.41
Sphericity	0.80	0.77
D50: μm	12	600
Permeability: m/s	3×10^{-7}	1.7×10^{-3}
ϕ_{crit}: degrees	37.5	37.5

Sample/Model preparation

On top of stress effects, the bearing capacity is also dependent on I_D. To simplify the validation exercise, the sand and silt beds were placed at their $e \approx e_{min}$. This also facilitated the achievement of uniform density in the models.

The dense sand was laid down by pluviation into the tub at a rate of 1.33 kg/min from a suspended conical hopper by way of a plastic hose 600 mm long, with 28 mm internal diameter, and at a constant drop height of 600 mm. The tub was 850 mm in diameter and 400 mm high. The sand was poured until the target height of the model was slightly passed. A modified vacuum cleaner was used to level the model back down to the target height. The tub was weighed to the nearest 1 kg before and after the sand was placed. When the volume was known, the bulk dry density could be estimated. With this method, e_{min} for the sand was found to be 0.60 ± 0.04.

When a saturated model was required, a steel lid was used to seal the top of the tub and the air inside was evacuated by a vacuum pump. Water was then permitted to percolate upwards through the bottom drainage layer at a flow rate slow enough to ensure piping would not occur during the saturation process (Lau, 1988).

The saturated silt was compacted by a vibrator bolted onto the wall of the tub and the slurry was introduced into the tub in layers with the vibrator switched on in order to let out any trapped air. When the required amount of slurry was in place and no more air bubbles could be seen escaping through the top of the slurry, which normally took roughly 24 hours to achieve. The top of the silt bed was covered with filter paper before a steel piston weighing 345 kg was put on top. Vibration continued until no further settlement of the piston could be observed: this normally took another 48 hours. During compaction, both top and bottom drainages were provided. In order to determine the bulk density, the tub was weighed to the nearest 1 kg before and after the compacted silt was in place. The volume of the silt was calculated by measuring the thickness of the model with the aid of a template at a regular grid of no less than 96 sampling points. With this method, e for the silt was found to be 0.59 ± 0.04.

Triaxial specimens of the sand were formed similarly by pluviation in a triaxial sample mould. For the silt, eighteen 100 mm cubes were excavated from the previously compacted bed. As soon as they were excavated, the samples were swiftly sealed with cling film with a view to retaining as much moisture as possible until triaxial tests could be carried out later. Immediately before the triaxial test, the sample was put on a lathe and trimmed carefully to the required diameter by a sharp cutting edge, guided by two vertical straight edges. It was then put on a cradle to be trimmed carefully to the required height before mounting on the platen in the triaxial cell. The volume was then measured. The dry weight was measured after each test when the specimen was oven-dried.

6.4.1. Triaxial Test Programme

It was decided to investigate the absolute particle size and stress effects by triaxial tests conducted over a wide range of effective confining pressures ranging from 10 to 10,000 kPa. More details on the extremely low and extremely high cell pressure equipment can be found in Lau (1988). In a triaxial test, the stress and strain fields are intended to be uniform; the formation of localised shear in rupture bands is not encouraged, at least before $Ø_{max}$ has been reahed.

The triaxial test results from 38 mm diameter sample tested with smooth platens and a height: diameter ratio of 2 are presented in Figs. (**6.32** and **6.33**) as the secant of the angle of friction and volumetric strain against axial strain. The unload-reload loops have been omitted from the plots of Ø against ε_a plots only for clarity. The secant of the angle of friction is obtained by dropping a tangent from the origin to a single Mohr circle of stress. The attainment of $Ø_{max}$ in general coincides with the maximum rate of dilation. $Ø_{crit}$ is not reached in low-pressure tests, as dilation was continuing even after an axial strain of 20%. $Ø_{max}$ is higher for silt than for sand, and the corresponding rate of dilation is also higher. One interesting observation is that ruptures were obviously formed in the silt samples after $Ø_{max}$ has been reached, resulting in a sudden decrease in both Ø and the rate of dilation (see Fig. **6.32**). It is obvious that the interference of ruptures was deferred when cell pressure was higher. Formation of ruptures was not obvious in the sand specimens.

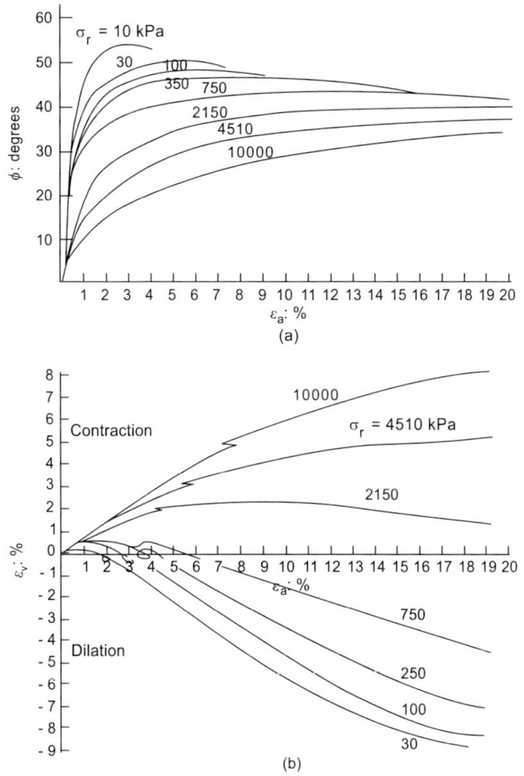

Fig. (6.32). Effects of effective cell pressure on the stress-strain behavior of sand in triaxial compression.

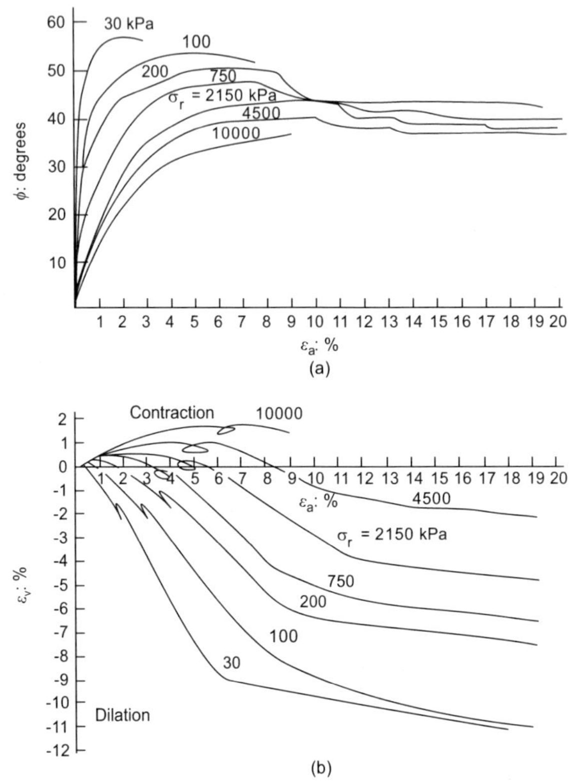

Fig. (6.33). Effects of effective cell pressure on the stress-strain behavior of silt in triaxial compression.

Fig. (**6.34**) summarises the pressure effects on sec ∅ for the two materials. Sec ∅ is $4^{\circ}-5^{\circ}$ higher for the silt than for the sand, but the trend of pressure effects is similar. In both cases, the secant of the angle of friction reduces linearly with the logarithm of mean stress. The results of the sand agree well with Bolton's (1986) empirical correlation for sands, also shown in Fig. (**6.34**): the deviation is smaller than $\pm 2^{\circ}$. Shifing of the grading curves shows that crushing occurred during the triaxial tests for sand under high pressure (see Fig. **6.35**). More crushing occurs when the confining pressure is increased. However, no crushing occurs when the sand is subjected to isotropic compression only. It is the crushing of grains that leads to reductions in dilatancy, and thereby to reductions in the angle of shearing resistance.

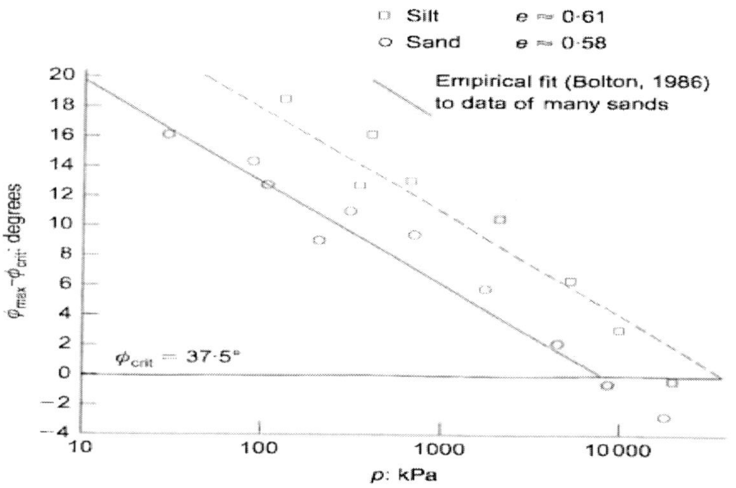

Fig. (6.34). $Ø_{max}$ plotted against logarithm of mean stress.

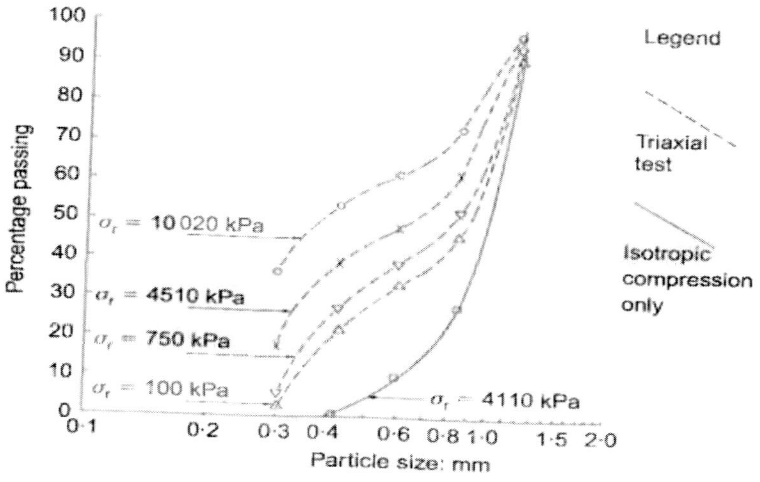

Fig. (6.35). Grading curves of sand samples after triaxial tests and isotropic compression.

The silt was able to maintain a higher peak angle of shearing and rate of dilation than the sand in the triaxial tests. At a similar voids ratio (close to minimum) and confining pressure, $\emptyset_{max}$ of the silt is 4° to 5° higher than that of the sand. From Fig. (**6.34**), it can equally be said that the silt is about four to five times less stress sensitive than the sand. This may be explained by the inevitable reduction in the size of flaws in the smaller particles. Griffith's (1921) criterion for facture proposes that rupture stress is inversely proportional to the square root of the size of a flaw. If the flaws in the silt particles had been scaled down on the same ratio as their diameter (factor 50 compared with the sand), they would have proved to be about seven times more resistant, rather than five times.

1-g Model Tests

The 1 **g** test series was designed to investigate scale effects in footing tests on weightless soil. A schematic layout of the test set-up is shown in Fig. (**6.36**). The boundary value problem chosen was a cylindrical bed of soil 850 mm in diameter and 350 mm high. The soil model was either fully submerged in water or dry. The side and bottom of the soil model were assumed to be supported by smooth and rigid boundaries. In order to simulate this smooth condition, the inside of the tub wall was lubricated with plumber's grease. The top surface of the soil model around the punch was subjected to a constant surcharge pressure ranging from 5 to 200 kPa. In this range of surcharge, the soil could practically be treated as weightless. The bottom of the model was a free-draining boundary. During a test, a centrally placed rough, rigid punch of diameter 100 mm (or 14.2 mm) was pushed axisymmetrically into the soil bed in a displacement-controlled manner in order to facilitate the investigation of post-peak behaviour. Throughout the indentation process, the average bearing pressure under the footing and the settlement of the punch were monitored.

The 1 **g** test programme is summarised in Table **6.5**. The three variables were B/d_{50}, σ_o and d_{50}. Apart from studying the scale effects, this series was also used to investigate the effect of settlement: diameter ratio (w/B) on N_q.

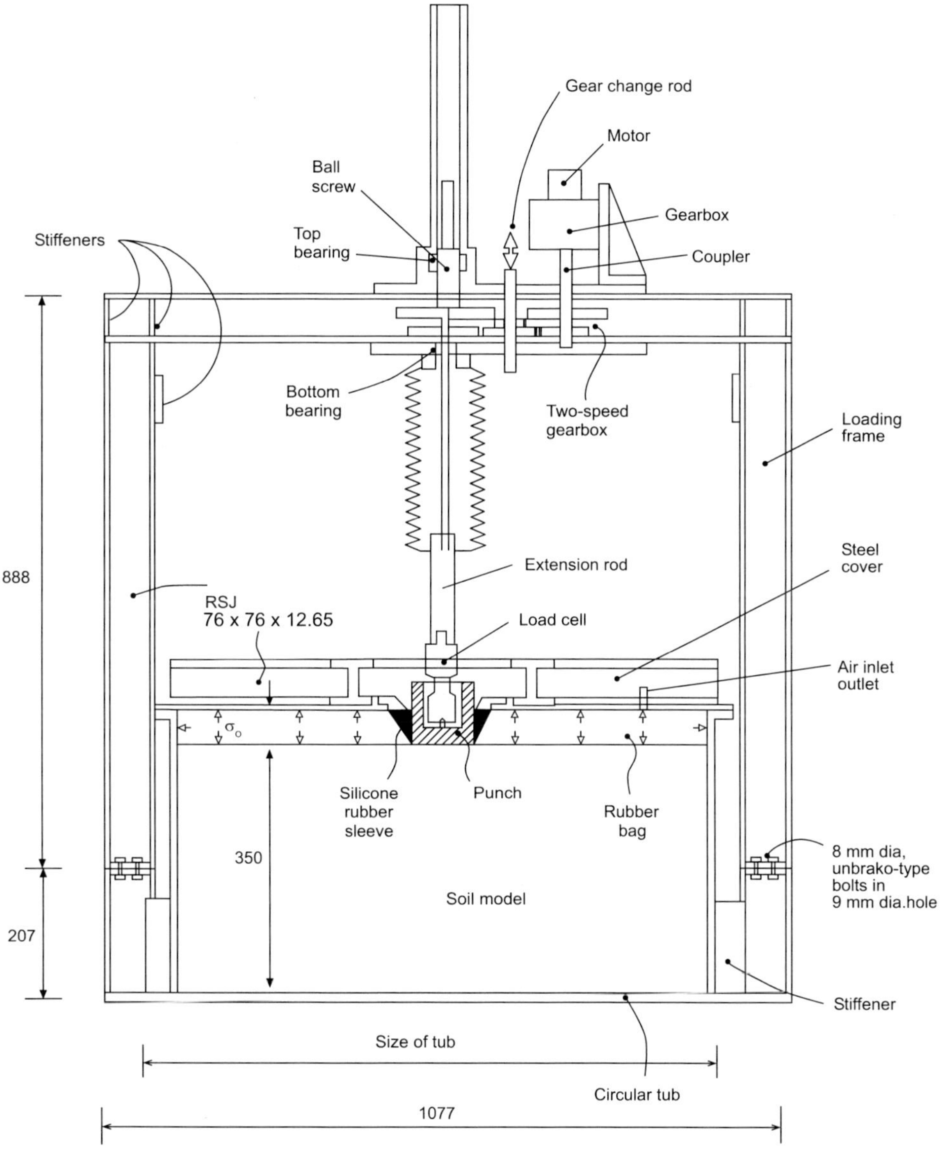

Fig. (6.36). General layout of test rig at 1g.

Table 6.5. Summary of 1g test programme.

Test no.	Material	*e*	*B*: mm	σ_o: kPa	d_{50}: μm	*B*/d_{50}	Condition
5	Silt	0.63	14.2	100	12	1183	Saturated
6	Silt	0.57	14.2	50	12	1183	Saturated
7	Silt	0.56	14.2	25	12	1183	Saturated
8	Silt	0.57	14.2	10	12	1183	Saturated
9	Silt	0.55	14.2	200	12	1183	Saturated
12	Sand	0.58	14.2	50	600	24	Dry
13	Sand	0.58	14.2	25	600	24	Dry
14	Sand	0.58	14.2	10	600	24	Dry
15	Sand	0.58	14.2	100	600	24	Dry
16	Sand	0.58	14.2	200	600	24	Dry
17	Sand	0.58	100	25	600	167	Dry
18	Sand	0.58	100	10	600	167	Dry
19	Sand	0.59	100	100	600	167	Dry
20	Sand	0.57	100	200	600	167	Dry
21	Sand	0.57	100	50	600	167	Dry
23	Sand	0.58	100	5	600	167	Saturated
24	Sand	0.63	14.2	25	600	24	Saturated
25	Sand	0.64	100	5	600	167	Saturated
26	Sand	0.56	100	5	600	167	Saturated
28	Silt	0.58	14.2	5	12	1183	Saturated

6.4.2 Penetration Effects

Fig. (**6.38**) shows the indentation response in test 14 (diameter B=14.2 mm) and test 18 (B=100 mm) plotted in terms of footing pressure σ_f against settlement w, and also against non-dimensional settlement ratio w/B, for a surcharge σ_o = 10kPa.

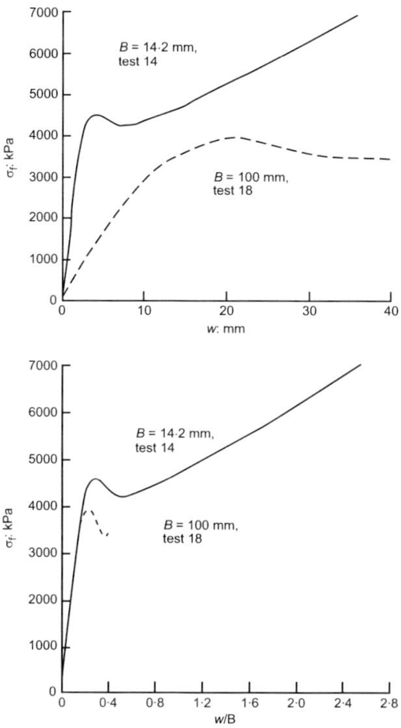

Fig. (6.37). Comparison between load-settlement response with load-relative settlement response of a footing on sand at 1g with σ_o =10 kPa.

Two features of Fig. (**6.37**) are noteworthy.

(a) It must be expected that, in early loading, soil strength should be mobilised as a function of strain, and also that strain must be a function of the only available non-dimensional measure of kinematics, namely w/B. This is well demonstrated in Fig. (**6.37**), in which scale effects on settlement prior to plastic indentation are correctly eliminated when settlement ratio w/B is used.

(b) Following plastic yielding, there is evidence of a steady enhancement of bearing pressure with settlement ratio for w/B > 0.5, particularly in test 14. Two possible explanations for this increase might be suggested:

(i) additional surcharge due to self-weight of the sand overburden now above the current elevation of the footing base

(ii) additional shear strength being mobilised in the soil involving the overburden

However, the overburden effect of 30 mm of sand is only 0.7 kPa, or 7% of σ_0, so the first explanation fails to tally with the observation of a 60% increase in test 14. The resistance must therefore be a mobilisation of the additional soil shear strength involving the overburden.

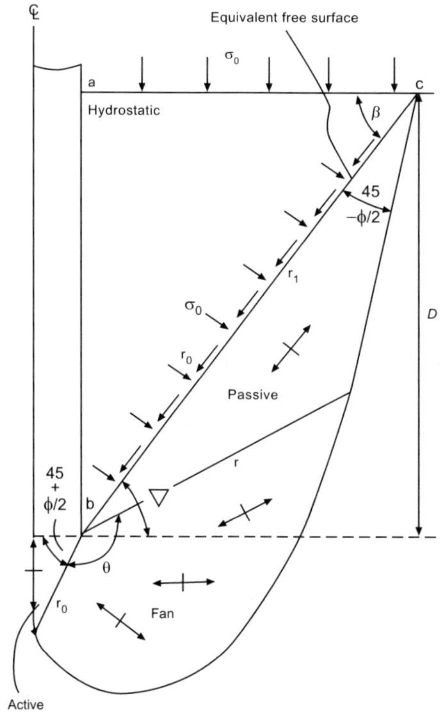

Fig. (6.38). Extended stress field to account for penetration.

The development of strength involving the overburden as understood by Meyerhof (1951) demands an extra rotation of the principal stress direction.

He proposed an extension of the work of Prandtl (1920) and Reissner (1924) for plane footings with initial burial: see (Fig. **6.38**). The soil within region abc of Fig. (**6.38**) is conservatively taken to act as a hydrostatic fluid, developing no shear resistance. The surcharge σ_o is therefore taken to act on plane bc sloping at angle ∇ to the horizontal. The rotation of the principal stress direction, θ, in the fan is simply $(\pi/2 + \nabla)$. The penetration effects of a strip footing on frictional weightless soil can now be incorporated in

$$\sigma_f = \sigma_o k_p e^{2\theta \tan\emptyset} \tag{6.38}$$

where $\theta = \pi/2 + \nabla$ and $0 \leq \nabla \leq \pi/2$. When $\nabla = 0$, the footing is at the surface and $\theta = \pi/2]$

$$\sigma_f = \sigma_o k_p e^{\pi \tan\emptyset} \tag{6.39}$$

When $\nabla = \pi/2$, the footing is at a critical depth D_c, that is, all additional rotation of principal stress has been exhausted and $\theta = \pi$

$$\sigma_f = \sigma_o k_p e^{2\pi \tan\emptyset} \tag{6.40}$$

It is now clear that, for $0 \leq D \leq D_c$, the penetration effects can be treated as a problem of gradual additional rotation of principal stress. The bearing capacity equation can now be treated as a simple geometry problem that relates penetration D to the additional rotation ∇.

It can be shown that

$$D = \frac{B \sin\nabla \cos\emptyset e^{\theta \tan\emptyset}}{2\cos(\pi/4+\emptyset/2)\sin(\pi/4-\emptyset/2)} \tag{6.39}$$

where $\theta = \frac{\pi}{2} + \nabla$. For $D < D_c$, ∇ can be calculated from eq. 6.39, and when it is substituted in eq. 6.38, σ_f can be found.

For $D < D_c$, the footing is at or below the critical depth D_c, where $\nabla = \pi/2$ and $\theta = \pi$, and any further increase in bearing capacity would be attributed to the additional surcharge of the overburden only.

$$D_c = \frac{B\cos\phi e^{\pi\tan\phi}}{2\cos(\pi/4+\phi/2)\sin(\pi/4-\phi/2)} \quad \textbf{(6.40)}$$

D_c may properly be taken as the demarcation between shallow and deep foundation. With this new definition, it can be said that the bearing capacity of a shallow foundation should be corrected for geometry effects, whereas for deep foundation the correction for overburden alone is sufficient, provided that the rotation in the shear fan is set at π instead of π/2.

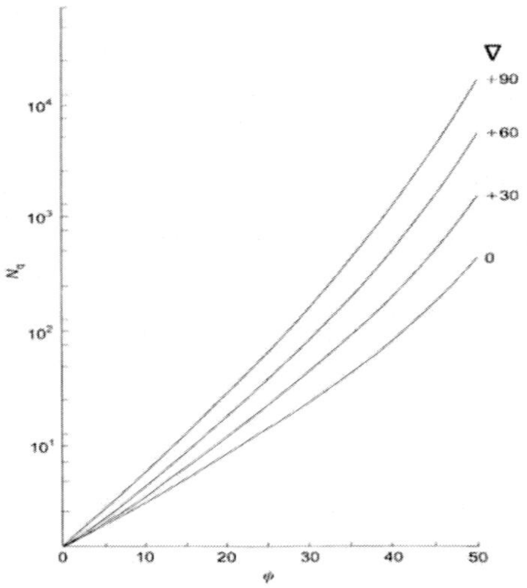

Fig. (6.39). Effects of shear stress on the equivalent free surface on N_q (after Meyerhof, 1951).

Meyerhof (1951) used Fig. (**6.39**) to show a typical 20-fold increase in bearing capacity factor for a footing placed at $D = D_c$ in sand developing

$\emptyset = 45^{\circ}$. In the current tests, the depth D is taken to be equal to the observed settlement w, and for a typical settlement w = 30 mm, in test 14 with B = 14.2 mm, we obtained *D*/B = 2.1 giving $\nabla = 9^{\circ}$and offering an enhancement factor of 1.5.

In order to interpret the data from the tests, the bearing capacity coefficient for a rough circular footing on the surface of soil with a given $\emptyset$ value was first calculated using the technique suggested by Cox (1962) as described earlier. This value was then enhanced by Meyerhof's factor, notwithstanding the change from plane strain to axisymmetry. Fig. (**6.40**) shows the resulting prediction for N_q against depth ratio *D*/B.

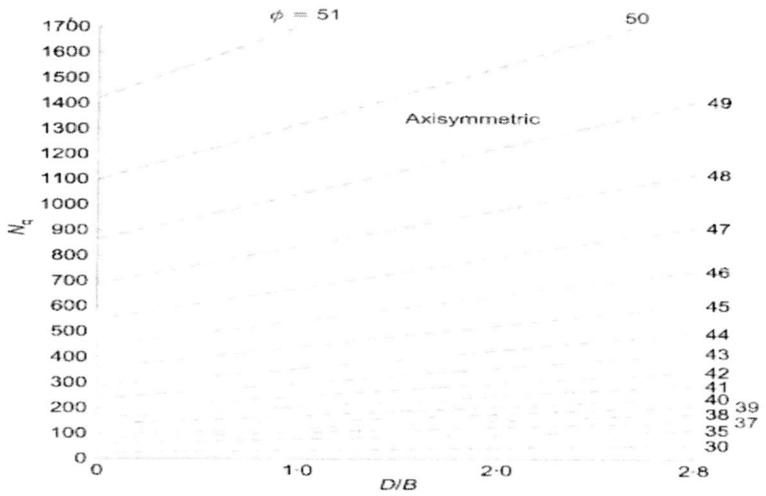

Fig. (6.40). Theoretical bearing capacity of shallow footing on weightless soil with overburden with shear strength.

1-g Tests: Discussion

Figs. (**6.41** and **6.42**) compare data with the theory for the 1 **g** test of the 14.2 mm footng on sand and silt respectively. It can be seen that the

Meyerhof correction for w/B leads to an interpretation in terms of ∅ remaining approximately constant for $0.5 < w/B < 2.5$. It may be concluded that the Meyerhof correction is useful in the back-analysis of these axisymmetric footings. Comparing the sand with the silt, it may be seen that the latter mobilised a slightly larger angle of shearing: a larger value was expected following the triaxial test results. Fig. (**6.43**) shows a similar pattern of penetration for the 100 mm footing on sand: those tests were curtailed at an earlier relative settlement. Fig. (**6.44**) shows the use of Meyerhof's w/B correction to obtain the equivalent value of N_q for w/B = 0. When a peak ∅ was observed (test m), this was used for the back-extrapolation; otherwise (test n) the asymptotic Meyerhof line was produced back to the N_q axis. Fig. (**46**) shows the exponential values of N_q *versus* overburden pressure σ_o , each on a logarithmic scale. A corresponding non-linear scale of ∅ is placed against the $\log_{10} N_q$ scale. In addition to the data points, the lines in Fig. (**6.45**) represent the results of the variable-∅ analyses reported in Fig. (**6.29a**) which were based on the best fit to the Ø - $\log_p$ data for the two soils.

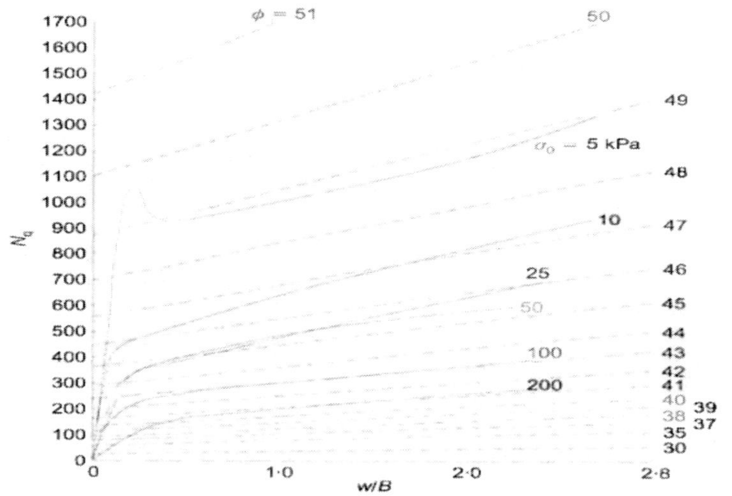

Fig. (6.41). Interpretation of 1g test data by superposing the normalized theoretical and experimental curves for silt, *B*=14.2mm.

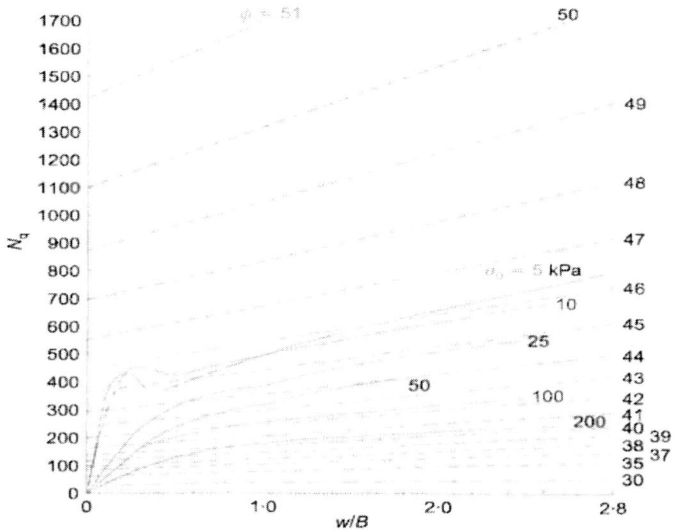

Fig. (6.42). Interpretation of 1g test data by superposing the normalized theoretical and experimental curves for sand, B=14.2mm.

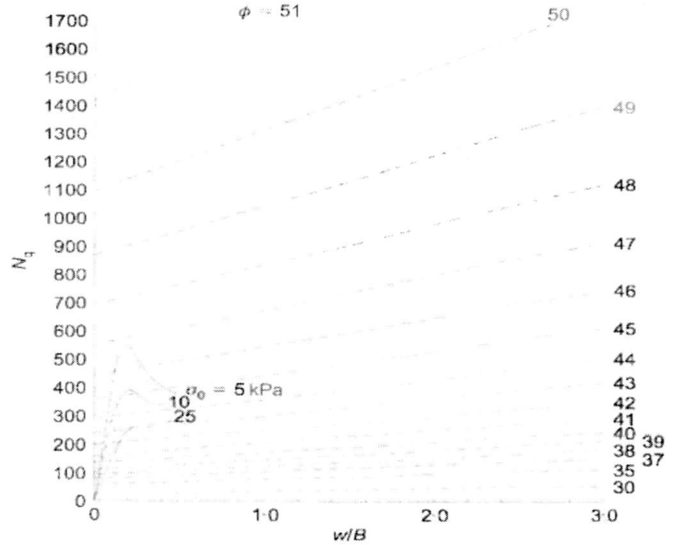

Fig. (6.43). Interpretation of 1g test data by superposing the normalized theoretical and experimental curves for sand, B=100mm.

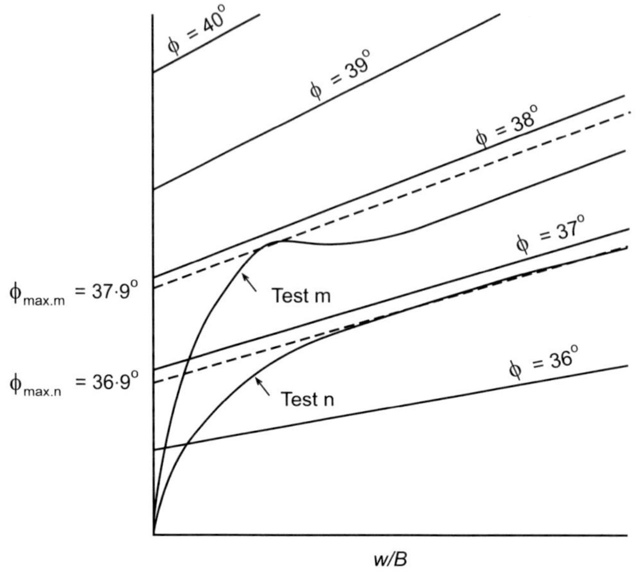

Fig. (6.44). Schematic diagram showing normalization of test curves to obtain equivalent N_q for w/B=0.

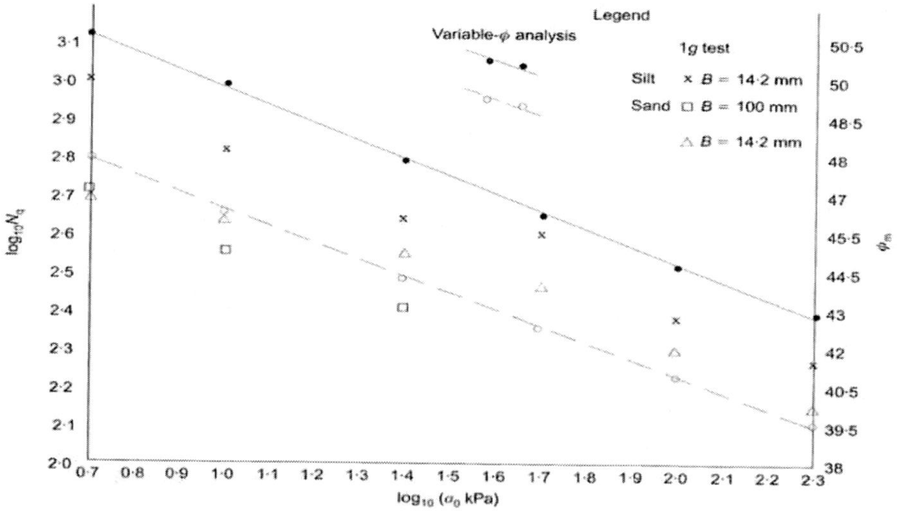

Fig. (6.45). Calculated and experimental results of the 1g tests.

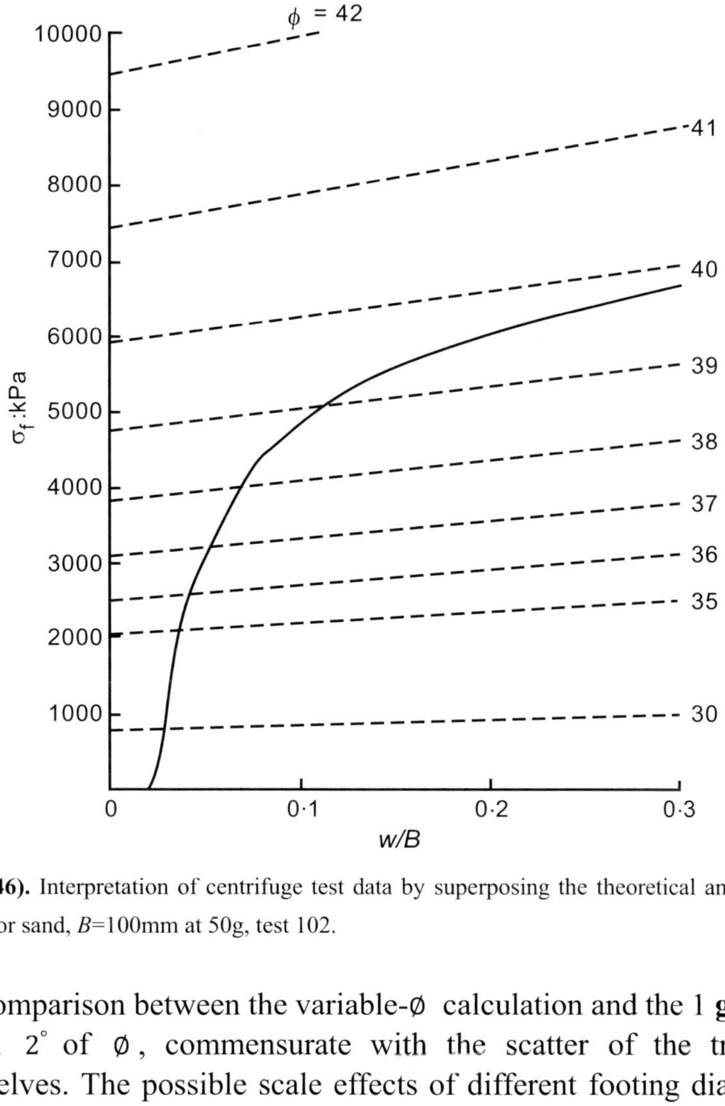

Fig. (6.46). Interpretation of centrifuge test data by superposing the theoretical and experimental curves for sand, *B*=100mm at 50g, test 102.

The comparison between the variable-∅ calculation and the 1 **g** test data is within 2° of ∅, commensurate with the scatter of the triaxial data themselves. The possible scale effects of different footing diameters and different soil particle sizes are seen to be eliminated when triaxial data of ∅ varying with p are used in the calculation of bearing capacity. Footing: particle size ratios between 24 and 1183, and container: footing diameter ratios between 8.5 and 60, create no measureable effect.

Centrifuge Model Tests

The earlier sections have demonstrated how 1 **g** tests, when subjected to a suitable range of surcharge pressures, can yield useful information on the scale effects in tests on footings. The main advantage of the 1 **g** tests is that they can mimic the behaviour of a weightless soil. This advantage, however, becomes its disadvantage when modelling a prototype shallow footing when self-weight of the soil become significant. The stresses at homologous points of a scaled-down model should replicate those of a prototype soil construction. This has been shown to be possible if the 1/n scale model is subjected to an acceleration field n times of that experienced by the prototype in a geotechnical centrifuge.

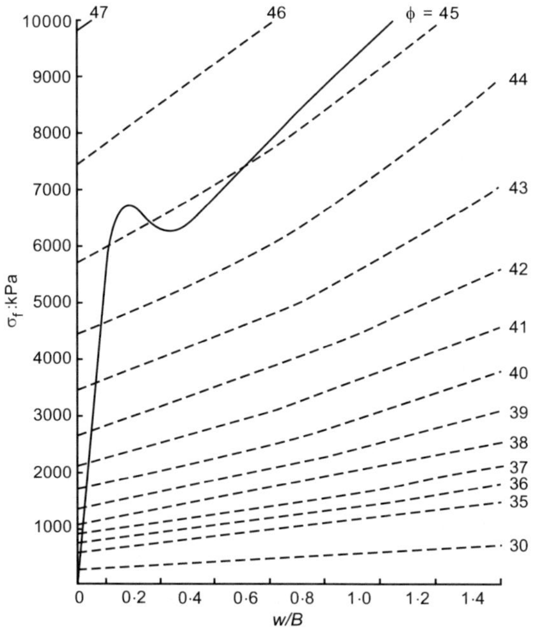

Fig. (6.47). Interpretation of centrifuge test data by superposing the theoretical and experimental curves for sand, B=14.2mm at 100g, test 104.

The same apparatus Fig. (**6.37**) was used for the centrifuge tests, except that the surcharge system was removed and replaced by a frame carrying an array of displacement transducers (Fig. **47**). The soil bed was fully submerged in water. The following tests at an elevated acceleration field of '50 **g**' (sand) and '14.2, 50/30 and 100 **g**' (silt) were conducted. The whole top surface of the soil model around the footing would be free of surcharge pressure at all times. As before, the average bearing pressure and settlement of the footing were monitored. The details of the tests are summarised in Table **6.6**.

Table 6.6. Summary of centrifuge model tests.

Test no.	Material	*e*	*B*: mm	*n*	d_{50}: μm	B/d_{50}	Condition
101	Silt	0.63	100	50g/30g	12	8333	Saturated
102	Sand	0.61	100	50g	600	166.7	Saturated
104	Silt	0.61	14.2	100g	12	1183.3	Saturated

6.4.3. Centrifuge Tests: Interpretation

For the purpose of interpretation, it was decided to make allowance for the settlement of the footing, as with the 1 **g** tests. In this case, however, the settlement w firstly produces a significant surcharge effect due to self-weight, enhancing the bearing capacity by a $\gamma' w N_q$ term, and secondly creates a strong w/B Meyerhof enhancement of the basic bearing capacity of a surface footing for which N_γ was calculated as before on the basis of a rough circular footing on constant- Ø soil.

Fig. (**6.46**) shows data from test 102 for a 5 m (prototype) footing on sand. Fig. (**6.47**) shows data for test 104 for a 1.42 m (prototype) footing on silt. As before, the w/B correction led to an understanding of the steady increase in bearing pressure with settlement, during plastic penetration. There is no evidence of any softening to $Ø < Ø_{max}$. Back-figuring to w/B = 0 permitted the estimation of N_γ in each case. A similar extrapolation procedure was used in test 101, but the interpretation was less secure, because the centrifuge acceleration had to be reduced halfway through the test in order to allow the loading system to be powerful enough to create a bearing failure.

Centrifuge Tests: Discussion

The centrifuge test data, back-analysed for N_γ in the way described above, are plotted against $\log_{10}$ B (metres prototype) in Fig. (**6.48**). The data are compared with the best-fit lines drawn from the variable- Ø analyses plotted in Fig. (**6.30a**). It will be seen that the error is within $\pm 1^\circ$, which is as small as it could be, considering the scatter in the original triaxial test data. It can therefore be concluded that there is no discernible relative particle size effects for B/d_{50} varying from 165 to 8333. This observation is consistent with the recommendation by Kusakabe (1995). There are also no measureable chamber effects when $\frac{Ø_{tub}}{B} \geq 8.5$ and $\frac{H_{tub}}{B} \geq 3.5$, which is in general agreement with the results reported by Bagge & Christensen (1977).

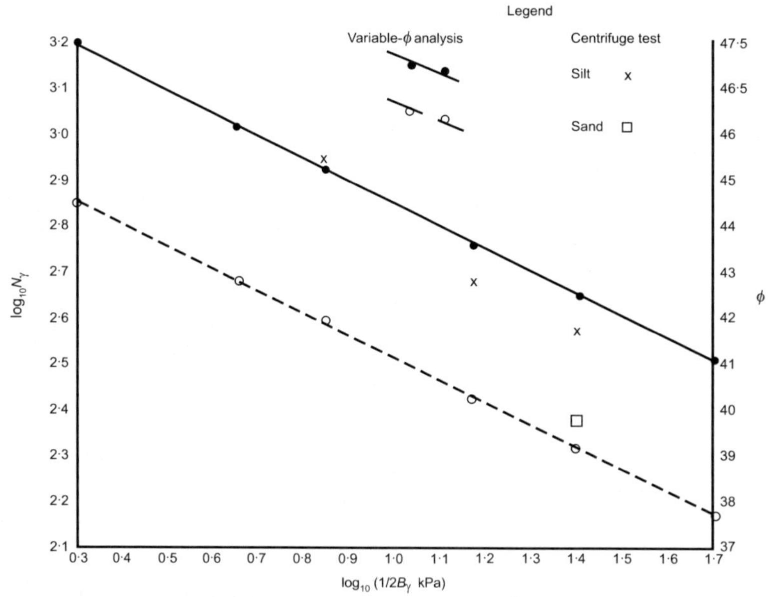

Fig. (6.48). Calculated and experimental results of the centrifuge tests.

CONCLUSION

Two issues are involved in the estimates of the drained bearing capacities of shallow circular footings. The first task is the establishment of a calculation procedure, based on soil strength parameters, which is capable of predicting bearing capacity. The second task is the establishment, through back-analysis of prototypes or models, of a means of defining representative values of soil strength to use in given circumstances.

A great deal of previous effort has gone into the comparison between strip and circular footings. Unfortunately, empirical investigations conflate the two issues referred to above: the use of different triaxial or plane soil strength parameters is confused with the possible existence of different bearing capacity factors. Other problems such as the nonlinearity of soil strength envelopes will inevitably require empirical treatment. In these circumstances, it is essential to remove empiricism from the calculation procedure wherever possible.

A coherent analysis of the collapse of plane and axisymmetric footings has been undertaken. Although it cannot be claimed that the assumptions which have been made are inevitable, they are at least consistent between the two cases. The method of characteristics has been used to confirm the utility of separating the N_q, N_γ, and N_c components of bearing. A table of bearing capacity factors has been produced for a wide range of ϕ values. Quite distinct values for plane and axisymmetric cases are listed; deep mechanisms beneath rough, rigid footings provide much larger self-weight components of bearing capacity than shallow mechanisms beneath smooth or compliant footings. Bearing capacity factors for a given angle ϕ are shown to vary over a factor of 4 from case to case. An error of 5° in ϕ produces an error of a factor of 2 to 3 in each factor. It is suggested that engineers find bearing capacity factors directly, rather than relying on shape factors, and should attempt to define strength envelopes appropriate to soil density, stress range and strain condition.

The stress-rotation equations used for constant-$\emptyset$ analysis need to be modified when $\emptyset$ is a function of mean stress p. These modifications have been derived and executed in a computer program that solves for equilibrium using the method of characteristics in the region of a known boundary stress condition.

Dilative granular materials generally exhibit a linear rate of reduction of sec $\emptyset$ with an increase in log p. In problems of plastic indentation, it has been shown that the variation in $\emptyset$ beneath a footing has a significant effect on the bearing capacity factor. Bearing capacity factors for dense sand and silt have been derived. The silt, which developed the same peak friction angle as the sand when it was subjected to five times greater confining pressure, produced bearing capacity factors approximately twice as large as the bearing capacity factors in sand. For each soil, a tenfold increase in resistive effective stress (σ_o or $0.5B\gamma'$) caused a threefold reduction in bearing capacity factor. Looser soils, which are closer to the critical state, must be expected to be less sensitive to stress-level effects.

If precise failure loads and plastic mechanisms are required, the new program VARPHI must be employed. However, if only the failure loads need to be estimated, an approximate treatment has been validated. The failure load with VARIPHI can be equated to that running an equivalent-constant angle of shearing $\emptyset_m$. The triaxial data of $\emptyset$ against p can be used to find the p_m at which $\emptyset = \emptyset_m$. Each class of problem will tend to display some functional linkage between p_m and the other imposed stresses. Two different functions have been suggested linking p_m to the harmonic mean vertical stress $\left(\sqrt{\sigma_f\sigma_o} \text{ or} \sqrt{\sigma_f 0.5B\gamma'}\right)$ in N_q or N_γ bearing problems.

Tests on small footings bearing on a sand and a silt with very similar, but differing in size by a factor of 50, mechanical characteristics have been carried out both at 1 **g** under surcharge, and in a centrifuge. Interpretation of the test data demanded that a correction be made for settlement effects. This was achieved using established techniques. The axisymmetic bearing capacity coefficients were calculated for zero penetration but constant-Ø following Cox (1962) and Meyerhof's (1951) construction was used to correct for geometry effects. This led to an empirical demonstration that bearing capacity coefficients reduce approximately linearly with stress level, on a $\log_{10}$–$\log_{10}$ plot.

Independent analytical estimates of bearing capacity were made based only on the variation of Ø with mean stress p in triaxial tests, and using the method of characteristics. The comparison between measurements and calculations was excellent, lying within $\pm 2^{\circ}$ scatter observed in the triaxial results themselves.

There is no strong evidence that anisotropy (strength reduction for shearing parallel to bedding planes) should have been taken into account. Tatsuoka *et al.,* (1991) reported up to 15% loss of Ø on plane shearing of bedding planes created after pluviation of dense sand samples. However, they also reported that this minimum plane-strain strength was similar to that measured in conventional triaxial compression tests with horizontal bedding. Apparently, the use of Ø from triaxial compression tests has been adequate in the present back-analysis of circular footings.

Similarly, there is no strong evidence that progressive failure has taken place: peak Ø values offer reasonable correlations with the peak loads observed in the model tests. Shear bands have previously been observed to propagate progressively downwards from the edge of strip footings on sand. Kimura *et al.,* (1985) associated 'scale effects' with these progressive distortions, while not accounting for the effects of Ø decreasing under increased confining stress. After some 1 **g** and centrifuge tests, the sand models that had horizontal colour marker beds were set by sugar syrup and sectioned for visual inspection. Likewise, the silt model inserted with vertical lead threads was X-rayed (Lau, 1988). No such shear band propagation was observed in the current tests on circular footings, but more tests should be carried out to confirm this observation.

A wide range of footing: container and footing: particle size ratios were investigated and no geometrical scale effect could be discerned. This suggests that the reduction in $\emptyset$ with increase in confining stress is the only significant 'size effect' determining the vertical bearing capacity of the model footings on sand and silt used in these experiments.

The VARIPHI program was used for the calculation of bearing capacity based on $\emptyset$ varying with p. This also demonstrated that an almost identical solution could be generated using constant-$\emptyset$ analysis, and therefore using existing bearing capacity factors, if appropriate value of $\emptyset_m$ is used. It was shown that, for surcharge effects only, $p_m = 2\sqrt{\sigma_f \sigma_o}$, whereas for self-weight effects only $p_m = 13\sqrt{\sigma_f 0.5B\gamma'}$. These semi-empirical rules could replace Meyerhof's suggestion of $p_m = \sigma_f/10$ for circular footings.

Notation

All stresses are effective stresses unless otherwise stated

A	Stress-induced strength reduction constant for sand
B	Width or diameter of footing
B/d_{50}	Footing : particle size ratio
c	Cohesion
D	Initial embedment of footing
D_c	Critical depth of footing
d_{50}	Mean particle size
e	Voids ratio
e_{min}	Minimum voids ratio
g	Acceleration due to earth's gravity
H_{tub}	Height of tub
I_D	Relative density
I_R	Relative dilatancy index
k_p	Coefficient of passive earth pressure
N_c	Bearing capacity factor (cohesion) ($=\sigma_f/c$)

Notation contd.....

N_q	Bearing capacity factor (surcharge) (=σ_f/σ_o)
N_γ	Bearing capacity factor (self-weight) (=$\sigma_f/0.5B\gamma$)
p	Mean stress (=$(\sigma_1+\sigma_2+\sigma_3)/3 \approx (\sigma_1+\sigma_3)/2$)
p_c	Intrinsic parameter related to the crushing of soil grains
p_m	Equivalent mean working stress
R	Normalised radius
r	Radius
s	Mean stress in the plane of shear $(=(\sigma_1+\sigma_3)/2)$
s_f	Mean stress in the active zone
s_o	Mean stress in the passive zone
w	Footing settlement
w/B	Relative footing settlement
Z	Normalised length
α, β	Pair of stress characteristics
∇	Equivalent relative settlement angle
γ	Bulk density of soil
γ'	Submerged density of soil
δ	Trapped cone/wedge-soil interface friction
θ	Principal stress rotation
σ	Stress
$\sigma_1, \sigma_2, \sigma_3$	Principal stresses
σ_f	Bearing capacity
σ_r	Triaxial cell pressure
σ_o	Surcharge
ε_a	Axial strain
ε_v	Volumetric strain
τ	Shear stress
τ_o	Shear stress acting on the free equivalent surface
$\emptyset$	Secant angle of friction
$\emptyset_{crit}$	Critical state angle
$\emptyset_{en}$	Angle of local strength envelope describing the stress characteristics

Notation contd.....

$\emptyset_m$	Mobilised angle of friction
$\emptyset_{max}$	Maximum angle of friction
$\emptyset_{op}$	Operative angle of friction
$\emptyset_{tub}$	Diameter of tub
φ	Orientation of principal stress direction
Ω	Superposition factor

REFERENCES

Bagge, G. and Christensen, S. N. 1977. Centrifuge testing on the bearing capacity of circular footings on the surface of sand. In Dialog 77, 20th anniversary, pp. 337-346. Lyngby: Denmarks Ingeniorakademi Bygningsafdelingen.

Bishop, A. W. 1961. Soil properties and their measurement: Discussion. In Proceedings, 5th International Conference on Soil Mechanics, vol. 3, p.97.

Bolton, M. D. 1986. The strength and dilatancy of sands. Geotechinque, 36(1): 65-78.

Bolton, M. D. & Lau, C. K. 1993. Vertical bearing capacity factors for circular and strip footings on Mohr Coulomb soil. Can. Geotech. J. 30. no. 6, 1024-1033.

Cerato, A. B. and Lutenegger, A. J. 2007. Scale effects of shallow foundation bearing capacity on granular material. J. Geotech. Geoenviron. Engng 133, no. 10, 1192-1202.

Chen, W. F. 1975. Limit analysis and soil plasticity. Elsevier. Amsterdam.

Corte, I. F. 1989. General report, model testing: geotechnical model tests. Proc. 12th Int. Conf. Soil Mech. Found. Engng Rio de Janeiro 4, 2553-2571.

Cox, A. D. 1962. Axially symmetry plastic deformation in soils-II. Indentation of ponderable soils. International Journal of Mechanical Sciences, 4: 371-380.

Cox, A. D., Eason, G. and Hopkins, H. G. 1961. Axially symmetric plastic deformation in soils. Philosophical Transactions of the Royal Society of London, Series A, 254: 1-45.

Davis, E. H. and Booker, J. R. 1971. The bearing capacity of strip footings from the standpoint of plasticity theory. In Proceedings, 1st Australian- New Zealand Conference in Geomechanics, Melbourne. pp. 276-282.

De Beer, E. E. 1965a. The scale effects on the phenomenon of progressive rupture in cohesionless soils. Proc. 6th Int. Conf. Soil Mech. Found. Engng, Montreal 2, 13-17.

De Beer, E. E. 1965b. Influence of the mean normal stress on the shearing strength of sand. Proceedings 6th Int. Conf. Soil Mech. Found. Engng. Montreal 1, 165-169.

Graham, J. and Hovan, J. M. 1986. Stress characteristics for bearing capacity in sand using a critical state model.

Can. Geotech. J. 23, no. 2, 195-202.

Graham, J. and Pollock, D. J. 1972. Scale dependent plasticity analysis for sand. Civ. Engng. Public Works Rev. 67, no. 788, 245-251.

Griffith, A. A. 1921. The phenomenon of rupture and flow in solids. Phil. Trans. R. Soc. London ser. A 221, no. 582-593, 163-197.

Griffiths, D. V. 1982. Computation of bearing capacity factors using finite elements. Geotechnique, 32(3):195-202.

Haar, A. and von Karman, T. 1909. Zur Theorie der Spannungs-zustaende in plastichen und sandartigen Medien. Nachrichten von der Koniglichen Gsellschaft der Wissenchaften, Gottingen, Math.-phys. Klasse, p. 204.

Hill, R. 1950. The mathematical theory of plasticity. Clarendon Press. Oxford.

Kimura, T. Kusakabe, O. and Saitoh, K. 1985. Geotechnical model tests of bearing capacity problems in a centrifuge. Geotechnique 35, no. 1, 33-45.

Ko, H. Y. and Davidson, L. W. 1973. Bearing capacity of footings in plane strain. ASCE Journal of the Soil Mechanics and Foundations Division, 99(SM1): 1-23.

Kumar, J. and Khatri, V. N. 2008. Effect of footing width on bearing capacity factor N_γ for smooth strip footings. J. Geotech. Geoenviron. Engng, 134, no. 9, 1299-1310.

Kutter, B. L., Abghari, A. and Cheney, J. A. 1988. Strength parameters for bearing capacity of sand. J. Geotech. Geoenviron. Engng, ASCE 114, no.4, 491-498.

Kusakabe, O. 1995. Foundations. In Geotechnical centrifuge technology (ed. R. N. Taylor), pp. 118-167. London: Blackie Academic & Professional.

Larkin, L. A. 1968. Theoretical bearing capacity of very shallow footings. ASCE Journal of Soil Mechanics and Foundations Division, 94(SM6): 1347-1357.

Lau, C. K. 1988. Scale effects in tests on footings. PhD thesis, University of Cambridge, U.K.

Lau, C. K. and Bolton, M. D. 2011a. The bearing capacity of footings on granular soils. I: Numerical analysis. Geotechnique 61, no.8, 627-638.

Lau, C. K. and Bolton, M. D. 2011b. The bearing capacity of footings on granular soils. II: Experimental evidence. Geotechnique 61, no.8, 639-650.

LeBlanc, L. 1981. Tracing the causes of rig mishaps. Offshore, March, 51-62.

Meyerhof, G. G. 1951. The ultimate bearing capacity of foundations. Geotechnique, 2(4): 301-332.

Meyerhof, G. G. 1961. Shallow foundations: Discussion. In Proceedings, 5th International Conference on Soil Mechanics, vol.3, p.193.

Meyerhof, G. G. 1963. Some recent research on the bearing capacity of foundations. Canadian Geotechnical Journal, 1: 16-26.

Muhs, H. 1965. Discussion: On the phenomenon of progressive rupture in connection with the failure behaviour of footings in sand. Proc. 6th Int. Conf. Soil Mech. Found. Engng. Montreal 3, 419-421.

Ovesen, N. K. 1975. Centrifuge testing applied to bearing capacity problems of footings on sand. Geotechnique 25, no. 2, 394-401.

Ovesen, N. K. 1979. Panel discussion: The scaling law relationship. Proc. 7th Eur. Conf. Soil Mech. and Found. Engng. Brighton 4, 319-323.

Prandtl, L. 1920. Uber die Harte plasticher Korper. Nachrichten von der Koniglichen Gesellschaft der Wissenschaften, Gottingen, Math.-phys. Klass, pp. 74-85.

Randolph, M., Cassidy, M. , Gourvenec, S. and Erbrich, C. 2005. Challenges of offshore geotechnical geotechnical engineering. Proc. 16th Int. Conf. Soil Mech. Geotech. Engng. Osaka 1, 123-176.

Reissner, H. 1924. Zum Erddruckproblem. In Proc. 1st Int. Conf. Applied Mechanics, Delft, 295-311.

Schofield, A. N. 1980. Cambridge geotechnical centrifuge operations. Geotechnique 30. No. 3, 227-268.

Shi, Q. 1988. Centrifugal modeling of surface footings subjected to combined loading. PhD thesis, University of Cambridge, U.K.

Shield, R. T. 1955. On the plastic flow of metals under conditions of axial symmetry. Proceedings of Royal Society of London, Series A, 233: 267-287.

Sokolovskii, V. V. 1960. Statics of soil media. Butterworths, London. (Translated from the 1942 Russian edition).

Tatsuoka, F., Okahara, M. Tanaka, T., Tani, K. Morimoto, T. and Siddiquee, M. S. A. 1991. Progressive failure and particle size effect in bearing capacity of footing on sand. Proc. Of the Geotech. Engng. Congress, Boulder, CO, ASCE Special Publ. no. 27, pp. 788-802.

Terzaghi, K. 1943. Theoretical soil mechanics. John Wiley & Sons, New York.

Terzaghi, K. and Peck, R. B. 1948. Soil mechanics in engineering practice. New York: John Wiley.

Ueno, K. Miura, K. and Maeda, Y. 1998. Prediction of ultimate bearing capacity of surface footing with regard to size effects. Soils Found. 38, no. 3, 165-178.

Ueno, K., Miura, K., Kusakabe, O. and Nishimura, M. 2001. Reappraisal of size effect of bearing capacity from plastic solution. J. Geotech. Geoenviron. Engng 127, no.3, 275-281.

Vesic, A. S. 1963. Bearing capacity of deep foundations in sand. Highway Research Record, no. 39, 112-153.

Vesic, A. S. and Clough, G. W. 1968. Behaviour of granular materials under high stresses. J. Soil Mech. Found. Div. ASCE 94, no. SM3, 661-688.

Yamaguchi, H, Kimura, T. and Fuji-I, N. 1976. On the influence of progressive failure on the bearing capacity of shallow foundations in dense sand. Soils Found. 16, no. 4, 11-22.

Yamaguchi, H, Kimura, T. and Fuji-I, N. 1977. On the scale effect of footings in dense sand. Proc. 9th Int. Conf. Soil Mech. Found. Engng. Tokyo 1, 759-798.

Zhu, F., Clark, J. I. and Phillips, R. 2001. Scale effect of strip and circular footings resting on dense sand. J. Geotech. Geoenviron. Engng 127, no. 7, 613-621.

SUBJECT INDEX

A

D

E

F

M

N

O

P

Made in United States
North Haven, CT
04 November 2026

21743224R00238